CONTEMPORARY HEALTH PHYSICS:
PROBLEMS AND SOLUTIONS

CONTEMPORARY HEALTH PHYSICS

Problems and Solutions

JOSEPH JOHN BEVELACQUA
Wisconsin Electric Power Company

A Wiley-Interscience Publication
JOHN WILEY & SONS
New York • Chichester • Brisbane • Toronto • Singapore

This text is printed on acid-free paper.

Copyright © 1995 by John Wiley & Sons, Inc.

All rights reserved. Published simultaneously in Canada.

Reproduction or translation of any part of this work beyond that permitted by Section 107 or 108 of the 1976 United States Copyright Act without the permission of the copyright owner is unlawful. Requests for permission or further information should be addressed to the Permissions Department, John Wiley & Sons, Inc., 605 Third Avenue, New York, NY 10158-0012.

Library of Congress Cataloging in Publication Data:
Bevelacqua, Joseph John, 1949–
    Contemporary health physics : problems and solutions / Joseph John Bevelacqua.
      p.   cm.
    "A Wiley-Interscience publication."
    Includes index.
    ISBN 0-471-01801-5 (cloth : alk. paper)
    1. Medical physics—Outlines, syllabi, etc. 2. Medical physics—Problems, exercises, etc. I. Title.
    [DNLM: 1. Health Physics—examination questions. WN 18 B571c 1995]
R895.b48   1995
616.9'897—dc20
DNLM/DLC
for Library of Congress   94-19851
                                                              CIP

Printed in the United States of America

10 9 8 7 6 5 4 3 2

This book is dedicated to my wife, Terry. Her love and understanding have been of great assistance to the completion of this text.

PREFACE

This book contains over 375 problems in health physics and discusses their practical applications. It assumes that the reader is familar with the science of radiation protection and is either an active participant in that field or interested in learning more about the health physics profession. In particular, this text is particularly useful to individuals preparing for the American Board of Health Physics Certification Examination.

The first part of this book provides an overview of the scientific basis for the field of health physics. The reader is provided with a comprehensive set of references supplemented by appendices that outline selected concepts required to fully appreciate the specialized Part II material. Over 130 problems and their solutions are provided to permit the reader to demonstrate a sound knowledge of health physics fundamentals. The problems are set within scenarios that are intended to enhance the reader's existing knowledge by demonstrating the basic principles in complex situations requiring a sound knowledge of both theoretical health physics principles and good judgment.

Part II provides the reader with examples of the concepts and calculations frequently encountered in the various fields of health physics. Chapter titles are selected to loosely conform to the various subfields of the health physics profession—that is, medical, university, fuel cycle, power reactor, environmental, and accelerator health physics. The problems are intended to illustrate general concepts within the framework of specific areas such as medical or power reactor health physics.

In addition to illustrating the fundamental concepts of health physics, the collection includes a large number of detailed problems that are often encountered by the radiation protection professional. Some of these problems involve considerable effort, whereas others are more simplistic and can be solved from traditional lectures in health physics. In addition, there are problems which address topics not usually covered in existing texts. These problems are not presented as isolated bits of health physics knowledge, but are introduced within a scenario that stimulates an integrated professional approach to the problem. Professional judgment and sound health physics principles are emphasized.

The third part of this book provides the solutions to the problems presented in the first and second parts. Many of these are worked in considerable detail to further illustrate and emphasize the concepts introduced in Parts I and II.

The present collection of problems is largely based upon the American Board of Health Physics Comprehensive Examination. The author was privileged to serve 4 years as a member, Vice-Chairman, and Chairman of the ABHP Comprehensive Panel of Examiners. The experience gained in the development of this examination and the weaknesses of candidates attempting this examination have affected the content of this work.

The author is deeply indebted to the members of the examination panels and the ABHP Board for their professional interaction which greatly expanded the author's own health physics knowledge. The opinions and interpretations reflected in this work are the author's and do not necessarily reflect those of his current or previous employers.

JOSEPH JOHN BEVELACQUA

Wisconsin Electric Power Company

A NOTE ON UNITS

In the United States many regulations, most reporting requirements, and a large portion of practicing health physicists utilize traditional units (Ci, R, rad, rem, etc.). The use of traditional units is currently in conflict with much of the international community and scientific publications which have adopted the SI system (Bq, C/kg, Gy, Sv, etc.).

This book adopts a set of units that are commonly utilized in practical health physics applications. These traditional units are selected because they are what the practicing health physicist will most frequently encounter in daily assignments and they can be easily related to their SI counterparts. Traditional units are also utilized to ensure that communications between the health physicist and the health physics technician are clearly understood. When SI units are utilized to convey ICRP dose limit recommendations or criteria, their traditional counterparts are also provided.

The conflict of units will remain until the United States adopts the SI system in its regulations. This should be done over a period of years in order to ensure that all health physicists are thoroughly familiar and comfortable with the SI units.

For those readers that feel more comfortable with the SI system, the following conversion factors are provided:

SI Unit	Traditional Unit
Bq	2.70×10^{-11} Ci
Gy	100 rad
C/kg of air	3881 R
Sv	100 rem

As the reader can quickly note the choice of units is more a matter of familiarity rather than scientific rigor. By using these simple factors, the reader should begin to feel more comfortable with either set of units.

CONTENTS

PART I	**BASIC CONCEPTS: THEORY AND PROBLEMS**	**1**
	1 Introduction	3
PART II	**SPECIALIZED AREAS: THEORY AND PROBLEMS**	**41**
	2 Medical Health Physics	43

        Historical Perspective, 43
        Medical Accelerator Physics, 44
        Diagnostic Nuclear Medicine, 46
        Therapeutic Nuclear Medicine, 48
        Facility Design, 51
        Shielding Design, 51
        X-Ray Shielding, 52
        NCRP-37 Exposure Recommendations, 56
        Ventilation Considerations, 56

    3 **University Health Physics**   68

        Research Utilizing Radionuclides, 68
        Engineering Considerations, 71
        Sample Counting, 72
        Intake of Radionuclides, 73
        Other Research Activities, 74

Agricultural/Environmental Research, 74
Research Reactors, 75
Particle Accelerators, 75
Materials Research via X-Ray Diffraction Techniques, 76
Fusion Energy Research, 76

4 Fuel Cycle Health Physics — 91

Radiation in Fuel Cycle Facilities, 91
Occupational Exposure, 92
Nuclear Fuel Cycle, 93
Uranium Ore and Chemical Processing, 95
Gaseous Diffusion, 95
Gas Centrifuge, 98
Laser Isotope Separation, 99
Spent Power Reactor Fuel, 101
Radioactive Waste, 102
Criticality, 104
Dispersion of Radioactive Gas from a Continuous Source, 107
Dispersion of Radioactive Particulates from a Continuous Source, 109
Fuel Cycle Facilities, 112

5 Power Reactor Health Physics — 126

Overview, 126
Health Physics Hazards, 127
Health Physics Program Elements, 135
Radioactive Waste, 137
Outages, 138
Radiological Considerations During Reactor Accidents, 139
Mitigation of Accident Consequences, 141

6 Environmental Health Physics — 158

Naturally Occurring Radioactive Material, 158
Radon, 159
Environmental Monitoring Programs, 164
Environmental Releases, 164
Regulatory Guidance for Effluent Pathways, 166
Doses from Liquid Effluent Pathways, 167
Doses from Gaseous Effluent Pathways, 171
Pathway Selection, 177
Model Parameters, 177

7 Accelerator Health Physics — 186

High-Energy Interactions, 186
Proton Accelerators, 187
Electron Accelerators, 188
Heavy-Ion Accelerators, 188
Residual Radioactivity, 190
Buildup of Radioactive and Toxic Gases in an Irradiation Cell, 192
Other Radiation Sources, 192
Shielding, 193
Accelerator Beam Containment, 196
Dose Equivalent from the Accelerator Target, 196
Beam Current, 197
Pulsed Radiation Fields, 198

PART III ANSWERS AND SOLUTIONS — 209

Solutions for Chapter 1 — 211

Solutions for Chapter 2 — 261

Solutions for Chapter 3 — 279

Solutions for Chapter 4 — 300

Solutions for Chapter 5 — 319

Solutions for Chapter 6 — 336

Solutions for Chapter 7 — 352

PART IV APPENDICES — 363

Appendix I Serial Decay Relationships — 365

Appendix II Basic Source Geometries and Attenuation Relationships — 368

Appendix III Neutron-Induced Gamma Radiation Sources — 376

Appendix IV Selected Topics in Internal Dosimetry — 380

Appendix V Radiation Risk and Risk Models — 408

INDEX — 417

PART I

BASIC CONCEPTS: THEORY AND PROBLEMS

1

INTRODUCTION

Health physics or radiation protection is the science dealing with the protection of radiation workers and the general public from the harmful effects of radiation. Health physicists work in a variety of environments, including medical facilities, universities, accelerator complexes, power reactors, and fuel cycle facilities. The health physicist is responsible for the radiological safety aspects of facility equipment and services. Radiological assessments of plant equipment, facility modifications, design changes, employee exposures, or the assessment of radiological effluents are key functions of a health physicist.

The fundamental tools of the health physicist include the fields of mechanics, electricity and magnetism, energy transfer, and quantum mechanics. Atomic and nuclear structure, radioactive transformations, and the interaction of radiation with matter are the cornerstones of health physics knowledge. Application of these fundamental tools permits the health physicist to measure, quantify, and control radiation exposures to affected groups.

Introductory health physics texts typically cover these topics in several hundred pages. Because the scope of this text builds upon these fundamental concepts, we will not repeat them herein. The reader is referred to the texts listed as references to this chapter for a discussion of health physics fundamentals. We will, however, provide several appendices that illustrate selected fundamental concepts. Also included is an extensive set of scenarios, including over 130 worked examples, that illustrate the fundamental concepts and permit the reader to assess his or her knowledge of these concepts. Because the fundamentals are needed to fully understand the remaining chapters in this text, a review of the scenarios in this chapter is recommended.

4 INTRODUCTION

SCENARIOS

Scenario 1

One of your neighbors, while digging up his back yard to build a pool, has discovered some old planks. Another neighbor, who has been investigating the possibility of the existence of a Viking settlement in the area, believes that the planks may be significant. He wishes to conduct an archeological expedition prior to any further construction. You offer to carbon date the wood to help settle the argument.

1.1. Carbon dating is possible because:
   a. The specific activity of carbon-14 in living organisms has changed over time, and one can identify the era of time the organism lived based on its current specific activity.
   b. Carbon-14 is in secular equilibrium with its daughter.
   c. The specific activity of carbon-14 in living organisms is relatively constant through time, but decays after the death of the organism.
   d. The specific activity of carbon-14 in wood increases over time due to shrinkage of the wood.

1.2. Calculate the approximate age of the wood given the following:

$$\text{C-14 } T_{1/2} = 5600 \text{ years}$$

Specific activity for C-14 in a nearby living tree
$$= 1.67 \times 10^{-1} \text{ Bq/g}$$

Specific activity for C-14 in the old wooden plank
$$= 1.50 \times 10^{-1} \text{ Bq/g}$$

Scenario 2

A nearby hospital has received a shipment of a Mo-99 generator. The shipment contained 1000 mCi of Mo-99 when manufactured. It arrived at the hospital 48 h after its production.

The decay scheme is illustrated in Fig. 1.1.

1.3. If the generator is milked exactly upon arrival at the hospital, how much Tc-99m will be obtained? Assume that 95% of the available Tc-99m is eluted.

1.4. If the generator is milked 24 hr after the initial milking, how much Tc-99m will be obtained?

Scenario 3

Consider a parent radioisotope A ($T_{1/2} = 10$ hr) that decays to a daughter radioisotope B ($T_{1/2} = 1$ hr).

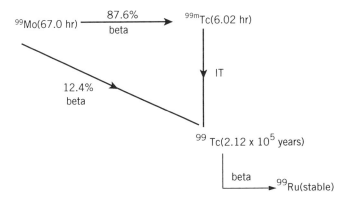

Fig. 1.1 Decay scheme for Mo-99.

1.5. Which of the following statements is true concerning these radioisotopes?
   a. Because $\lambda_A > \lambda_B$, the parent and daughter will eventually reach the condition of transient equilibrium.
   b. Because $\lambda_A \gg \lambda_B$, the parent and daughter will eventually reach the condition of secular equilibrium.
   c. Because $\lambda_A = \lambda_B$, no state of equilibrium can ever exist between the parent and daughter.
   d. Because $\lambda_B > \lambda_A$, the parent and daughter will eventually reach the condition of transient equilibrium.
   e. Because $\lambda_B \gg \lambda_A$, the parent and daughter will eventually reach the condition of secular equilibrium.

1.6. Assuming that the activity of the daughter is zero at time zero, at what time (t) will the daughter reach its maximum activity?

Scenario 4

The plant manager at your facility has requested that you review the following questions and provide the best solution. These questions will be used to assess the qualification of health physics candidates for entry-level positions in your facility's radiological controls department.

1.7. Tissue dose from thermal neutrons arises principally as a result of:
   a. (n, γ) reactions with hydrogen
   b. (n, γ) reactions with hydrogen and (n, p) reactions with nitrogen
   c. (n, p) reactions with carbon
   d. (n, α) reactions with carbon
   e. (n, α) reactions with carbon and (n, γ) reactions with hydrogen

6 INTRODUCTION

1.8. Tissue dose from fast neutrons (0.1 to 14 MeV) is due principally to:
   a. Resonance scattering with nuclei
   b. Inelastic scattering with nuclei
   c. Coulomb scattering with nuclei
   d. Nuclear capture and spallation
   e. Elastic scattering with nuclei

1.9. The most probable process for energy deposition by a 1-MeV photon in tissue is:
   a. Photoelectric absorption
   b. Pair production
   c. Compton scattering
   d. Photonuclear absorption
   e. Bremsstrahlung

1.10. The principal mechanism of dose deposition by a 5-MeV alpha particle that stops in tissue is:
   a. Inelastic scattering by atomic electrons
   b. Elastic scattering by atomic electrons
   c. Elastic scattering by atomic nuclei
   d. Inelastic scattering by atomic nuclei
   e. Nuclear spallation

1.11. The principal mechanism of dose deposition by a 100-keV beta particle that stops in tissue is:
   a. Elastic scattering by atomic electrons
   b. Elastic scattering by atomic nuclei
   c. Inelastic scattering by atomic nuclei
   d. Inelastic scattering by atomic electrons
   e. Bremsstrahlung

1.12. The average number of ion pairs produced by a 100-keV beta particle that stops in air is approximately:
   a. 300
   b. 30
   c. 30,000
   d. 3000
   e. 300,000.

1.13. The average number of ion pairs produced by a 100-keV beta particle that stops in a germanium semiconductor is:
   a. 30,000
   b. 30
   c. 300

    d. 3000
    e. 300,000
1.14 A nuclide that undergoes orbital electron capture:
    a. Emits an electron, a neutrino, and the characteristic x-rays of the daughter.
    b. Emits a neutrino and the characteristic x-rays of the daughter.
    c. Also decays by positron emission.
    d. Also emits internal conversion electrons.
    e. Makes an isomeric transition.
1.15. The specific gamma-ray emission rate for Cs-137 in units of R hr^{-1} Ci^{-1} m^2 is approximately:
    a. 1.3
    b. 0.12
    c. 0.33
    d. 0.05
    e. 0.77
1.16. An example of an organ or tissue for which the Annual Limit on Intake (ALI) is determined by the limit for nonstochastic effects is the:
    a. Red bone marrow
    b. Gonads
    c. Lung
    d. Breast
    e. Thyroid

Scenario 5

The radioisotope I-126 (atomic number 53) can decay into stable Te-126 (atomic number 52) by orbital electron capture (EC) or by positron emission. It can, alternatively, decay by negative beta emission into stable Xe-126 (atomic number 54). The fractions of the transformations that take place via these modes are: EC 55%, positron emission 1%, and beta decay 44%. An I-126 source also emits gamma photons of energy 386 keV and 667 keV as well as characteristic x-rays of Te. The energy equivalents (Δ) of the mass excesses of the atoms involved in these transformations are (Δ = atomic mass − atomic number):

Atom	Δ (MeV)
Te-126	−90.05
I-126	−87.90
Xe-126	−89.15

8 INTRODUCTION

The energy equivalent of the electron rest mass is 0.511 MeV, and the binding energy of the K-shell electron in I-126 is 32 keV.
For the following questions, choose the best answer.

1.17. The energy released (Q-value) by the decay of I-126 via capture of a K-shell electron, going directly to the ground state of Te-126, is:
  a. 0.03 MeV
  b. 1.13 MeV
  c. 2.12 MeV
  d. 2.15 MeV
  e. 2.18 MeV

1.18. The energy released (Q-value) by the decay of I-126 via position emission to the ground state of Te-126 is:
  a. 0.51 MeV
  b. 1.02 MeV
  c. 1.13 MeV
  d. 1.64 MeV
  e. 2.15 MeV

1.19. The energy released in the decay of I-126 to the ground state of Xe-126 by beta emission is:
  a. 0.20 MeV
  b. 0.23 MeV
  c. 0.90 MeV
  d. 0.74 MeV
  e. 1.25 MeV

1.20. Of the following kinds of radiation emitted from I-126, which is the single least significant potential contributor to internal dose?
  a. Annihilation photons
  b. Bremsstrahlung
  c. Internal-conversion electrons
  d. Auger electrons
  e. Antineutrino

  How would your answer change if external dose contributions were under consideration?

1.21. Why are the 32-keV Te x-rays present with an I-126 source?
  a. The nucleus of Te-126 has excess energy after the EC event. This excess energy is released by Te-126 as x-rays.
  b. Stable Te-126 has excess energy after the positron emission. This excess energy is released by Te-126 as x-rays.
  c. Electrons rearranging between the L and M shells produce x-rays.

d. Te x-rays are released when the EC event creates a vacancy in the inner shells, and electrons from outer shells fill the vacancy.

e. Te x-rays are equivalent to the bremsstrahlung radiation emitted by I-126.

Scenario 6

The nuclide Sr-90 (atomic number 38) decays by beta emission into Y-90 (atomic number 39), which then decays by beta emission into Zr-90 (atomic number 40), with the half-lives noted below:

$$\text{Sr-90} \xrightarrow{27.7 \text{ years}} \text{Y-90} \xrightarrow{64.2 \text{ hr}} \text{Zr-90}$$

1.22. What is the mean, or average, lifetime of a Y-90 atom?
   a. 31.1 hr
   b. 44.5 hr
   c. 77.04 hr
   d. 92.6 hr
   e. 128.4 hr

1.23. What is the specific activity of Y-90 in SI units?
   a. 5.42×10^5 Bq/kg
   b. 7.22×10^{16} Bq/kg
   c. 2.01×10^{19} Bq/kg
   d. 7.22×10^{19} Bq/kg
   e. 6.49×10^{21} Bq/kg

1.24. Starting with a pure Sr-90 sample at time $t = 0$, a researcher finds that the Y-90 activity is 3.4 mCi at $t = 72.0$ hours. What was the activity of the Sr-90 at $t = 0$?
   a. 1.84 mCi
   b. 3.40 mCi
   c. 4.37 mCi
   d. 6.29 mCi
   e. 7.39 mCi

Scenario 7

You have been asked to assist in the technical evaluation of an ionization chamber and environmental sampling results. Your boss has requested answers to the following questions. Assume the density of air at STP = 1.293×10^{-6} kg/cm^3.

10 INTRODUCTION

1.25. A free air ionization chamber shows a flow of electrical charge of 1×10^{-9} A. The chamber has a sensitive volume of 4 cm³. The reading is taken at 10°C and 755 mm Hg. Find the exposure rate in R/sec based on STP conditions.

1.26. You are asked to provide immediate, on-site measurement results for a series of environmental samples that are being collected every 100 min. It has been requested that each sample count be preceded by a background count. From past experience, you estimate that the net sample and background counting rates should be approximately 2400 and 300 cpm, respectively. Assuming that each sample must be analyzed before the next one is received, how long would you count the sample to minimize the standard deviation estimate for the sample's net activity?

1.27. A water sample that was counted for 10 min yielded 600 counts. A 40-min background count yielded a background rate of 56 cpm. At a 95% confidence level (one-tail test), determine whether or not there was any net activity in the sample.

Scenario 8

You are responsible for operating the counting room at a nuclear facility. You need to minimize the counting time required for air samples because of the heavy workload and a need to streamline operations in the court room. The bulk of your air sample workload is counting I-131. The following parameters are applicable to your operation:

>   Counting efficiency = 20%
>   Background count time = sample count time
>   Background count rate = 50 cpm
>   Sampling flow rate = 5 liters/min
>   Sample collection time = 10 min
>   Iodine collection efficiency = 70%
>   MPC for iodine = 1×10^{-9} μCi/cm³

1.28. Calculate the minimum sample and background counting time required to ensure an LLD at the 95% confidence level less than or equal to 0.10 MPC for I-131.

1.29. List methods that could be used in the field or in the counting room to reduce the time required to process I-131 samples. Explain how each method reduces processing time.

Scenario 9

As a health physicist at a nuclear facility, you are asked to develop a program to characterize the radioactive particulate emissions through the facility's main ventilation stack. The following questions relate to various aspects of this assignment.

1.30. In designing the sampling system, you have determined that the stack internal diameter is 0.5 m and the volumetric flow through the stack is 20 m^3/min. You want to use a vacuum source which will provide a constant volumetric flow of 200 liters/min through your sampling train. Assuming laminar flow, what should the internal diameter of the sampling nozzle be to ensure isokinetic sampling conditions?

1.31. To ensure that your sample is representative of laminar flow conditions (nonturbulent, constant velocity) within the stack, discuss factors that you should consider relative to the location of your sampling nozzle within the stack.

1.32. You have decided to use filtration techniques to capture your sample and are evaluating three types of media (cellulose, glass-fiber, and membrane filters). List advantages and disadvantages of each.

Scenario 10

You are responsible for a high-volume air sampler located downwind from a Department of Energy (DOE) facility following a suspected release of Pu-239. The air sampler has a calibrated volumetric flow rate of 55 SCFM, and the filter has an alpha self-absorption factor and filter collection efficiency of 0.4 and 0.8, respectively. The air sampler is operated at this flow rate for 1 hr, and the filter surface is measured with a gross alpha probe detector having an active detection area of 60 cm^2 and a background count of 20 counts in 100 min. The detector efficiency for alpha is 0.3 cpm/dpm, and the active filter area is 500 cm^2. Assume that the filter face velocity is uniform.

Data

Half-Life for Pu-239 = 24,390 years

Alpha yield = 100%

LLD(95%) = 4.66 sb (where sb is the standard deviation

of the background)

1.33. The initial filter-face alpha count immediately after the 1 hr sampling period was 2000 for a 10-min counting interval. Forty-eight hours later, the same filter is measured again with the same detector, and the count was 220 in 100 min. Explain why the count rate is lower 48 hr later.

1.34. What is Pu-239 airborne activity (in dpm/m^3) and the standard deviation for this measured quantity?

1.35. What is the lower limit of detection (LLD) at the 95% confidence level for this air sampling and detection system (in dpm/m^3) for the same sampling conditions?

Scenario 11

This scenario deals with the working-level unit.

With the passage of the Radon Control Act of 1988, the Environmental Protection Agency (EPA) is now instructed by the Congress to assess public risks of radon exposure in public buildings (including schools) throughout the nation. Regarding the measurement, detection, and health physics of radon-222 and its daughter products, answer the following questions:

1.36. Historically, an operational definition of the working-level exposure unit (WL) for radon-222 daughters has been 100 pCi/liter of each short-lived daughter product in secular equilibrium. Using this definition and the data provided derive the total alpha energy per liter of air (MeV/liter) associated with a concentration of one working level. Radon and its short-lived daughters include:

Nuclide	Alpha Energy (MeV)	Half-life
Radon-222	5.49	3.82 days
Polonium-218	6.00	3.05 min
Lead-214	0	26.8 min
Bismuth-214	0	19.7 min
Polonium-214	7.68	1×10^{-6} min

1.37. Using the data provided, calculate the concentration of radon-222 gas in air determined from a single-count, filter collection method for radon daughters. Assume a 50% equilibrium between radon-222 and its daughters. Neglect special considerations for radioactive growth and decay during sampling and counting. The following data are provided:

Sample collection period = 5 min

Counting time = 1 min

Total alpha counts = 230

Counting efficiency = 0.3

Pump flow rate = 10 liters/min

Conversion factor = 150 dpm alpha liter^{-1} WL^{-1}

1.38. List common methods for the detection and measurement of radon and/or its daughters for use in assessing public exposure in building structures.

Scenario 12

A common type of portable beta–gamma survey instrument uses an air ionization chamber vented to atmospheric pressure. The cylindrical detector is 3 in. high and 3 in. in diameter with a 7-mg/cm^2 beta window and a 400-mg/cm^2 beta shield. The side walls are 600 mg/cm^2. Answer the following questions with respect to the instrument's response versus the "true" dose rates specifically associated with the following conditions.

1.39. Briefly describe a potential source of error associated with measuring gamma and beta dose rates while moving in and out of a noble gas environment.

1.40. List and briefly explain two harsh environmental conditions which could have an adverse effect on the accuracy of the instrument response while in the area.

1.41. Briefly describe the most significant source of error associated with measuring true beta and gamma surface dose rates from contact measurements of small sources.

1.42. Briefly explain a source of error associated with measuring beta dose rates from large-area sources, with each source comprised of a different radionuclide.

1.43. Briefly describe a source of error associated with measuring beta dose rates from high-energy beta sources using open minus closed window readings.

Scenario 13

ANSI N13.11-1983, "American National Standard for Dosimetry—Personal Dosimetry Performance Criteria for Testing," is used as a basis for testing the performance of suppliers of dosimetry services. This standard provides criteria for testing personnel dosimetry performance for any type of dosimeter whose reading is used to provide a lifetime cumulative personal radiation record. The test procedure in this standard evaluates the absorbed dose and dose equivalent at two irradiation depths (0.007 cm and 1.0 cm). The radiation sources used for the performance tests are Cs-137, Sr-90/Y-90, heavy water moderated Cf-252, and an x-ray machine. The x-ray machine is used to generate several photon beams with average energies between 20 keV and 70 keV. Choose the single answer which is most correct.

14 INTRODUCTION

1.44. The provisions of this standard apply:
   a. to neither pocket dosimeters nor extremity dosimeters.
   b. to pocket dosimeters but not to extremity dosimeters.
   c. only to beta and gamma radiation.
   d. to extremity dosimeters but not to pocket dosimeters.
   e. to film badges but not to thermoluminescent dosimeters (TLDs).

1.45. Because of the particular irradiation depths chosen for the tests, a dosimetry system which is calibrated with the standard tests may be reporting doses which are different than the actual dose received. For which of the following tissues (red bone marrow, skin, gonads, lens of the eye, or whole body) is this difference most significant?

1.46. Because of the particular radiation sources specified, the standard least adequately tests for radiations emitted by:
   a. C-14, power reactor leakage neutrons
   b. P-32, Cf-252
   c. Y-90/Sr-90, Am–Be source
   d. Co-60, Ni-65
   e. Uranium slab, Cf-252

1.47. A dosimeter of a processor who has passed the test category for:
   a. beta radiation is appropriate for measuring low-energy photons.
   b. beta radiation is not appropriate for measuring beta radiation from all sources.
   c. low-energy photons can be used to pass the performance test for beta radiation.
   d. high-energy photons and the category for low-energy photons can be assumed to pass the test for mixtures of high-energy and low-energy photons.
   e. neutrons is appropriate for measuring neutron radiation from any source.

1.48. This standard:
   a. forms the basis for the National Voluntary Laboratory Accreditation Program for dosimetry processors.
   b. provides guidance for individual variability from reference man.
   c. provides guidance for summing the internal and external dose.
   d. is applicable to the entire range of gamma energies.
   e. is not required to be implemented by 10 CFR 20.

Scenario 14

For each of the situations below (1.49 to 1.53), select the personnel dosimeter which is most suitable for the purpose of establishing primary dose records. In each case substantiate your choice of dosimeter. Limit your choice of dosimeter to the following:

1. A common film badge with 300 mg/cm^2 plastic filtration over all areas except for the 14-mg/cm^2 mylar window.
2. A TLD albedo containing both Li-6 and Li-7 elements.
3. A TLD albedo containing both Li-6 and B-11 elements.
4. A calcium sulfate, manganese-activated TLD element in a tissue equivalent holder.
5. A proton recoil film badge.
6. A four-element TLD with lithium borate phosphors, 300-mg/cm^2 plastic filtration over two elements, aluminum over the third element, and lead over the fourth element.
7. A four-element TLD with lithium borate phosphors, a thin mylar filter over one element, plastic filters over two elements, and an aluminum filter over the fourth element.
8. A natural LiF TLD element.
9. A calcium sulfate, dysprosium-activated TLD element in a tissue equivalent holder.
10. A two-element TLD with lithium borate phosphors and 300-mg/cm^2 plastic filters.

1.49. An accelerator facility using tritiated targets with 14-MeV deuteron beams.
1.50. A mixed neutron and gamma field where gamma dose predominates.
1.51. A radiographer using a 320-kVp x-ray machine.
1.52. A field of high-energy, 6-MeV photons.
1.53. A field of mixed beta (average energy of 200 keV) and gamma (average energy of 800 keV) radiation.

Scenario 15

You supervise an in-house TLD system for occupationally exposed workers. The TLD badge consists of two LiF chips of 235-mg/cm^2 thickness. Chip 1 is covered by 7 mg/cm^2 of plastic, and Chip 2 is shielded by 850 mg/cm^2 of lead and 150 mg/cm^2 of plastic. The TLD system is calibrated by exposing badges to known quantities of beta and gamma radiations and plotting the TL reader output versus mrem dose equivalent. Both the gamma and beta calibration curves are linear and pass through the origin (0, 0) on the graph. The gamma calibration curve indicates that 6000 TL units equals 500 mrem of gamma dose equivalent, and the beta curve yields 750 TL units per 1000 mrem of beta dose equivalent.

16 INTRODUCTION

The following data are provided:

1. The control dosimeter reads 120 TL units on both Chips 1 and 2. (Both Chips have the same gamma sensitivity.)
2. Chip 1 = 12,270 TL units
   Chip 2 = 11,520 TL units
3. The beta calibration curve for other tissue depths includes the following:

Tissue Depth (mg/cm^2)	Percentage of dose equivalent at 7 mg/cm^2
100	50
300	25
500	10
1000	1

4. The gamma dose equivalent remains constant at tissue depths from 7 to 1000 mg/cm^2.

1.54. Calculate the skin and whole-body dose equivalents for the exposed TLDs noted above in item 2.
1.55. Calculate the dose to the lens of the eye.
1.56. Explain if any dose limits were exceeded. Justify your answer by stating the limits and identifying the source of the limits that you applied.

Scenario 16

A facility is in the process of setting up a neutron dosimetry program. You have been asked to consult on this matter. The following dosimeters are under consideration: TLD, recoil track-etch, neutron track type A (NTA) film, and bubble detectors. A final option is to use stay time calculations based on survey results from a "rem-ball" that has been calibrated using D_2O-moderated Cf-252.

1.57. Which one of the following statements is incorrect?
   a. There is no neutron dosimetry system in use today that is adequate ($\pm 50\%$ of the true dose equivalent) for all situations where neutron dosimetry is required.
   b. The neutron quality factor between 0 and 20 MeV is relatively constant at a value of about 10.
   c. Neutron energies can span nine decades in some monitoring situations.

d. Neutron monitoring is usually performed in a mixed field of neutron and gamma radiation.

e. Stay-time calculations, though often used, may be unreliable due to variations of neutron dose rates and energies in a given neutron radiation area.

1.58. In a field of mixed neutron and gamma radiation, the gamma dose measured on a phantom is:

a. greater than the gamma dose measured in air due to the $H(n, \gamma)$-D reaction in the phantom.

b. less than the dose measured in air due to the moderation of neutrons in the phantom.

c. the same as the measured dose in air because phantoms do not influence gamma irradiation.

d. less than the dose measured in air because some incident gamma rays are absorbed in the phantom.

e. not a quantity of interest in a dosimetry program.

1.59. If no corrections are made to the dosimeter response for neutron energy, TLD albedo dosimeters calibrated with a bare Cf-252 source will:

a. give accurate indications ($\pm 50\%$) of neutron dose equivalent in soft (thermal or epithermal) spectra.

b. underestimate the neutron dose equivalent by as much as a factor of 2 in soft (thermal or epithermal) spectra.

c. overestimate the neutron dose equivalent regardless of the incident spectrum.

d. underestimate the neutron dose equivalent regardless of the incident spectrum.

e. overestimate the neutron dose equivalent in soft (thermal or epithermal) spectra.

1.60. Which one of the following statements is true regarding neutron bubble detectors?

a. They are insensitive to intermediate-energy neutrons.

b. They are accurate within $\pm 30\%$ in neutron dose rates of over 1000 rad/hr.

c. They are affected by temperature.

d. They cannot measure the total integrated dose.

e. They are not yet commercially available.

1.61. Which of the following choices would most accurately measure the neutron dose equivalent for commercial power reactor containment entries?

a. A TLD albedo dosimetry system calibrated to D_2O-moderated Cf-252.

b. A TLD albedo dosimetry system calibrated to AmBe.
c. A proton-recoil dosimetry system calibrated to D_2O-moderated Cf-252.
d. A proton-recoil dosimetry system calibrated to AmBe.
e. An NTA film dosimetry system calibrated to D_2O-moderated Cf-252.

Scenario 17

A large community hospital wishes you to set up a personnel monitoring program. The following organization information is provided:

Department A: The nuclear medicine department is a well-equipped department using technetium-99m for all its studies. The Tc-99m is milked from a generator, and the radiopharmaceuticals are prepared within the nuclear medicine department. The department has sealed sources of cobalt-57, cesium-137, and barium-133 for calibrating the dose calibrator.

Department B: The x-ray department is an active group using fluoroscopic procedures, general diagnostic x-ray procedures, and some special procedures.

Department C: The radiation therapy department is an active group using a Co-60 teletherapy device and a 4.0-MeV linear accelerator, but no brachytherapy.

Department D: The research department is a fairly active department using only hydrogen-3 and carbon-14.

1.62. What departments will require personnel monitoring for photons?
1.63. What department will require personnel monitoring for neutrons?
1.64. What departments will benefit from both a personnel monitor at the belt (under leaded apron) and one at the collar?
1.65. What department would need ring badges?
1.66. In what department would the assessment of skin dose be important?
1.67. What department might require bioassay?
1.68 List positive characteristics of film dosimeters for personnel monitoring.
1.69. List negative characteristics of film dosimeters for personnel monitoring.
1.70. List positive characteristics of TLDs for personnel monitoring.
1.71. List negative characteristics of TLDs for personnel monitoring.

Scenario 18

This scenario involves the properties of gas-filled detectors.

Data

$$\text{Air density} = 1.29 \text{ kg/m}^3 \text{ at STP}$$
$$1 \text{ torr} = 1 \text{ mm Hg at } 0°C$$

1.72. Consider two cylindrical gas ionization chambers, A and B. The chamber of detector A has the dimensions 0.5 cm in diameter and 5 cm in height. Detector B has the dimensions 1.0 cm in diameter and 5 cm in height. Both detectors have the same chamber wall material and thickness, fill gas, and chamber pressure. If detector A shows an output current of 1.0×10^{-10} A when placed in an isotropic gamma field, what theoretical response should be given by detector B when placed in the same field? Neglect detector end effects.
  a. 2.5×10^{-11} A
  b. 4.0×10^{-10} A
  c. 2.0×10^{-10} A
  d. 5.0×10^{-11} A
  e. 1.0×10^{-10} A

1.73. The gas fill pressure in detector A is 7600 torr, and the detector sensitivity is 1.2×10^{-10} A-hr/R. What would the detector sensitivity be if the gas fill pressure was increased to 11,400 torr?

1.74. Assuming a chamber pressure of 7600 torr, a chamber volume of 100 cm^3, and a temperature of 20°C, calculate the exposure rate in R/hr for an air-equivalent wall ion chamber if the saturated ion current is 9.0×10^{-14} A.

1.75. An ambient-pressure air ion chamber is calibrated at 7000-feet altitude in New Mexico at 20°C, 591.6-torr air pressure to read correctly under those conditions. What exposure will it indicate in a 100-mR/hr field at sea level in the Marshall Islands at 36°C, 760.0-torr air pressure?
  a. 74 mR/hr
  b. 82 mR/hr
  c. 100 mR/hr
  d. 122 mR/hr
  e. 136 mR/hr

Scenario 19

You are the station health physicist at a nuclear power station. The Chemistry Manager has asked you to review a purchase requisition for an N-16 calibration source. The source generates N-16 via an (α, p) reaction involving 160 mCi of curium-244 and carbon-13. The source gamma emission strength is 2.2 ×

10^6 gammas/sec, and the neutron emission strength is 2.0×10^5 neutrons/sec. Assume a gamma energy of 6.1 MeV and an average neutron energy of 2.5 MeV. The following information is provided for your evaluation:

Physical Quantity (6.1 MeV)	Water	Air	Muscle	Lead
Density (g/cm³)	1.00	0.001293	1.0400	11.35
Mass-energy absorption coefficient (cm²/g)	0.0180	0.0163	0.0178	—
Mass-attenuation coefficient (cm²/g)	0.0277	0.0252	0.0274	0.0435

Point Source Dose Buildup Factors in Lead (ux)

Energy = 6.0 MeV	1	2	4	7	10	15	20
	1.18	1.40	1.97	3.34	5.69	13.8	32.7

The neutron flux to dose equivalent (k) at 2.5 MeV is $k = 20$ n/cm²-sec = 2.5 mrem/hr.

1.76. Calculate the total gamma dose equivalent rate at 1 ft. Assume a 100% emission rate from the principal gamma peak.

1.77. Calculate the total neutron dose equivalent rate at 1 ft.

1.78. Lead and polyethylene are available to shield the source. How would you arrange these materials to yield the lowest overall dose rate?

   a. Polyethylene followed by lead
   b. Lead only
   c. Polyethylene only
   d. Lead followed by polyethylene
   e. No shielding is necessary because the 12-in. air gap will sufficiently scatter/attenuate the neutrons.

1.79. What is the shielding requirement (cm of lead) to reduce the gamma dose rate at 1 ft by a factor of 5?

Scenario 20

As the HP supervisor at a reactor decommissioning project, the project engineer has asked you to assist in the evaluation of methods to reduce radiation levels emanating from a neutron-activated concrete shield to meet release limits for unrestricted use. The preferred method requires you to predict the depth to which a slab of neutron-activated concrete must be excavated to allow free release. Other methods she has asked to be evaluated include delayed decommissioning and the addition of shielding.

Assume that the neutron relaxation length in concrete is 15 cm. The current exposure rate is 20 μR/hr 1 m from the slab. The only applicable release limit is 5 μR/hr 1 m from the surface. The concrete source term, based upon a single concrete core sample 1 in. deep, is as follows:

Activation Data for Concrete Source Term

Nuclide	Decay Mode	Radiations Energy (MeV)	$T_{1/2}$	Specific Activity (pCi/g)	Gamma Constant (R/Ci-hr @ 1 m)
H-3	Beta	0.0186	12.3 years	1000	—
C-14	Beta	0.156	5730 years	500	—
Mn-54	Gamma	0.835	312 days	2500	0.47
Co-60	Gamma	1.332	5.27 years	2500	1.32
	Gamma	1.173			
	Beta	0.314			

1.80. For each of the following three methods for meeting the release limit, list two advantages and two disadvantages:
1. Time to allow decay
2. Immediate removal
3. Add shielding

1.81. Based upon the data provided, estimate the depth of the excavation required to allow free release.

1.82. Assuming that no concrete removal occurs, predict the time necessary to allow the principal radionuclide of interest, Co-60, to decay to the release limit. Assume that the 20 μR/hr exposure rate is due solely to the Co-60.

1.83. How much more concrete shielding would be needed to reduce the exposure rate at 1 m to the release limit? Neglect the geometry considerations. The mass attenuation coefficient is 0.06 cm^2/g, and the density of concrete is 2.5 g/cm^3. Buildup is assumed to be a constant factor of 2.

Scenario 21

You are involved in an assessment of the results of an activation experiment that produced Na-24. The buildup and decay of this source and the resultant dose rates require your attention. Answer the following questions regarding the shielding and activation of the Na-24 source.

1.84. What is the flux in particles per square centimeter per second which will produce 4.0×10^7 Bq of Na-24 at saturation in an aluminum target of 1-cm^2 cross section and 1-g weight? Assume that the production cross section is 20 mb.

Data

$$\text{Atomic weight of aluminum} = 27$$
$$\text{Avogadro's number} = 6.02 \times 10^{23}$$
$$1 \text{ barn} = 1.0 \times 10^{-24} \text{ cm}^2$$

1.85. Immediately after an irradiation time of 30 hr, what would be the amount of Na-24 present? Assume no initial activity. The half-life of Na-24 is 15 hr.

1.86. What is the dose equivalent rate to a person standing 1 m from the unshielded Na-24 source in air. The following information should be considered in your answer:

$$\text{Gamma 1} = 1.4 \text{ MeV @ } 100\%$$
$$\text{Gamma 2} = 2.8 \text{ MeV @ } 100\%$$
$$\text{Air density} = 0.00129 \text{ g/cm}^3$$
$$\text{Air attenuation coefficient} = 2.3 \times 10^{-5} \text{ cm}^{-1}$$
$$\text{Assume 1 mrem} = 1 \text{ mrad}$$
$$1 \text{ MeV} = 1.6 \times 10^{-6} \text{ erg}$$

Scenario 22

You are a consulting health physicist. A client plans to build a 10-MCi Co-60 irradiation facility to sterilize surgical equipment. You may assume that the activity is in the form of a point source. The following data may be useful:

$$\text{Co-60 gamma constant} = 13.2 \text{ R-cm}^2 \text{ hr}^{-1} \text{ mCi}^{-1}$$

Table of Linear Attenuation Coefficients and Fluence Buildup Factors for Concrete

Energy (MeV)	u (1/cm)	B ux = 7	ux = 10	ux = 15	ux = 20
0.5	0.204	16.6	29.0	58.1	98.3
1.0	0.149	11.7	18.7	33.1	50.6
1.17	0.140	11.0	17.5	30.6	46.4
1.25	0.135	10.7	16.9	29.4	44.4
1.33	0.130	10.4	16.3	28.2	42.4
1.5	0.121	9.7	15.0	25.7	38.2

1.87. What would be the exposure rate at a distance of 3 m from the unshielded 10-MCi Co-60 source?

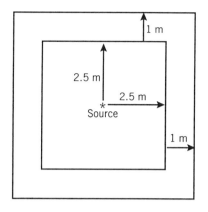

Fig. 1.2 Proposed irradiation facility floor plan.

1.88. When exposed, the source will be in the center of a room having internal dimensions of 5 m × 5 m × 5 m with walls of 1-m-thick concrete. The room layout is illustrated in Fig. 1.2. Based upon this information, calculate the maximum photon fluence rate in photons cm^{-2} sec^{-1} at a point on the exterior surface of the shield wall. State any assumptions used. Ignore scatter off air or walls other than the wall between the source and the reference point.

1.89 The buildup factor should only be used:
   a. for photons of energy below 3 MeV.
   b. for photons of energy above 0.5 MeV.
   c. in cases where the shield thickness exceeds 3 relaxation lengths.
   d. for situations involving "broad-beam" or "poor" geometry.
   e. for situations involving "narrow-beam" or "good" geometry.

Scenario 23

Consider both broad and narrow beams of 1-MeV photons, illustrated in Fig. 1.3, that are normally incident on different thicknesses of uranium slabs. The measured radiation levels for three different thicknesses (x) are given below for both the broad-beam and narrow-beam situations. The following data are provided:

$$\text{Density of uranium} = 18.9 \text{ g/cm}^3$$

Measured Radiation Levels for Various Thicknesses of Uranium

Slab Thickness (cm)	Broad Beam (mR/hr)	Narrow Beam (mR/hr)
0.0	127.0	127.0
1.0	43.1	29.5
2.0	13.0	7.7
3.0	4.0	1.9

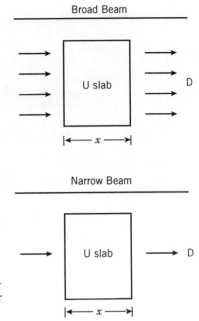

Fig. 1.3 Geometry for broad and narrow photon beam scattering experiments. The detector position is indicated by the letter "D."

From these data, determine:

1.90. The linear attenuation coefficient of uranium for the narrow beam of 1-MeV photons.
1.91. The buildup factor for the broad beam with a slab thickness of 2.5 cm. Assume the linear attenuation coefficient is 2/cm.
1.92. What is the mass-attenuation coefficient of uranium for 1-MeV photons if the linear attenuation coefficient is 2/cm?

Scenario 24

The International Commission on Radiation Protection (ICRP) publishes various reports on basic radiation protection policy, practices, and research. Two such reports, Report Number 23 and Report Number 26, are of interest in this question.

1.93. ICRP Report Number 23 describes reference man as containing 140 g of potassium. The following data apply:

Mass of reference man = 70,000 g
0.012% of the potassium is K-40
K-40 decays by emitting a beta particle with a 90% probability

Maximum beta energy = 1.3 MeV
Half-life of K-40 = 1.2×10^9 years
Avogadro's number = 6.023×10^{23}
1.6021×10^{-6} erg/MeV

What is the average beta dose rate in rad/week to the whole body from K-40?

1.94. Which statement is most accurate?
   a. It is hard to identify K-40 in the presence of 10 nCi of Co-60.
   b. The quantity of K-40 does not vary by more than ±5% from individual to individual.
   c. K-40 has no regulatory significance in the whole-body counting program but serves as an important qualitative system check.
   d. K-40 should be omitted from the radionuclide library for the whole-body counting because it is of no regulatory interest.
   e. A multidetector counter will typically not identify K-40.

1.95. Strict adherence to ICRP Report Number 26 would allow:
   a. plutonium internal doses to be regulated using annual dose equivalent rather than committed dose equivalent.
   b. deletion of record keeping for internal doses less than 50% of the allowable dose limit.
   c. consideration of internal and external dose limits separately.
   d. the worker to choose the type of respiratory protection device if use is required.
   e. use of air samples and stay-time calculations instead of respirator usage, if it is deemed to be ALARA.

1.96. The assumption of electronic equilibrium for a Co-60 source at 1-m distance is least likely to be correct at the:
   a. surface of the skin.
   b. center of a large muscle mass.
   c. bone–tissue interface.
   d. center of a large bone mass.
   e. internal surface of the lung.

Scenario 25

A worker at your facility received a diagnostic administration of I-131 as NaI for a thyroid function test. Your radiation protection program restricts workers to 0.1 times the ICRP-10 investigation level from certain work even if the exposure resulted from a medical procedure.

You are asked to estimate how long he must be placed on a restricted status.

26 INTRODUCTION

He is also concerned about the dose that he will receive from this diagnostic procedure. Use the following data to answer the questions for this scenario:

Administered activity = 1.0 μCi

Adminstrative limit (0.1 times the ICRP-10 investigation level) = 30 nCi

Biological half-life (thyroid) = 74 days

Biological half-life (whole body) = 0.4 days

Physical half-life = 8.08 days

$f_2 = 0.3$

Thyroid mass = 20 g

$S(T \leftarrow S)$ for the thyroid = 2.2×10^{-2} rad/μCi-hr (MIRD-11)

1.97. Calculate the thyroid dose received from this procedure.
1.98. Based on the thyroid retention, how many days must pass until the worker can be released from restricted status?
1.99. Which one of the following statements is incorrect?
   a. Because a thyroid abnormality is suspected, these calculations are only an estimate of the organ dose.
   b. The most accurate method of assessing the actual dose is to obtain *in vivo* bioassay data and calculate an organ retention function.
   c. The values of S take beta dose within the organ of interest into account, but do not consider beta doses between organs except for organs with walls and bone and bone marrow.
   d. Because of the reciprocal dose theorem, the dose to testes from the thyroid is equal to the dose to the thyroid that would be produced if the same activity were in the testes.
   e. The ICRP-10-derived investigation level for short-lived transportable radionuclides is based on one-quarter of the maximum permissible quarterly intake for short-lived transportable radionuclides.

Scenario 26

You are responsible for the health physics input to the design of a new laboratory which will be handling tritium. One room in the lab is 30 ft long, 20 ft wide, and 10 ft high and contains the primary tritium handling glove box. The glove box will contain a maximum of 10 Ci of tritium, all of which could be released into the room should an accident occur. You are concerned about the doses that could be received by an operator in the room and by an individual standing downwind at the site boundary, 1 mile away. The following data should be considered:

Breathing rate of both individuals = 3.5×10^{-4} m³/sec
Atmospheric diffusion factor at 1 mile = 1.0×10^{-4} sec/m³
Dose conversion factor for tritium (including absorption through the skin) = 158 rem/Ci inhaled
Time the operator remains in the room after the accident without respiratory protection = 30 min
Assume that the position of the operator in the room does not affect the dose received.

1.100. Determine the maximum dose that could be delivered to the operator.
1.101. Calculate the maximum dose that could be delivered to the person at the site boundary. Assume that all tritium is released to the environment in 30 min and that the person stays at the boundary for the entire release.
1.102. The ventilation system design criteria call for three complete air changes per hour in the lab. If the ventilation system works as designed, what is the maximum dose the operator could receive?
1.103. If the ventilation system works as designed, what is the maximum dose that could be delivered to the person at the site boundary?
1.104. Your design goal for the dose equivalent delivered to the operator during the tritium accident is 500 mrem. How many air changes per hour will be required to limit the operator's dose equivalent to this value?

Scenario 27

1.105. Calculate the committed dose equivalent (CDE), the committed effective dose equivalent (CEDE), the annual limit on intake, and the derived air concentration for the inhalation of Cs-137. The Cs-137 decay scheme is illustrated in Fig. 1.4. The following data are given:

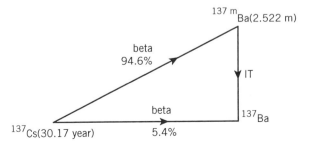

Fig. 1.4 Decay scheme for Cs-137.

Specific Effective Energy (MeV per gram per transformation) of Cs-137

Targets	Sources	
	Lungs	Total Body
Gonads	0.0	2.7×10^{-6}
Breast	0.0	2.7×10^{-6}
Red Marrow	0.0	2.7×10^{-6}
Lungs	1.9×10^{-4}	2.7×10^{-6}
Thyroid	0.0	2.7×10^{-6}
Bone surface	0.0	2.7×10^{-6}
SI wall	0.0	2.7×10^{-6}
ULI wall	0.0	2.7×10^{-6}
LLI wall	0.0	2.7×10^{-6}
Uterus	0.0	2.7×10^{-6}
Adrenals	0.0	2.7×10^{-6}

Specific Effective Energy (MeV per gram per transformation) of Ba-137m

Targets	Sources	
	Lungs	Total Body
Gonads	5.7×10^{-8}	4.7×10^{-6}
Breast	2.7×10^{-6}	3.9×10^{-6}
Red marrow	2.5×10^{-6}	4.3×10^{-6}
Lungs	9.5×10^{-5}	4.0×10^{-6}
Thyroid	2.6×10^{-6}	3.9×10^{-6}
Bone surface	2.0×10^{-6}	4.0×10^{-6}
SI wall	5.9×10^{-7}	4.9×10^{-6}
ULI wall	8.0×10^{-7}	4.8×10^{-6}
LLI wall	1.7×10^{-7}	4.9×10^{-6}
Uterus	2.1×10^{-7}	4.9×10^{-6}
Adrenals	4.9×10^{-6}	5.2×10^{-6}

Number of Nuclear Transformations Over 50 Years in Source Organs or Tissues per Unit Intake of Activity (Transformations/Bq) of Cs-137 (U_S)

Organ	Isotope	Oral	Inhalation (Class D)
	Cs-137	$f_1 = 1.0$	$f_1 = 1.0$
	Ba-137m	$f_1 = 1.0$	$f_1 = 1.0$
Lungs	Cs-137		1.9×10^4
	Ba-137m		1.8×10^4
Other tissue	Cs-137	1.2×10^7	7.7×10^6
(whole body 70,000 g)	Ba-137m	1.2×10^7	7.3×10^6

f_1 is the fraction of a stable element reaching the body fluids following ingestion or inhalation.

Scenario 28

Assuming that Tc-99m acts as an insoluble compound, calculate the following for an uptake of 1 μCi of Tc-99m into the stomach:

1.106. The cumulated activity of Tc-99m in μCi-hr in each segment of the gastrointestinal (GI) tract.
1.107. The dose equivalent in rem to the walls of each segment of the GI tract.
1.108. The maximum permissible uptake rate in μCi/hr and the maximum allowed concentration in water for occupational exposure in μCi/ml. Assume that an intake of 1100 ml/day of contaminated water is consumed during the work day and that the maximum allowed organ dose permitted at your facility is 15 rem.

GI Tract Parameters

Section of GI Tract	Mass of walls (g)	Mass of contents (g)	Mean Residence Time (day)	λ (1/day)
Stomach (ST)	150	250	1/24	24
Small intestine (SI)	640	400	4/24	6
Upper large intestine (ULI)	210	220	13/24	1.8
Lower large intestine (LLI)	160	135	24/24	1

Absorbed Dose per Unit Cumulated Activity (rad/μCi-hr) for Tc-99m with a Half-Life of 6.03 hr

| Target Organs | Source Organs | | | |
	Stomach Contents	SI Contents	ULI Contents	LLI Contents
GI stomach wall	1.3×10^{-4}	3.7×10^{-6}	3.8×10^{-6}	1.8×10^{-6}
GI SI wall	2.7×10^{-6}	7.8×10^{-5}	1.7×10^{-5}	9.4×10^{-6}
GI ULI wall	3.5×10^{-6}	2.4×10^{-5}	1.3×10^{-4}	4.2×10^{-6}
GI LLI wall	1.2×10^{-6}	7.3×10^{-6}	3.2×10^{-6}	1.9×10^{-4}

Scenario 29

Two possible approaches for estimating the risk of cancer induction from exposure to low levels of ionizing radiation are ICRP-26 and the Probability of Causation (PC) Tables published by the Department of Health and Human

30 INTRODUCTION

Services (HHS). Answer the following questions to demonstrate your understanding of these reports.

1.109. Based upon ICRP risk estimates, what is the probability of developing a radiation-induced fatal cancer over a lifetime for an average occupationally exposed radiation worker who has received 100,000 mrem of uniform, whole-body external exposure?

1.110. Assuming a normal cancer fatality rate of 20%, what would be the total probability of developing a fatal cancer for a group of occupationally exposed workers with a 3 in 1000 probability of contracting a radiation-induced fatal cancer?

1.111. The ICRP-26 risk model for cancer is based on:
   a. An absolute risk model
   b. A relative risk model
   c. An absolute and relative risk model
   d. A stochastic model
   e. A linear stochastic model

1.112. The PC Tables are based on:
   a. An absolute risk model
   b. A relative risk model
   c. An absolute and relative risk model
   d. A stochastic model
   e. A linear stochastic model

1.113. Which statement is not true regarding the PC tables?
   a. The formulation of these tables was mandated by Congress.
   b. Smoking history is not considered when using the tables to estimate risk.
   c. Source of data for the table include: rodent data, *in vitro* cell studies, and human data.
   d. The tables were published to provide scientific evidence to resolve radiation litigation cases.
   e. Prior medical x-ray exposure is not considered when using the tables to estimate risk.

Scenario 30

The biological effects of ionizing radiation encompass a broad range of topics. The following questions are designed to indicate your general understanding of this area.

1.114. Equal amounts of tritium as tritiated water and tritiated thymidine (a basic component of DNA) are incorporated into a large volume of cells. Which statement best describes the biological effectiveness of these compounds?

a. Tritiated water will cause more biological damage to the cell because the cell is principally made up of water.
b. Tritiated water will cause more biological damage to the cell because tritiated thymidine is quickly metabolized by the cell.
c. Tritiated thymidine will cause a greater biological effect than tritiated water because it is incorporated into the cell's nucleus.
d. Both compounds will deliver the same biological effect because they are distributed in equal activities.
e. The biological effect will be the same for both compounds because both emit the same low-energy beta radiation and are equal in activity.

1.115. Match the following inhaled radionuclides with the adult critical organ. The critical organ may be used more than once.
   a. Lung   _____ Strontium-90 (soluble)
   b. Bone   _____ Cesium-137
   c. Total body   _____ Plutonium-239 (soluble)
   d. Liver   _____ Uranium-238 (insoluble)
   e. Kidney   _____ Radon-222

1.116. Based on the law of Bergonie and Tribondeau, order the following cells from most to least radiosensitive:
   a. Mature lymphocytes
   b. Intestinal crypt cells
   c. Mature spermatocytes
   d. Erythrocytes (red blood cells)
   e. Nerve cells

1.117. Ionizing radiation has been directly associated with cataract formation. Select the statement that is incorrect.
   a. The cataractogenic dose response is considered a threshold effect.
   b. Fast neutrons are more effective at producing cataracts than are other forms of radiation.
   c. The cataract effect is dependent on age at the time of irradiation.
   d. Occupational exposure to x-rays accounts for approximately 1% of the cataracts observed in x-ray technicians.
   e. Radiogenic cataracts are distinct in that they originate on the anterior epithelium of the lens.

1.118. Figure 1.5 can be used to express cell survival under a number of different irradiation circumstances. Which of the following statements is not true?
   a. Curve A best represents the response of a cell system to a high dose rate, whereas curve B best represents the response to a low dose rate.

32 INTRODUCTION

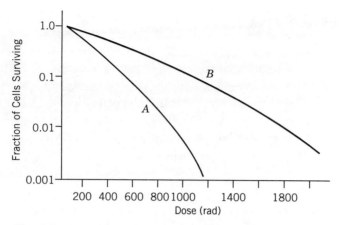

Fig. 1.5 Fractional cell survival curves as a function of dose.

- b. Curve B best represents a multitarget cell system response, whereas curve A best represents a single-target system.
- c. Curve A best represents the effect in a cell system that is irradiated under hypoxic conditions, whereas curve B best represents the response of the same system under aerated conditions.
- d. Curve B best represents the response of a cell system to low LET radiation, whereas curve A best represents the response of the same system to high LET radiation.
- e. Curve B best represents the response of a cell system when a radioprotective compound is used, whereas curve A best represents the response of that cell system without a radioprotective compound.

Scenario 31

The following series of questions relates to dosimetry and dose limits.

1.119. The major pathway by which soluble radioactive material is removed from the body is
   a. Perspiration
   b. Feces
   c. Respiration
   d. Exhalation
   e. Urine

1.120. Prior to January 1993, the internal dose assessment methodology used to meet regulatory requirements in Title 10 CFR Part 20 is:
   a. NCRP-84
   b. NCRP-91
   c. ICRP-30

d. ICRP-2

e. ICRP-26

1.121. The ICRP (ICRP-26) determined that to prevent nonstochastic effects, the annual dose equivalent limit which must not be exceeded for all tissues except the lens of the eye is:
   a. 30.0 rem
   b. 5.0 rem
   c. 5.0 Sv
   d. 0.5 Sv
   e. 30 rad

1.122. In order to limit stochastic effects, the dose limit recommended in ICRP 26 is based on the principle that the risk associated with uniform irradiation of the whole body is
   a. equal to
   b. greater than
   c. less than
   d. related to
   e. not related to
   the risk associated with nonuniform radiation.

1.123. ICRP 26 replaces the ICRP 2 concept of the critical organ with the concept of:
   a. Genetic region
   b. Source region
   c. Organ equivalent
   d. Tissue region or target tissue
   e. Weighted critical organ

1.124. Most large data sets of measurable occupational annual dose equivalents have been found to fit a:
   a. Poisson distribution
   b. Normal distribution
   c. Log-normal distribution
   d. Binomial distribution
   e. Weibull distribution

1.125. In 1980, the ICRP reviewed its annual dose limitation recommendations on the lens of the eye. They decided to:
   a. change its recommendations from 0.3 Sv to 0.15 Sv.
   b. change its recommendations from 0.3 Sv to 0.50 Sv.
   c. leave the number unchanged.
   d. drop its recommendations for the lens of the eye.
   e. make the eye limitations 0.50% of that for other tissues.

34 INTRODUCTION

1.126. NCRP 91 recommends that the cumulative limit for occupational dose equivalent given by $5 \times (N - 18)$ rem be:
   a. changed to $1 \times N$ rem (N is in years).
   b. changed to $2 \times N$ rem (N is in years).
   c. increased by 50%.
   d. decreased by 50%.
   e. changes to $10 \times (N - 18)$ rem (N is in years).

1.127. Assuming an average annual dose equivalent of 500 mrem and employing BEIR III methodology, the annual risk estimate (fatal cancer) for occupational radiation workers is considered to be a nominal value of about:
   a. 1×10^{-2}
   b. 1×10^{-3}
   c. 1×10^{-4}
   d. 1×10^{-5}
   e. 1×10^{-6}

   How would your answer change if BEIR V were the basis of the risk coefficient?

1.128. The average annual fatal accident rate in safe industries in the United States is approximately:
   a. 1×10^{-3}
   b. 1×10^{-4}
   c. 1×10^{-5}
   d. 1×10^{-6}
   e. 1×10^{-7}

Scenario 32

The biological effects of ionizing radiation depend upon the tissues involved and the nature of the radiation impinging upon the cellular structures. One of the more radioresilient tissues is the skin of the whole body. This scenario addresses radiation effects on skin.

1.129. Which one of the following lists the skin response to acute radiation exposure in correct chronological order?
   a. Dry desquamation, moist desquamation, erythema, recovery
   b. Moist desquamation, dry desquamation, erythema, recovery
   c. Dry desquamation, moist desquamation, recovery, erythema
   d. Erythema, moist desquamation, dry desquamation, recovery
   e. Erythema, dry desquamation, moist desquamation, recovery

1.130. Which one of the following factors does not affect the severity of the skin's reaction to radiation?
   a. Skin pigmentation
   b. Fractionation of dose
   c. Charged particle equilibrium at the basal cell layer
   d. Dose rate
   e. LET

1.131. The radiosensitivity of skin is based on the sensitivity of which tissue?
   a. Epidermal layer
   b. Basal cell layer
   c. Horny layer
   d. Hair follicle
   e. Fat cells

1.132. If the skin were contaminated by an isotope with a half-life of 8 days, and assuming an exponential turnover time of the skin of 50% in 5 days, calculate the time to reduce the contaminant to 10% of the initial level. Assume that decontamination has been ineffective.

1.133 ICRP recommends a weighting factor of 0.01 for assessing stochastic risk to the skin. This means that:
   a. Radiogenic skin cancer is a low risk.
   b. Radiogenic skin cancer is a high risk.
   c. Spontaneous skin cancer is a low risk.
   d. Radiogenic skin cancer exceeds spontaneous skin cancer as a risk.
   e. Dose equivalent to the whole body is 100 times dose to the skin.

Scenario 33

The BEIR V report gives an analysis for cancer risk assessment from exposure to low levels of ionizing radiation. This report was preceded by a number of studies that addressed human risk estimation. The following questions address biological risk from ionizing radiation.

1.134. There are several important areas for which human data are inadequate for risk estimation. Provide the types of information from animal studies that can be useful in human risk estimation.

1.135. Initiation, promotion, and progression are three distinct stages of experimental radiation-induced carcinogenesis. Identify the radiobiological factors which affect either the onset or the development of malignant tumors in experimental animals.

1.136. Based upon available evidence, the weighted average risk of death from cancer following an acute dose equivalent of 0.1 Sv of low-LET radiation to all body organs is estimated to be:
   a. 1×10^{-4}
   b. 8×10^{-3}
   c. 2×10^{-3}
   d. 8×10^{-4}
   e. 1×10^{-2}

1.137. Two competing functional forms have been used for describing fatal cancer risks from radiation exposures. Choose the best statement:
   a. The BEIR V multiplicative risk model multiplies the dose by a constant to determine cancer risk.
   b. The BEIR IV additive risk model adds a constant times the underlying risk of cancer to the age- and sex-dependent radiation dose to determine cancer risk.
   c. The BEIR V relative risk model computes fatal cancer risk for individuals by an age- and sex-dependent factor times the cancer risk in the victim's relatives.
   d. The additive risk model has been dropped by the BEIR V Committee in favor of the multiplicative risk model.
   e. The BEIR IV absolute risk model holds that fatal cancer risk is a linear function of the absolute value of the radiation dose.

1.138. Consider a general form of a cancer risk estimate from BEIR V:

$$r(d) = r_0[1 + f(d)g(B)]$$

where

$r(d)$ = total risk

r_0 = background risk

$f(d)$ = function depending on the dose d

$g(B)$ = function of dose-modifying parameters B

Which of the following statements is not correct?
   a. The constant 1 ensures positive values of the excess risk estimate.
   b. $g(B)$ may include components which depend on sex and age.
   c. $f(d)$ can be a linear or linear-quadratic function.
   d. r_0 can vary significantly for different populations at risk.
   e. r_0 is not specifically modeled by the Committee.

REFERENCES

H. L. Andrews, *Radiation Biophysics*, Prentice-Hall, Englewood Cliffs, NJ (1962).

F. H. Attix, editor, *Topics in Radiation Dosimetry*, Academic Press, New York (1972).

F. H. Attix and W. C. Roesch, editors, *Radiation Dosimetry, Volume II: Instrumentation*, 2nd edition, Academic Press, New York (1966).

C. C. Brown and K. C. Chu, *Approaches to Epidemiologic Analysis of Prospective and Retrospective Studies, Epidemiology: Risk Assessment*, SIAM, Philadelphia, PA (1982).

A. P. Casarett, *Radiation Biology*, Prentice-Hall, Inc., Englewood Cliffs, NJ (1968).

H. Cember, *Introduction to Health Physics*, 2nd edition, Pergamon Press, Oxford, England (1983).

B. L. Cohen and I. Lee, A Catalog of Risks, *Health Physics* **36,** 707 (1979).

W. H. Ellett, Editor, *An Assessment of the New Dosimetry for A-Bomb Survivors*, National Research Council, National Academy Press, Washington, DC (1987).

R. D. Evans, *The Atomic Nucleus*, McGraw-Hill, New York (1970).

J. I. Fabrikant, *Radiobiology*, Year Book Medical Publishers, Chicago (1972).

E. F. Gloyna and J. O. Ledbetter, *Principles of Radiological Health*, Dekker, New York (1969).

H. Goldstein, *Fundamental Aspects of Reactor Shielding*, Addison-Wesley, Reading, MA (1959).

D. A. Gollnick, *Basic Radiation Protection Technology*, Pacific Radiation Press, Temple City, CA (1984).

D. S. Grosch, *Biological Effects of Radiation*, Blaisdell, New York (1965).

E. J. Hall, *Radiobiology for the Radiologist*, Harper & Row, 2nd edition, New York (1978).

G. J. Hine and G. L. Brownell, editors, *Radiation Dosimetry*, Academic Press, New York (1956).

ICRP Publication 2, *Permissible Dose for Internal Radiation*, Pergamon Press, Oxford, England (1959).

ICRP Publication 10, *Evaluation of Radiation Doses to Body Tissues from Internal Contamination Due to Occupational Exposure*, Pergamon Press, Oxford, England (1968).

ICRP Publication 10A, *The Assessment of Internal Contamination Resulting from Recurrent or Prolonged Uptakes*, Pergamon Press, Oxford, England (1971).

ICRP Publication 23, *Reference Man: Anatomical, Physiological, and Metabolic Characteristics*, Pergamon Press, Oxford, England (1975).

ICRP Publication 26, *Recommendations of the International Commission on Radiological Protection*, Pergamon Press, Oxford, England (1977).

ICRP Publication 30, *Limits for Intakes of Radionuclides by Workers*, Pergamon Press, Oxford, England (1979).

ICRP Publication 38, *Radionuclide Transformations: Energy and Intensity of Emissions*, Pergamon Press, Oxford, England (1983).

ICRP Publication 41, *Non-Stochastic Effects of Ionizing Radiation Pergamon Press*, Oxford, England (1984).

ICRP Publication 42, *A Compilation of the Major Concepts & Quantities in Use by ICRP*, Pergamon Press, Oxford, England (1984).

ICRP Publication 45, *Quantitative Bases for Developing a Unified Index of Harm*, Pergamon Press, Oxford, England (1986).

ICRP Publication 49, *Development Effects of Irradiation on the Brain of the Embryo & Fetus*, Pergamon Press, Oxford, England (1987).

ICRP Publication 51, *Data for Use in Protection Against External Radiation*, Pergamon Press, Oxford, England (1988).

ICRP Publication 54, *Individual Monitoring for Intakes of Radionuclides by Workers: Design and Interpretation*, Pergamon Press, Oxford, England (1988).

ICRP Publication 58, *RBE for Deterministic Effects*, Pergamon Press, Oxford, England (1990).

ICRP Publication 60, *1990 Recommendations of the ICRP*, Pergamon Press, Oxford, England (1991).

ICRP Publication 61, *Annual Limits on Intake of Radionuclides by Workers Based on the 1990 Recommendations*, Pergamon Press, Oxford, England (1991).

ICRU Report 20, *Radiation Protection Instrumentation and Its Application*, ICRU Publications, Bethesda, MD (1971).

ICRU Report 22, *Measurement of Low-Level Radioactivity*, ICRU Publications, Bethesda, MD (1972).

R. G. Jaeger, editor, *Engineering Compendium on Radiation Shielding*, Springer-Verlag, New York (1968).

R. L. Kathern, *Radiation Protection*, Adam Hilger, Ltd., Bristol, England (1985).

G. F. Knoll, *Radiation Detection and Measurement*, John Wiley & Sons, New York (1979).

D. C. Kocher, *Radioactive Decay Data Tables: A Handbook of Decay Data for Application to Radiation Dosimetry and Radiological Assessments*, USDOE Report DOE/TIC 11026, US Department of Energy, Springfield, VA (1981).

R. E. Lapp and H. L. Andrews, *Nuclear Radiation Physics*, Prentice–Hall, New York (1972).

R. Loevinger, T. F. Budinger, and E. E. Watson, *MIRD Primer for Absorbed Dose Calculations*, The Society of Nuclear Medicine, New York (1988).

K. Z. Morgan and J. E. Turner, *Principles of Radiation Protection*, John Wiley & Sons, New York (1967).

National Research Council, Committee on the Biological Effects of Ionizing Radiation, *The Effects on Populations of Exposures to Low Levels of Ionizing Radiation (BEIR III)*, National Academy Press, Washington, DC (1980).

National Research Council, *The Health Effects of Exposure to Low Levels of Ionizing Radiation, BEIR V*, National Academy Press, Washington, DC (1990).

NCRP Report No. 30, *Safe Handling of Radioactive Materials*, NCRP Publications, Bethesda, MD (1964).

NCRP Report No. 53, *Review of NCRP Radiation Dose Limit for Embryo and Fetus in Occupationally Exposed Women*, NCRP Publications, Bethesda, MD (1977).

NCRP Report No. 57, *Instrumentation and Monitoring Methods for Radiation Protection*, NCRP Publications, Bethesda, MD (1978).

NCRP Report No. 58, *A Handbook of Radioactive Measurement Procedures*, NCRP Publications, Bethesda, MD (1978).

NCRP Report No. 64, *Influence of Dose and Its Distribution in Time on Dose–Response Relationships for Low-LET Radiations*, NCRP Publications, Bethesda, MD (1980).

NCRP Report No. 65, *Management of Persons Accidentally Contaminated with Radionuclides*, NCRP Publications, Bethesda, MD (1980).

NCRP Report No. 80, *Induction of Thyroid Cancer by Ionizing Radiation*, NCRP Publications, Bethesda, MD (1985).

NCRP Report No. 83, *The Experimental Basis for Absorbed-Dose Calculations in Medical Uses of Radionuclides*, NCRP Publications, Bethesda, MD (1985).

NCRP Report No. 84, *General Concepts for the Dosimetry of Internally Deposited Radionuclides*, NCRP Publications, Bethesda, MD (1985).

NCRP Report No. 87, *Use of Bioassay Procedures for Assessment of Internal Radionuclide Deposition*, NCRP Publications, Bethesda, MD (1987).

NCRP Report No. 91, *Recommendations on Limits for Exposure to Ionizing Radiation*, NCRP Publications, Bethesda, MD (1987).

Oversight Committee on Radioepidemiological Tables, *Assigned Share for Radiation as a Cause of Cancer—Review of Radioepidemiological Tables Assigning Probabilities of Causation (Final Report)*, National Academy Press, Washington, DC (1984).

G. Paic, editor, *Ionizing Radiation: Protection and Dosimetry*, CRC Press, Boca Raton, FL (1988).

E. Pochin, *Nuclear Radiation: Risks and Benefits*, Clarendon Press, Oxford, England (1983).

D. L. Preston and D. A. Pierce, *The Effects of Changes in Dosimetry on Cancer Mortality Risk Estimates in Atomic Bomb Survivors*, RERF TR 9-87, Radiation Effects Research Foundation, Hiroshima, Japan (1987).

B. T. Price, C. C. Horton, and K. T. Spinney, *Radiation Shielding*, Pergamon Press, Elmsford, NY (1957).

W. J. Price, *Nuclear Radiation Detection*, 2nd edition, McGraw-Hill, New York (1964).

T. Rockwell, editor, *Reactor Shielding Design Manual*, D. van Nostrand, Princeton, NJ (1956).

C. L. Sanders and R. L. Kathren, *Ionizing Radiation: Tumorigenic and Tumoricidal Effects*, Battelle Press, Columbus, OH (1983).

N. M. Schaeffer, editor, *Reactor Shielding for Nuclear Engineers*, TID-25951, NTIS, US Department of Commerce, Springfield, VA (1973).

J. Shapiro, *Radiation Protection*, 2nd edition, Harvard University Press, Cambridge, MA (1981).

Y. Shimizu, *Life Span Study Report 11, Part II: Cancer Mortality in the Years 1959–1985 Based on the Recently Revised Doses (DS86)*, RERF TR 5-88, Radiation Effects Research Foundation, Hiroshima, Japan (1988).

Y. Shimizu, H. Kato, and W. J. Schull, *Life Span Study Report 11, Part I: Comparison of Risk Coefficients for Site-Specific Cancer Mortality Based on DS86 and T65DR*

Shielded Kerma and Organ Doses, RERF-TR-12-87, Radiation Effects Research Foundation, Hiroshima, Japan (1987).

B. Shleien, *Supplement 1 (1986): The Health Physics and Radiological Health Handbook*, Scinta, Inc., Silver Spring, MD (1986).

B. Shleien and M. S. Terpilak, editors, *The Health Physics and Radiological Health Handbook*, Nuclear Lectern Associates, Olney, MD (1984).

J. E. Turner, *Atoms, Radiation, and Radiation Protection*, Pergamon Press, Oxford, England (1986).

UNSCEAR, *Sources, Effects and Risks of Ionizing Radiation*, United Nations Committee on the Effects of Atomic Radiations, 1988 Report to the General Assembly, United Nations, New York (1988).

USNRC Regulatory Guide 8.29, *Instruction Concerning Risks from Occupational Radiation Exposure*, Washington, DC (1981).

R. J. Vetter, editor, The Biological Effects of Low-Dose Radiation: A Workshop, *Health Physics* **59**, No. 1 (1990).

O. J. Wallace, *WAPD-TM-1453: Analytical Flux Formulas and Tables of Shielding Functions*, Bettis Atomic Power Laboratory, West Mifflin, PA (1981).

R. E. Yoder, *Course 1B: An Overview of BEIR V*, 1992 Health Physics Society Meeting, Columbus, OH (1992).

PART II

SPECIALIZED AREAS: THEORY AND PROBLEMS

2

MEDICAL HEALTH PHYSICS

The medical use of radiation is the major contributor to man-made radiation exposures to the public. This exposure is deliberate because it is intended to benefit the individual. Other types of public exposures, such as nuclear power plant releases, are generally acceptable because of their low levels and because they benefit society in general.

The Medical Health Physicist is responsible for maintaining the radiation exposures of the hospital staff, patients, and general public to values that are as low as reasonably achievable. In addition, the medical health physicist has radiation safety program management and regulatory compliance responsibilities. He or she will also be involved in facility design and modifications, training, and emergency response. As part of these duties, he or she must assess medical related radiation exposures that arise from a variety of sources.

Medical uses of radiation include diagnostic procedures, radiotherapy, and biomedical research. Diagnostic techniques include x-ray and nuclear medicine procedures that are used to determine noninvasively the absence, presence, and extent of disease. Radiotherapy includes external beam therapy, brachytherapy, and radionuclide therapy which utilize radiation to deposit large amounts of energy in specific tissues.

Brachytherapy, which involves implants of sources into patient tissue, provides an external radiation hazard. Other external hazards include x-ray or imaging procedures. Internal radiation hazards are derived from medical administrations of radioisotopes.

HISTORICAL PERSPECTIVE

Within months of Wilhelm Roentgen's discovery of x-rays in November 1895, they were being used for medical diagnosis. In conjunction with this diagnostic use, biological effects were observed. By the turn of the century, the destructive effects of radium on skin were noted. However, radiation was found to have a more destructive effect on diseased cells than on normal cells. By fractionating the exposure, researchers discovered that the difference was quite significant and that radiation exposures provided a curative effect.

Radiotherapy is based on the differences in cell characteristics between diseased and healthy cells. Normal cells have active repair mechanisms, whereas cancer cells have degraded repair characteristics. Radiotherapy utilizes this difference by applying the radiation in fractionated doses over an extended period of time. During the time between exposures, the normal cells repair themselves and undergo cell division, where the cancer cells either die or are significantly reduced. Radiotherapy fails when the radiation tolerance of normal tissue is reached before all the cancer cells are destroyed.

Radiation therapy is possible because limited areas of the body can tolerate at least 10 times the lethal whole-body radiation dose. The lethal dose is relatively low because certain organs are radiosensitive. In humans, bone marrow is the first system to be destroyed by radiation and it determines the maximum survivable radiotherapy exposures.

The choice of using radiotherapy involves a careful consideration of risks. A cancer patient will normally view the radiation risks as small compared to the risks of not being treated. Often the radiation risks are smaller than the risks of other medical options, including surgery and chemotherapy.

Although radiation is a valuable medical tool, it is also a recognized carcinogen which requires that its use be restricted and controlled. Radiation is almost entirely reserved for the treatment of cancer. For that reason, the term *radiotherapy* is often referred to as *radiation oncology*, which recognizes the broader range of care provided to the cancer patient.

MEDICAL ACCELERATOR PHYSICS

Radioactive implants are used in treating about 5% of cancer cases, but most treatments are administered by external radiation sources. Over the last 20 years, the electron linear accelerator has become a dominant tool replacing the betatrons and Co-60 devices. Treatment utilizes both photon and electron beams in the 4- to 20-MeV energy range. Linear accelerators are utilized by university medical centers and hospitals as a routine treatment device.

Although accelerators will be discussed in a subsequent chapter, it is appropriate to briefly address the physics of radiotherapy as it relates to accelerator-produced radiations. Photon or x-ray beams have the widest application (80%) in radiotherapy, and electron beams are used in about 10–15% of the cases. Radiotherapy doses are often as large as 60 Gy delivered in 30 or more sessions over 6 or more weeks. The biological effect of the incident high-energy photons or electrons is to create large electron densities in the irradiated tissue.

For photon beams, divergence and attenuation reduce the photon fluence as a function of the depth in tissue. However, the electron density builds to an equilibrium value inside the tissue. The combination of these two effects produces a depth dose curve that rises to a maximum and then decreases with increasing depth in the tissue. Electron backscatter increases the surface dose to a value between 15% and 100% of the maximum dose. The depth of the maximum dose increases with increasing beam energy. For example, the depth dose curve for 15-MeV photons is illustrated in Figure 2.1. The depth dose curve peaks at 2.7-cm depth, and clinically useful radiation is available beyond 10-cm tissue depth.

Curves similar to that in Fig. 2.1 can be developed for electron beams. In electron beams, the primary electrons slow down in tissue and produce high ionizations per unit length as they reach their maximum range. For tissue depths beyond the maximum range, the electron dose decreases very rapidly to a value of only a few percent of the maximum dose. The energy loss for high-energy electrons is about 2 MeV/cm in tissue and about twice this value in bone.

For energies below 1 MeV, the maximum dose occurs near the skin surface. Because most lesions are below the surface of the skin, it is useful to use higher-energy beams, which have a larger dose in the tissue. Figures 2.2 and 2.3 provide depth dose curves for 6-MeV and 18-MeV electrons.

Electron beams are useful when the tumor volume is near the skin surface. By properly selecting the beam energy, the tumor is attacked while the underlying tissue is spared. For example, a chest wall tumor must be treated without damaging the underlying lung tissue. As the electron beam energy increases from 4 to 20 MeV, the shape of the depth dose curve shifts from a surface peak to a broader plateau extending into tissue. Beyond 20 MeV, the plateau expands, and the advantage of sparing healthy tissue at depth is lost. In general, the useful electron energy range is between 4 and 20 MeV.

Proton and heavy-ion beams produce a relatively low constant depth dose that terminates in a narrow peak at the end of the depth dose curve. The dose can be highly localized, which produces high tumor energy deposition and

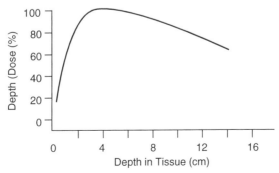

Fig. 2.1 Typical depth dose curve as a function of tissue depth for a beam of 15-MeV x-rays filtered for uniformity over a 35-cm × 35-cm area. (From Ford 1993.)

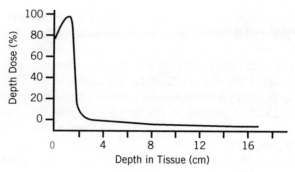

Fig. 2.2 Typical depth dose curve as a function of tissue depth for 6-MeV electrons spread by a system of scattering foils to provide for uniformity over a 35-cm × 35-cm area. (From Ford 1993.)

lower tissue doses. Dose localization may have some disadvantages when the tumor geometry is considered. A tumor may be approximated as a central mass with numerous protrusions or microextensions extending radially outward in random directions. In order to destroy the tumor, both the central mass and the microextensions must be destroyed. A highly localized beam could destroy the central tumor mass, but leave the microextensions relatively intact and capable of further growth. Therefore, some spreading in the depth dose profile is desirable.

DIAGNOSTIC NUCLEAR MEDICINE

Diagnostic techniques have been used to determine the presence and location or absence of infections, blood clots, myocardial infarctions (heart attacks), pulmonary emboli (blood clots in the pulmonary lung), occult bone fractures, and cancer. These techniques also have the ability to assess whether the organ

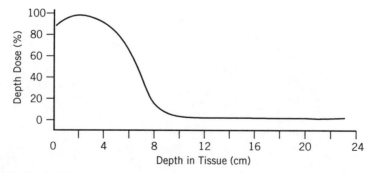

Fig. 2.3 Typical depth dose curve as a function of tissue depth for 18-MeV electrons spread by a system of scattering foils to provide for uniformity over a 35-cm × 35-cm area. (From Ford 1993.)

of interest is functioning properly without disturbing the organ under observation.

To accomplish this, a photon-emitting radionuclide is attached to a biologically active chemical to form a radiopharmaceutical which is administered to a patient by intravenous injection, oral ingestion, or inhalation. Following administration, the radiopharmaceutical is monitored over time using a gamma camera to detect photons escaping from the organ being irradiated. These photons create an image of the tissue distribution of the radionuclide via a technique called *scintigraphy*, which utilizes a gamma camera (sodium iodide crystal) to detect photons emitted from the radiopharmaceutical.

X-Rays

Diagnostic x-ray procedures include radiography, fluoroscopy, and mammography. Typical radiography involves the exposure of the body to a beam of x-rays. Attenuation of the x-ray beam depends on the intervening tissue which leads to variable darkening (contrast) of the photographic film. In diagnostic procedures, the attenuation of x-rays by the photoelectric effect is the most important contributor to the film contrast. Contrast materials with high atomic numbers provide for clearer images. Iodine ($Z = 53$) and barium ($Z = 56$) are commonly used contrast materials.

Fluoroscopy procedures take the x-ray beam, after it traverses the patient, and sends it into an imaging system that produces a video output. The video output permits real-time imaging and visual information that exceeds that obtained from snapshot x-ray images.

Nuclear Medicine

Nuclear medicine includes uptake studies, tracer studies, planar imaging, single photon emission computed tomography, and positron emission tomography. Commonly used diagnostic radionuclides are summarized in Table 2.1.

Computed Tomography

Diagnostic procedures include imaging the body's cross section via a technique known as *tomography*. When compared with planar imaging, tomography has the advantage of being able to resolve underlying organs with the aid of computational techniques. Tomography has been used to investigate abnormal physical shapes or functional characteristics in organs such as the liver, brain, or heart. It has also been used to determine the location and size of tumors in the body.

Tomography is commonly manifested in two forms: single photon emission computed tomography (SPECT) and positron emission tomography (PET). The SPECT method uses a rotating gamma camera and the same radionuclides used in conventional planar imaging. SPECT has been effective in the diagnosis of heart, liver, and brain disorders.

MEDICAL HEALTH PHYSICS

Table 2.1 Diagnostic Radionuclides Used in Nuclear Medicine

Radionuclide	Organ	Half-life	Gamma Energy (keV)
Tc-99m	Heart and bone	6.0 hr	141
In-111	Labeled blood products	2.8 days	171 245
Tl-201	Heart	73.1 hr	167
I-123	Thyroid	13.2 hr	159
I-131	Thyroid	8.0 days	365
Ga-67	Tumor agent	3.3 days	185 300
Xe-133	Lung	5.2 days	81
C-11	Heart	20.4 min	511
N-13	Heart	10.0 min	511
O-15	Brain	2.0 min	511
F-18	Brain	109.8 min	511

In contrast to the use of a rotating gamma camera, PET cameras are a stationary array of detectors that detect the two 511-keV photons produced by the annihilation of a positron–electron pair. PET scanners rely on coincidence detectors located 180° apart or time-of-flight electronic circuitry to determine the spatial coordinates of the annihilation event, which is assumed to be at the same location as that of the positron emission. The coincidence events are assembled to produce images of the scanned organs. Currently PET scans are performed on heart patients, but brain and cancer work is also significant.

Table 2.1 provides a listing of commonly used PET diagnostic radionuclides. These radionuclides typically have short half-lives (<2 hr) and have low atomic numbers. A cyclotron produces the positron emitter and the radionuclide is attached to a chemical compound such as $^{13}NH_3$, $H_2^{15}O$, ^{11}C-labeled amino acids, or ^{18}F glucose analogs.

Tracer Studies and Radioisotope Administration

The most commonly used diagnostic radionuclide is Tc-99m because its photon energy of 141 keV is optimal for imaging. Its half-life of 6.0 hr is long enough for the completion of most imaging procedures, but minimizes patient radiation exposures.

THERAPEUTIC NUCLEAR MEDICINE

Therapy applications include external beam therapy via linear accelerators or Co-60 teletherapy units. Brachytherapy using encapsulated sources implanted in tissue is another common therapy application. A final class of therapy procedures includes radionuclide administrations.

Table 2.2 Therapeutic Radionuclides Used in Nuclear Medicine

Radionuclide	Half-life	Average Beta Energy (keV)	Disease Treated
P-32	14.3 days	695	Leukemia
Y-90	64.0 hr	935	Cancer
I-131	8.0 days	192	Hyperthyroidism Thyroid Cancer
Sm-153	46.7 hr	220 and 226	Bone cancer pain
Re-186	90.6 hr	309 and 362	Bone cancer pain
Re-188	17.0 hr	728 and 795	Bone cancer pain
Au-198	2.7 days	315	Limit spread of ovarian cancer

Source: Stubbs and Wilson (1991).

Radionuclide Administration

Therapeutic radionuclides are used primarily in (a) the suppression of hyperthyroidism, (b) cancer treatment, and (c) reduction of pain levels associated with metastic cancer. Table 2.2 summarizes common therapeutic radionuclides and their uses.

Therapeutic doses are significantly larger than those associated with diagnostic radionuclide administrations. In the treatment of thyroid cancer, administrations of up to hundreds of millicuries of I-131 are given to the patient. This quantity of I-131 is very effective in nonsurgically destroying the thyroid gland via the deposition of local thyroid doses of more than 50,000 rad. In contrast, typical diagnostic doses are less than 10 rad to any organ.

External Beam Therapy

High doses may be applied to localized areas by utilizing either accelerators or teletherapy units. Accelerator applications were discussed earlier in this chapter. Another way of delivering high therapeutic doses is through the use of Co-60 teletherapy units. External beam therapy may be utilized to supplement brachytherapy.

Brachytherapy

Brachytherapy involves the placement of radioactive material in direct contact with tumors. It is reserved for tumors that are accessible via natural body cavities (intracavitary applications) or on body surfaces (interstitial applications). I-125, Cs-137, Ir-192, and Au-198 are commonly used brachytherapy materials.

Brachytherapy presents an external radiation hazard to hospital staff, patients, and the public. The unshielded exposure rate to patient rooms in the vicinity to the brachytherapy suite is given by

$$D_0 = m_{eqRa} G_{Ra}(1 - P)/r^2 \qquad (2.1)$$

where

D_0 = unshielded exposure rate to adjacent patient (mR/hr)
m_{eqRa} = radium equivalent activity in the brachytherapy patient (mg)
G_{Ra} = 8.25 mR-ft^2/mg-hr
P = radiation attenuation by the brachytherapy patient
r = distance between brachytherapy patient and an adjacent patient (ft)

The use of brachytherapy also exposes the patient to a potential internal radiation exposure as a result of using seeds composed of radioisotopes that can be absorbed into the body as a result of seed leaks or breaches in the seed capsules. For this type of situation, ICRP-26 methodology can be utilized to determine the dose consequence of the uptake:

$$D_T = BtS \qquad (2.2)$$

where

D_T = dose to the patient's organ (rad)
B = organ burden (μCi)
T_P = physical half-life of the radionuclide (hr)
T_B = biological half-life of the radionuclide (hr)
T_{eff} = effective half-life (hr)
   = $(T_P T_B)/(T_P + T_B)$
t = mean life of the radionuclide (hr)
   = $T_{eff}/\ln(2)$
S = mean dose per unit accumulated activity (rad/μCi-hr)

The total exposure, which is the sum of the external radiation exposure and the weighted internal exposure, is often of interest because it characterizes the total bodily insult from the brachytherapy. The total committed effective dose equivalent (D) from a brachytherapy source can be defined as

$$D = D_{whole\,body} + \sum_T w_T D_T \qquad (2.3)$$

where the sum includes all organs or tissues that have absorbed the radionuclide leaking from the source, and w_T is the organ weighting factor.

In order to perform the calculation dictated by the second term, a variety of approaches may be utilized to calculate the absorbed dose to the organ. One approach utilizes the Medical Internal Radiation Dose (MIRD)-type methodology to calculate the absorbed dose. The MIRD methodology is described in Appendix IV.

FACILITY DESIGN

The facility design depends on the characteristics of the radiation sources employed by the facility. These sources include the primary radiation source or beam, scatter radiation, leakage radiation, and radionuclide sources. Primary radiation is that radiation which exits the source after containment by a beam restriction device or shielding structure. Scattered radiation results from the interaction of the primary radiation and the scattering media, and it has an energy distribution similar to that of the primary radiation. Leakage radiation is radiation exiting the source in directions other than the intended beam direction. For x-ray tubes, where the output is highly filtered as it passes through the beam housing, leakage radiation has a higher energy than that of the primary or scattering radiation.

The primary as well as the scatter or leakage radiation must be shielded to meet the facility design requirements. Shielding requirements depend upon the beam quality, workload, distance from source to target individual, occupancy factor, and utilization.

SHIELDING DESIGN

Hospital or medical research environments utilize x-ray or imaging equipment. In order to reduce radiation exposures, shielding becomes an important consideration. In the shield design evaluation a number of input parameters are required to design a shielding configuration that is cost-effective but that meets the basic needs of the facility. One key design parameter is the facility workload (W) given by

$$W = EN_v N_p k \qquad (2.4)$$

where

W = weekly workload (mA-min/week)
E = exposure/view (mAs/view)
N_v = number of views per patient
N_p = number of patients per week
k = conversion factor (1 min/60 sec)

A second key design parameter is the use factor (U_x) for structure x. The structure may be any occupied area such as a control room, office, or waiting room.

$$U_x = V_x / N_v \qquad (2.5)$$

where V_x is the number of views directed toward structure x.

A third key parameter is the occupancy factor (T) which is the factor that multiplies the workload in order to correct for the degree or type of occupancy for the area under design evaluation. When historical occupancy data are not available, best estimate values of T are used. A value of $T = 1$ is assumed for full occupancy areas which include wards, work areas, wide corridors large enough for personnel occupancy at a desk or workstation, restrooms used by radiation workers, play areas, living spaces, and occupied areas in adjacent buildings. Partial occupancy ($T = 1/4$) may be assumed for narrow corridors, utility rooms, restrooms not normally used by radiation workers, elevators requiring an operator, and adjacent public parking lots. Occasional occupancy areas are assigned a $T = 1/16$ value and include stairways, automatic elevators, and closets too small for work-space utilization.

These factors are important design parameters that influence the calculation of the unshielded facility dose rate:

$$\dot{X}_u(r) = o_{PB}(T)(U)(W)k(r_0/r)^2 \tag{2.6}$$

where

$\dot{X}_u(r)$ = unshielded exposure rate at the location of interest (r) (mR/week)
r_0 = location of the primary beam output measurement.
o_{PB} = primary beam output @ r_0
T = occupancy factor
U = use factor
W = workload = mA-min/week
k = conversion factor (1 min/60 sec)
r = distance from x-ray source to point of interest

For locations not directly exposed to the primary beam, scattered radiation must be considered. The unshielded dose rate due to scattered radiation is

$$\dot{X}_u(r)^{scatt} = o_{PB}(W)kf_{scatt}(T)(r_0/r)^2 \tag{2.7}$$

where

$\dot{X}_u(r)^{scatt}$ = unshielded exposure rate at the location of interest (r) in mR/week due to the scattered radiation
f_{scatt} = ratio of scattered beam to incident beam at 1 meter

The reader should note that for scattered radiation, $U = 1$.

X-RAY SHIELDING

A common x-ray procedure involves a collimated beam of x-rays directed toward a patient being radiographed. After passing through the patient, it is attenuated to a design value by the primary protective barrier before irradiating other individuals in neighboring offices, hallways, or waiting areas. Leakage and scattered radiations are attenuated by a secondary protective barrier before irradiating other individuals. The following discussion will outline the procedure for calculating the primary and secondary barrier shielding.

The shielding requirements depend upon the maximum beam current and voltage of the x-ray tube, and the workload, use factor, and occupancy factor for the tube and shield structure. The workload measures the use of the x-ray machine, and it is normally measured in units of mA-min/week. The fraction of useful beam time during which the beam is pointed at the shield in question is defined as the use factor. The occupancy factor corrects for the degree of occupancy of the area in question.

Primary Barrier

The maximum exposure rate at an occupied location a distance d from the target in an x-ray tube in R/week is

$$\dot{X}_m = P/T \qquad (2.8)$$

where P is the maximum permissible weekly exposure rate (R/week) and T is the occupancy factor. NCRP-49 specifies 0.1 R/week for controlled areas and 0.01 R/week for uncontrolled areas. The maximum exposure may be used to obtain the exposure at other distances if the source geometry is known. Normally, a point source approximation is applicable.

For point source conditions, the exposure rate at 1 meter ($\dot{X}_1$) from the x-ray tube target is

$$\dot{X}_1 = d^2 \dot{X}_m = d^2 P/T \qquad (2.9)$$

If it is assumed that the exposure is due to a workload W and use factor U, the ratio K may be defined:

$$K = \frac{\dot{X}_1}{WU} = \frac{d^2 P}{WUT} \qquad (2.10)$$

where K is the primary beam ratio or transmission factor in units of R-m^2/mA-min, and W is the workload in mA-min/week.

Values of the transmission factor K have been measured for various x-ray energies and shielding thicknesses. Once the transmission factor is known, the required barrier thickness may be obtained from tabulated K versus shield thickness curves. NCRP-49 provides transmission versus shield thickness curves for common shielding materials such as concrete and lead.

Secondary Barrier

Scattered and leakage radiation are attenuated by the shielding of the secondary barrier. The required shield thicknesses to attenuate scattered and leakage radiation are calculated separately. NCRP-49 provides guidance for determining the required shielding. If the required thicknesses are about the same, an additional half-value layer is added to the larger thickness. If the difference between the calculated shielding thicknesses for scattered and leakage radiation is at least one-tenth-value layer, then the larger of the two values should be selected.

The intensity of the scattered radiation is a function of a number of parameters such as the scattering angle, primary beam energy, and the geometry of the scattering area.

The secondary barrier transmission factor relationship is based on the assumption that the point source approximation is applicable and that the exposure rate from the scattered radiation is directly proportional to the scattering area.

The transmission relationship for scattered radiation is written in terms of a scattering transmission factor (K_{ux}) with units of R-m^2/mA-min:

$$K_{ux} = \frac{400P(d_{scat})^2(d_{sec})^2}{aWTFf} \tag{2.11}$$

where P is the maximum weekly exposure (R), d_{scat} is the distance between the x-ray tube's target and the object scattering the x-rays (normally the patient undergoing the procedure), d_{sec} is the distance from the scatterer to the point of interest that is shielded by the secondary barrier, a is the ratio of scattered to incident radiation, and F is the scattering field size (in centimeters squared). NCRP-49 tabulates values of the ratio a based on $F = 400$ cm^2. The F factor is the actual scattering field size. The final factor appearing in Eq. (2.11) is f, which is a factor accounting for the fact that the x-ray output increases with voltage; that is, smaller K_{ux} values require larger shield thicknesses. Values assigned to f are summarized in Table 2.3.

Leakage Radiation

The protective tube housing limits the leakage radiation. Once a measured leakage value is determined at a fixed distance, usually 1 m, it is possible to determine the required barrier thickness at other distances if the energy of the

Table 2.3 Values of the factor f

Voltage (kV)	f
≤ 500	1
1000	20
2000	300
> 2000	700

leakage radiation is known. The leakage radiation is filtered and hardened in traversing the tube housing and emerges in a narrow, essentially monochromatic, energy region. This leakage property suggests that the half-value layer is strongly dependent on the voltage across the x-ray tube.

For a diagnostic x-ray tube, the leakage is restricted to 0.1 R in an hour at a distance of 1 m from the tube when it is operating at its maximum voltage and current rating. The leakage attenuation factor (B_{Lx}) from a diagnostic x-ray tube is

$$B_{Lx} = \frac{600Pd^2I}{WT} \quad \text{(diagnostic)} \qquad (2.12)$$

where d is the distance from the x-ray tube target to the point of interest and I is the x-ray tube electron beam current. The other factors appearing in this equation were previously defined.

The leakage radiation relationship for therapeutic exposures depend on the operating tube voltage. For a therapeutic tube with an operating voltage less than or equal to 500 keV, the leakage is limited to 1 R in an hour at 1 m from the target of the x-ray tube. The barrier attenuation relationship for voltages of 500 keV or less is given by

$$B_{Lx} = \frac{60Pd^2I}{WT} \quad \text{(therapeutic, } V \leq 500 \text{ keV)} \qquad (2.13)$$

This relationship is altered for potentials greater than 500 keV. For therapeutic x-ray tubes with potentials greater than 500 keV, the leakage is limited to 0.1% of the intensity of the useful beam at 1 m. The attenuation factor for this case is

$$B_{Lx} = \frac{1000Pd^2}{WT}(1/\dot{X}_n) \quad \text{(therapeutic, } V > 500 \text{ keV)} \qquad (2.14)$$

where $\dot{X}_n$ is the exposure rate at 1 meter from the therapeutic x-ray tube when the electron beam current is 1 mA.

The number of half-value layers (n) required to obtain the desired attenuation (B_{Lx}) is obtained from the relationship

$$B_{Lx} = (1/2)^n \qquad (2.15)$$

The barrier thickness as a function of the barrier attenuation is tabulated in NCRP-49 for common materials such as lead and concrete as a function of peak voltage.

NCRP-37 EXPOSURE RECOMMENDATIONS

NCRP-37 recommends that patients in a hospital receive no more than 100 mR per hospital stay from other patients. Because a patient could occupy an adjacent room for 168 hr each week, exposures to adjacent patients must not exceed 0.6 mR/hr. The shielding required to meet NCRP-37 recommendations can be determined in general in terms of a half-value layer equation:

$$(1/2)^N = D_{NCRP37}/D_0 \qquad (2.16)$$

where

N = number of half-value layers of shielding needed to meet the NCRP-37 dose recommendations
D_{NCRP37} = 0.6 mR/week (NCRP-37 recommendation)
D_0 = unshielded exposure rate (mR/week)

Once N is known, the required shielding thickness (t) is readily obtained from the definition of the half-value layer:

$$t = Nt_{HVL} \qquad (2.17)$$

where

t = thickness of shielding required to meet the NCRP-37 recommendations (cm)
t_{HVL} = half-value thickness (cm)

VENTILATION CONSIDERATIONS

Another interesting aspect of exposing a patient to an internal administration of radionuclides is that the patient will begin to excrete the material and becomes a source of exposure to other hospital occupants. Because the patient is normally confined to a room, the room ventilation parameters play a key role in the determination of the extent of this hazard.

With a single pass system and the assumption of uniform mixing of the air, the air concentration may be determined by considering (a) the total volume of air handled in the room in a specified time period such as 1 year and (b) the total activity excreted by the patient. The volume of air passing through the treatment room in a year (ml/year) ($\dot{V}$) is given by

$$\dot{V} = Vrk \qquad (2.18)$$

where

V = room volume (ft³)
r = number of room turnover's per hour (hr⁻¹)
k = conversion factor
  = (24 hr/day)(365 day/year)(2.832 × 10⁴ ml/ft³)

The resulting activity released into the room per year ($\dot{A}$) is

$$\dot{A} = ANf \quad (2.19)$$

where

N = number of subjects treated/year
f = loss factor or the fraction of material injected into the patient that is released into the room.
A = average activity burden of the patient (μCi/subject)

The average room concentration ($\overline{C}$) may be determined from the information derived above:

$$\overline{C} = \dot{A}/\dot{V} \quad (2.20)$$

The average room concentration may be compared to applicable standards or used as the basis for an internal dose assessment.

SCENARIOS

Scenario 34

Your hospital has decided to dedicate a lead-shielded room for brachytherapy patients to reduce exposure from this treatment to staff, adjacent patients, and visitors. Typical treatments for gynecological implants are Cs-137 sealed sources at a maximum treatment time of 50 hr each. Assume that the adjacent patient could remain in her bed for an entire 7-day period during her hospital stay. Also assume that the brachytherapy room is continuously occupied for at least 100 hr in a week by a patient implanted with Cs-137. A diagram is provided in Fig. 2.4.

Data

Specific gamma ray constant for radium = 8.25 mR-ft²/mg-hr

Patient attenuation = 30%

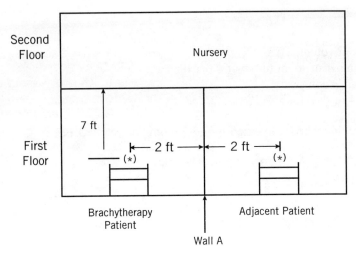

Fig. 2.4 Hospital cross section including the brachytherapy suite and its location relative to the nursery and an adjoining patient room. The symbol "(*)" denotes the patient locations.

Brachytherapy patient is: 2 ft from wall A;
4 ft from adjacent patient's bed;
7 ft from a nursery above;
able to move freely about the room

HVL for Cs-137 is lead = 0.65 cm

Mass attenuation coefficient for concrete = 0.06 cm^2/g

Density of concrete = 2.4 g/cm^3

Physical half-life of I-125 = 60 days

Biological half-life of I-125 = 138 days

S for I-125 = 3.0 × 10^{-3} rad/μCi-hr

w_T = 0.03 for the thyroid

2.1. What are the NCRP-37 dose recommendations regarding the dose a patient receives from other patients during a hospital stay?

2.2. Based on a maximum activity of 70-mg Ra equivalent, calculate the amount of lead in wall A necessary to comply with the NCRP-37 recommendations for adjacent patient exposure per visit. Assume that the dose at the walls controls implementation of the NCRP recommendations.

2.3. Using the same information as in the preceding question, calculate how much concrete is needed for the ceiling above.

2.4. List differences related to radiation protection for Cs-137 and I-125 brachytherapies.

2.5. A patient who received I-125 brachytherapy using seeds to treat prostate cancer had a confirmed I-125 thyroid burden of 300 nCi detected 3 days after the procedure. Calculate the dose equivalent from the measured uptake to the thyroid using ICRP-26 methodology. Ignore back-decay.

2.6. A week after the initial 300-nCi measurement, a second measurement indicated 330 nCi. As a health physicist, how would you evaluate this situation?

Scenario 35

A physician in your hospital wants to assess liver metabolism in diabetics using a C-14-labeled glucose compound. He wants to study a total of 20 patients. Assuming that all of the activity is instantaneously absorbed by the liver, glucose is metabolized into CO_2 with a 67-hr half-life. $^{14}CO_2$ is released from the whole body via the lungs with a 1.2-hr half-life. All 20 subjects will be studied in the same hospital room. The patients will be done one at a time over a 1-year period. Each patient will be in the room for 6 hr post injection so that blood samples may be obtained. The following information should be evaluated:

Activity administered to each patient = 200 μCi

$S_{\text{liver-liver}} = 5.8 \times 10^{-5}$ rad/μCi-hr

$S_{\text{whole body-whole body}} = 1.5 \times 10^{-5}$ rad/μCi-hr

ICRP weighting factor for the liver = 0.06

Hospital's administrative limit for $^{14}CO_2$ = 1 $\times$ 10^{-6} μCi/ml

1 ft^3 = 2.832 $\times$ 10^4 ml

2.7. Based on this information, calculate the committed effective dose equivalent to the patient.

2.8. Given that the room is 18 ft $\times$ 20 ft $\times$ 8 ft, uses single-pass air, and has three room-air changes per hour, assess the likelihood that the hospital's administrative C-14 limit for air will be exceeded over the year period in which the project is scheduled. Assume uniform mixing and that 20% of the administered activity will be exhaled into the room as $^{14}CO_2$.

2.9. Would recirculation of the room air affect your answer to question 2.8? Assume that the room represents 1/50 of the total air volume handled, that all of the areas served are turned over three times per hour, and that there is 33% fresh-air makeup.

2.10. Using the following equation, calculate the quantity of $^{14}CO_2$ released into the room per patient. Assume a 6-hr release period.

$$N_G \xrightarrow{\lambda_G} N_{CO_2} \xrightarrow{\lambda_C} N_{Lung}$$

$$N_{CO_2}(t) = \frac{\lambda_G}{\lambda_C - \lambda_G} N_G(0)[\exp(-\lambda_G t) - \exp(-\lambda_C t)]$$

where G and C label the glucose compound and CO_2, respectively.

Scenario 36

You are asked to recommend shielding for the mammographic suite, illustrated in Fig. 2.5. It can be assumed that the average kVp for all the views taken is 30 and that the average film requires 120 mA-sec in order to get an appropriate density. On average, 40 women per week will be radiographed. Each breast will be imaged once in the cephalocaudal orientation (primary beam directed straight down) and once in the mediolateral view (primary beam directed toward the control booth wall or the outside wall)

2.11. Calculate the weekly workload for this room.

2.12. Based upon the guidance give in NCRP-49, the cost of shielding will not increase significantly if the design exposure for the operator's position is based on:
a. 2 mR/week
b. 1 mR/week

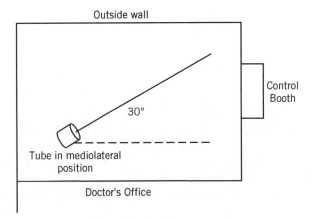

Fig. 2.5 Geometry of the mammographic suite and its relationship to the doctor's office, an outside wall, and the control booth.

c. 100 mR/week
d. 10 mR/week
e. 5 mR/week

2.13. Calculate the use factor for the control booth.

2.14. Based upon the recommendations of NCRP-49, the assumed occupancy factor (OF) for the control booth would be:
a. 0.10
b. 0.25
c. 0.50
d. 0.75
e. 1.00

2.15. Based upon the original layout, calculate the required thickness of gypsum board needed to ensure that the operator in the control booth receives no more than 5 mR/week from the primary beam. Consider the following in performing your estimate:

Data

Primary beam output = 1.0 mR/mA-sec at 100 cm

Occupancy factor = 0.5

Use factor = 0.5

Workload = 160 mA-min/week

Distance from x-ray source to point of interest

(mediolateral position) = 200 cm

A transmission factor of 8.3×10^{-3} corresponds

to 1.1 cm of gypsum board.

2.16. Based upon the original layout, calculate the required thickness of gypsum board needed to ensure that an individual in the doctor's office receives no more than 5 mR/week from the scattered radiation. Consider the following in performing your estimate:

Data

Scatter to incident ratio at 1 m = 0.0003

Occupancy factor = 0.6

Workload = 160 mA-min/week

Distance from point of scattering (0.5 m from

x-ray source) to point of interest = 150 cm

2.17. Assuming a leakage rate of 100 mR/hr at 100 cm when the unit is operated at 30 kVp and 7 mA continuously, calculate the required thickness of gypsum board to ensure that an individual in the doctor's office receives no more than 5 mR/week from leakage radiation. Consider the following in formulating your response:

Data

HVL (gypsum) for 30 kVp = 0.5 cm

Occupancy factor = 0.5

Workload = 640 mA-min/week

Distance from point of x-ray source to point of interest = 200 cm

Scenario 37

Consider an Mo-99/Tc-99m generator and radiopharmaceutical production. Answer the following questions after considering the information provided:

Data

1.

Isotope	Half-life
Mo-99	66 hr
Tc-99m	6 hr

2. Linear attenuation coefficient for Tc-99m gammas in lead is 3.25/cm.
3. Gamma constant for Tc-99m = 0.56 R/mCi/hr at 1 cm.

2.18. On day 1 at 8 a.m. an Mo-99/Tc-99m generator is milked of all its Tc-99m. The amount of Mo-99 present on day 1 at 8 a.m. is 1000 mCi. How much Tc-99m will be available for milking at 8 a.m. on day 2? Assume that all Mo-99 decays result in Tc-99m.

2.19. A pharmaceutical kit for a sulfur colloid contains 5 cm^3 of appropriate liquid. Assuming that the specific activity of the Tc-99m is 80 mCi/cm^3 at 8 a.m. on day 2, how many cubic centimeters of the Tc-99m milking must be placed with the kit's 5 cm^3 so that the specific activity of the sulfur colloid at 4 p.m. of day 2 is 10 mCi/cm^3?

2.20. If a lead syringe shield of 5-mm thickness is used on a syringe containing 50 mCi, what will the exposure rate be at 50 cm?

It is now 8 a.m. on day 90. This time is applicable for questions 2.21 and 2.22.

2.21. Assuming that the activity on day 1 is 1000 mCi, what activity of Mo-99 will be present?

2.22. A recent manufacturer's bulletin warns that the generator might be contaminated with Ru-106. You survey the spent generator column with a pancake GM probe, and you observe 3700 dpm in contact with the column. Explain this observation.

Scenario 38

A 30-keV photon beam with 4.0×10^{11} photons per second is to be used for angiography. The beam is 0.5 mm high by 123 mm wide. The patient is positioned in a chair that moves vertically through the beam.

Data

$$u/p \text{ for Al at 30 keV} = 1.12 \text{ cm}^2/\text{g}$$
$$u_{en}/p \text{ for Al at 30 keV} = 0.87 \text{ cm}^2/\text{g}$$
$$u_{en}/p \text{ for air at 30 keV} = 0.15 \text{ cm}^2/\text{g}$$
$$\text{Aluminum density} = 2.7 \text{ g/cm}^3$$
$$W_{air} = 34 \text{ eV/ion pair} = 34 \text{ J/coulomb}$$

2.23. Calculate the beamline exposure rate.

2.24. Calculate the thickness of an aluminum beam stopper to reduce the exposure rate to 2.0 mR/hr. The stopper is located downstream of the collimator. Assume that the collimator is very thick.

2.25. Calculate the patient's beamline exposure for a constant vertical chair movement of 60 mm/sec.

2.26. Discuss the use of a transmission ionization chamber located in front of the patient to provide a direct measurement of the surface dose.

2.27. List parameters that must be monitored to ensure that the patient dose is maintained below 1.0 rad for a single procedure.

Scenario 39

During a radiographic study of a small child, it was necessary to restrain the patient. The child's mother was asked to do this by standing alongside the x-ray table while holding the child motionless with her hands. The mother was supplied with a lead apron and lead gloves. No part of the mother's body was in the direct x-ray beam. Subsequently, the woman discovered that she was pregnant. Her physican is concerned about possible effects of the radiation exposure to the woman and fetus.

You are at home and about to leave for a 2-week vacation when the physican calls and tells you of the aforementioned situation. He requests that you quickly

64 MEDICAL HEALTH PHYSICS

assess the relevant radiation doses. During the subsequent 30-min phone conversation you will be asked the questions in this scenario. Use your best judgment to assess any information that the physician does not have available.

Data

> Voltage = 70 kVp
>
> 70 kVp yields 4 mR/mA-sec at 40 in.
>
> Filtration = 2.5 mm Al
>
> Time/current = 25 mA-sec/view
>
> Film size = 8 in. × 10 in.
>
> Source image detector distance = 40 in.
>
> Number of views = 5
>
> Scatter radiation at 1 m = 0.1% of the primary beam
>
> Distance between the primary beam and the abdomen = 18 in.

2.18. Estimate the dose to the skin of the woman's abdomen.

2.29. Estimate the dose to the fetus. Assumptions regarding x-ray penetration through the lead apron and percent depth dose at the uterus should be stated.

2.30. What guidelines exist regarding fetal x-ray exposure?

2.31. What other information should you provide to the physician?

2.32. Based upon your calculated dose to the fetus, discuss the practice of not allowing fertile females to hold patients during x-ray procedures.

Scenario 40

A patient is to be given a 200-mCi I-131 oral therapeutic dose as an iodide for an inoperable thyroid metastasis. The I-131 is administered in solution form. Thyroid surgery was unsuccessful during a previous hospitalization. This scenario addresses the health physics aspects of the I-131 administration and the following hospitalization of the patient.

Data for I-131

> Thyroid dose commitment = 6.6 rem/μCi in thyroid
>
> Gamma constant = 2.2 R hr^{-1} mCi^{-1} cm^2
>
> Radiological half-life = 8.05 days

2.33. Quantify the internal and external radiation hazards associated with I-131.

2.34. What is the contact (1 cm) exposure rate and the exposure rate at 1 m for the container holding the 200-mCi administration? Assume no shielding and that the activity may be treated as a point source.

2.35. What exposure control and health physics practices should be utilized as part of the I-131 administration and subsequent hospitalization of the patient?

REFERENCES

ANSI Standard N43.2-1977, *Radiation Safety for X-Ray Diffraction and Fluorescence Analysis Equipment*, NBS Handbook 111, US Government Printing Office, Washington, DC (1978).

F. H. Attix, *Introduction to Radiological Physics and Radiation Dosimetry*, Wiley, New York (1986).

FDA 76-8027, *The Use of Electron Linear Accelerators in Medical Radiation Therapy: Physical Characteristics*, Public Health Service, Rockville, MD (1976).

FDA 82-8181, *A Primer on Theory and Operation of Linear Accelerators in Radiation Therapy*, Public Health Service, Rockville, MD (1982).

J. Ford, SLAC Beam Line, *Little LINACS Fight Cancer*, **23,** 6 (1993).

A. M. Friedman, R. H. Seevers, Jr., and R. P. Spencer, *Radionuclides in Therapy*, CRC Press, Boca Raton, FL (1987).

W. R. Hendee, *Medical Radiation Physics*, 2nd edition, Year Book Medical Publishers, Chicago (1979).

W. R. Hendee, *Radiation Therapy Physics*, Year Book Medical Publishers, Chicago (1981).

ICRP Publication 44, *Protection of the Patient in Radiation Therapy*, Pergamon Press, Oxford, England (1985).

ICRP Publication 52, *Protection of the Patient in Nuclear Medicine*, ICRP Publications, Pergamon Press, Oxford, England (1987).

ICRP Publication 53, *Radiation Dose to Patients from Radiopharmaceuticals*, ICRP Publications, Pergamon Press, Oxford, England (1988).

ICRP Publication 57, *Radiological Protection of the Worker in Medicine and Dentistry*, ICRP Publications, Pergamon Press, Oxford, England (1990).

ICRU Report No. 24, *Determination of Absorbed Dose in a Patient Irradiated by Beams of X- or Gamma-Rays in Radiotherapy Procedures*, ICRU Publications, Bethesda, MD (1976).

ICRU Report No. 26, *Neutron Dosimetry for Biology and Medicine*, ICRU Publications, Bethesda, MD (1977).

H. E. Johns and J. R. Cunningham, *The Physics of Radiology*, 4th edition, Charles C Thomas, Springfield, IL (1983).

R. Loevinger, T. F. Budinger, and E. E. Watson, *MIRD Primer of Adsorbed Dose Calculations*, Society of Nuclear Medicine, New York (1988).

M. Mandelkern, SLAC Beam Line, *Positron Emission Tomography*, **23,** 15 (1993).

T. F. McAinsh, editor, *Physics in Medicine and Biology Encyclopedia*, Pergamon Press, Oxford, England (1986).

A. L. McKenzie, J. E. Shaw, S. K. Stephenson, and P. C. R. Turner, *Radiation Protection in Radiotherapy*, Institute of Physical Sciences in Medicine, London (1986).

K. L. Miller, Medical Doses—Clinical and Occupational, *Radiation Protection Management*, **7**, 30 (1990).

W. Minder and S. B. Osborne, *Manual on Radiation Protection in Hospitals and General Practice*, Volume 5, Personnel Monitoring Services, World Health Organization, Geneva (1980).

NBS Handbook 138, *Medical Physics Data Book*, US Department of Commerce, US Government Printing Office, Washington, DC (1982).

NCRP Report No. 33, *Medical X-Ray and Gamma-Ray Protection for Energies up to 10 MeV*, NCRP Publications, Bethesda, MD (1968).

NCRP Report No. 37, *Precautions in the Management of Patients Who Have Received Therapeutic Amounts of Radionuclides*, NCRP Publications, Bethesda, MD (1978).

NCRP Report No. 40, *Protection Against Radiation from Brachytherapy Sources*, NCRP Publications, Bethesda, MD (1972).

NCRP Report No. 48, *Radiation Protection for Medical and Allied Health Personnel*, NCRP Publications, Bethesda, MD (1976).

NCRP Report No. 49, *Structural Shielding Design and Evaluation for Medical Use of X-Rays and Gamma Rays of Energies Up to 10 MeV*, NCRP Publications, Bethesda, MD (1976).

NCRP Report No. 54, *Medical Radiation Exposure of Pregnant and Potentially Pregnant Women*, NCRP Publications, Bethesda, MD (1977).

NCRP Report No. 69, *Dosimetry of X-Ray and Gamma-Ray Beams for Radiation Therapy in the Energy Range 10 keV to 50 MeV*, NCRP Publications, Bethesda, MD (1981).

NCRP Report No. 70, *Nuclear Medicine Factors Influencing the Choice and Use of Radionuclides in Diagnosis and Therapy*, NCRP Publications, Bethesda, MD (1982).

NCRP Report No. 79, *Neutron Contamination from Medical Electron Accelerators*, NCRP Publications, Bethesda, MD (1984).

NCRP Report No. 83, *The Experimental Basis for Absorbed-Dose Calculations in Medical Uses of Radionuclides*, NCRP Publications, Bethesda, MD (1985).

NCRP Report No. 85, *Mammography—A User's Guide*, NCRP Publications, Bethesda, MD (1987).

NCRP Report No. 99, *Quality Assurance of Diagnostic Imaging Equipment*, NCRP Publications, Bethesda, MD (1988).

NCRP Report No. 100, *Exposure of the US Population from Diagnostic Medical Radiation*, NCRP Publications, Bethesda, MD (1989).

NCRP Report No. 105, *Radiation Protection for Medical and Allied Health Personnel*, NCRP Publications, Bethesda, MD (1989).

NCRP Report No. 107, *Implementation of the Principle of ALARA for Medical and Dental Personnel*, NCRP Publications, Bethesda, MD (1990).

NCRP Report No. 111, *Developing Radiation Emergency Plans for Academic, Medical, or Industrial Facilities*, NCRP Publications, Bethesda, MD (1991).

M. E. Noz and G. Q. Maguire, *Radiation Protection in the Radiologic and Health Sciences*, 2nd edition, Lea & Febiger, Philadelphia (1985).

J. Selman, *The Fundamentals of X-Ray and Radium Physics*, 6th edition, Charles C Thomas, Springfield, IL (1976).

E. M. Smith, General Considerations in Calculation of the Absorbed Dose of Radiopharmaceuticals Used in Nuclear Medicine, in *Medical Radionuclides: Radiation Dose and Effects*, edited by R. J. Cloutier, C. L. Edwards, and W. S. Snyder, CONF-691212, NTIS, Springfield, VA (1970).

J. A. Sorenson and M. E. Phelps, *Physics in Nuclear Medicine*, 2nd edition, Grune & Stratton, New York (1987).

J. B. Stubbs and L. A. Wilson, *Nuclear News*, **May,** 50 (1991).

M. M. Ter-Pogossian, *The Physical Aspects of Diagnostic Radiology*, Harper & Row, New York (1967).

3

UNIVERSITY HEALTH PHYSICS

A university health physicist is required to demonstrate many of the concepts discussed in the previous chapters. This professional may be required to perform internal dose assessments involving the wide variety of isotopes used within the particular research environment or to evaluate exposures from university medical or accelerator facilities. The university health physicist is also faced with dealing with assessments of effluent releases and for managing radioactive waste.

In a sense, the health physics concerns at a university encompass the wealth of the field. The limits of the scope of university health physics are only imposed by the imagination of the various research faculty. Therefore, the scope is quite broad. For example, university research activities utilize a variety of radiation sources. These include large fixed gamma sources, x-ray machines, nuclear reactors, particle accelerators, neutron sources, unsealed radioisotopes used in biomedical or chemical applications, and biomedical tracer studies. Because of this breadth, we will address a variety of topics that are representative of the concerns and challenges of the university health physicist.

RESEARCH UTILIZING RADIONUCLIDES

The initial part of this chapter will focus upon research activities involving specific radioisotopes, their hazards, and the engineering controls that limit their impact upon the researcher. The consequences of the failure of these controls, the resulting release of radioactive material to the work environment, the quantification of this release, and its impact upon the worker will also be addressed.

Table 3.1 Common University Research Radioisotopes

Isotope	Use
H-3	Biomedical tracer studies
C-14	Biomedical tracer studies
P-32	Biochemical labeling experiments
Co-60	Large sealed sources used in radiation damage studies
I-125/I-131	Medical research/therapeutic treatments
Cf-252	Neutron studies

Research involving radionuclides span a variety of areas, including the biomedical sciences, engineering, physics, chemistry, biology, and geology. A summary of common university radioisotopes is presented in Table 3.1. A summary of the characteristics of these research isotopes and their health physics considerations are outlined below. In all cases, proper contamination controls practices must be utilized.

H-3

Tritium enters the body by inhalation or absorption through the skin and equilibrates with the body's water. H-3, entering the body as tritiated water, will equilibrate with total body water after inhalation and transfer from the lung to the blood. Skin absorption follows transfer to the lymph system and then to the blood. Equilibration may take 2–4 hr.

Tritium uptakes are best detected by urinalysis. Urine samples should be counted for the 0.0186-MeV maximum beta energy using liquid scintillation counting.

C-14

Breath analysis is the most sensitive technique for analyzing intakes of C-14 via the excretion of CO_2 by exhalation. Urinalysis with a liquid scintillation counter may be more practical and is nearly as sensitive as breath analysis.

Engineering controls mitigate the uptake of C-14. The use of local ventilation and fume hoods are reasonable controls for handling C-14. The chemical form of the C-14 material also impacts its uptake.

P-32

P-32 is often administered as a biochemical label in an aqueous solution. The isotope is a pure beta emitter with a maximum energy of 1.71 MeV and a half-life of 14.3 days. The health physics considerations are potential contamination problems and direct beta dose to the researcher. Exposures may be minimized with proper shielding. The use of P-32 is normally performed in a ventilation hood which minimizes the airborne hazard and also protects the workers eyes from the beta particles emitted by P-32. P-32 will not volatilize under normal circumstances, but a spill or misapplication could lead to P-32 vapor.

Good housekeeping practices and laboratory procedures minimize the potential for contaminating other equipment. Finger rings or wrist badges should be utilized to assess the beta dose to the hands. Eye protection may be necessary depending upon the geometry and the research application. Gloves, shielded vests, shadow shielding, and short tongs could be employed to further minimize the worker's radiation exposure.

Uptakes of P-32 could be detected by either urinalysis or whole-body counting. Whole-body counting would detect the bremsstrahlung radiation produced by the 1.71-MeV beta particles.

Co-60

Radiation damage studies or food irradiation experiments utilize kilocurie sources of gamma isotopes such as Co-60. The irradiation chamber's door and shutter should be interlocked to prevent entry when the source is exposed.

The health physicist's major concern is direct exposure and appropriate training and surveillance of interlocks and warning systems. The source shield and the shutter should be labeled with radiation warning signs. The door should also have appropriate signs, warning lights, and audible alarms that provide an indication that the source is exposed. A whole-body thermoluminescent dosimeter (TLD) or film badge would be the recommended dosimetry.

I-125/I-131

The retention of iodine by the thyroid makes thyroid counting the most sensitive and practical form of bioassay for iodine radioisotopes. Both I-125 and I-131 are airborne hazards and should be utilized under controlled conditions such as in a fume hood or in a well-ventilated area.

I-125 has a 60-day half-life, and it is detected through the 35-keV x-ray emitted as a result of electron capture. Its emissions are weaker than those of I-131 and include x-rays and conversion electrons.

Iodine-131 poses both an external and internal radiation hazard due to the beta and gamma radiation associated with its decay governed by its 8-day half-life. Research studying thyroid dysfunction must consider the impact of radiation because 6.6-rem/μCi uptake occurs in the thyroid. The gamma constant for I-131 is 2.2 R-cm^2/hr-mCi, so that 100 mCi of unshielded I-131 leads to an exposure rate of about 22 mR/hr at 1 meter and 440 R/hr at 1 cm.

The administration of radioiodine is normally in a basic (pH > 7) solution to minimize the quantity of iodine that volatilizes. Shielded vials, protective gloves, and protective clothing should be utilized to minimize the technician's exposure. Oral administration should be followed by other fluids to wash the mouth and esophagus by removing residual radioiodine and thus lower the exposure to these tissues. These additional fluids will also serve to reduce the stomach and gastrointestinal (GI) tract exposures. Intravenous administration would eliminate these exposure concerns.

Cf-252

The radiation hazard from Cf-252 is primarily due to neutrons emitted following spontaneous fission. For Cf-252, the associated gamma dose equivalent is only about 5% of that due to the neutrons. The Cf-252 dose equivalent rate at 1 meter is about 2.4 rem/hr-μg, which suggests that personnel monitoring for both neutrons and gammas should be provided.

ENGINEERING CONSIDERATIONS

A university health physicist will be required to minimize the radiological impact of experiments involving radionuclides that have the potential to become an airborne hazard. These experiments often involve fume hoods or glove boxes that protect the researcher from the airborne hazard.

The university health physicist will become involved in the development of engineering controls to minimize releases of radioactive material into laboratory workspaces. These control measures are not always completely effective, due to researcher errors or equipment malfunctions, and releases may occur. The release has the potential to lead to an uptake of radioactive material, which also requires the health physicist to determine the radiological consequences of the release.

Engineering Controls

The uptakes noted in the previous section may be mitigated through the use of engineering controls. University research environments frequently utilize ventilation systems and other engineering controls (hoods and glove boxes) to minimize the uptakes of volatile radionuclides. The design of containment systems for volatile research material should consider the following features:

1. Supply air should enter near the ceiling and exhaust near the floor. This will sweep contamination that leaks from the hood or glove box toward the floor and away from the respirable air.
2. Components should be readily accessible for testing and component changeout.
3. Individual laboratory rooms should have a dedicated air supply and exhaust system.
4. Backup power should be provided to ventilation systems and radiation air monitors.
5. Components should be constructed of stainless steel or other nonreactive materials.

6. Sharp bends, nonlaminar transitions, and long pipe runs should be eliminated for efficient system operation.

The system design should be periodically tested to ensure that the intake, exhaust, and recirculating air flows meet the system specifications and that the system is properly balanced. Room air-flow patterns can be investigated using $TiCl_4$ smoke sticks. The $TiCl_4$ combines with the humidity in the air to produce a dense white smoke (TiO_2). The TiO_2 is carried by the air flow to produce a visible indication of air flow. Air flows around intake and exhaust ducts and in the vicinity of hoods/glove boxes are readily evaluated using the $TiCl_4$ method.

System filters should be tested with DOP (dioctylphthalate) to verify particulate collection efficiency. The standard test using 0.3-μm particles should lead to a collection efficiency of at least 99.95%. DOP testing can also be used to perform leakage tests of various system components by introducing the DOP particles into the duct, glove box, or hood during system operation. The DOP sampling probe can be used to measure DOP concentrations at areas of concern.

Should the engineering controls fail or degrade, it will be necessary to assess the quantity of radioactive material released as well as the radiological impact upon affected workers. The next section addresses the determination of activity from air sample analysis.

SAMPLE COUNTING

The concept of counting samples for gamma radionuclides is common to many fields of health physics, including the university environment. Sample counting involves a bookkeeping exercise in which an unknown spectrum is compared to radionuclide emissions having known energy emissions. The comparison of known radionuclide to unknown emission is made as a function of energy.

The concept of counting efficiency is essential to a quantitative assessment of the sample's activity. The counting efficiency is given by

$$e(E) = N(E)/S \times t \tag{3.1}$$

where

$e(E)$ = counting efficiency as a function of gamma ray energy
E = gamma ray energy
$N(E)$ = total detector counts at energy E
t = counting time (sec)
S = strength of the calibration source (gammas/sec) at energy E

The counting laboratory will have the necessary data to determine the activity of the unknown sample. This information will include a curve of counting efficiency versus gamma ray energy, a listing of peak energies and net detector counts, and a list of possible isotopes and their gamma energies and abundances. With these data, we can calculate the activity of the isotopes in the sample from the relationship

$$A_i = C(E)k/(t)[e(E)][Y(E)] \qquad (3.2)$$

where

A_i = activity of isotope i (μCi)
k = conversion factor $(2.22 \times 10^6 \text{ dpm}/\mu\text{Ci})^{-1}$
t = counting time (min)
$C(E)$ = counts at energy E
$Y(E)$ = gamma yield for the gamma ray of energy E

In estimating the activity, unique peaks are used to obtain the activity of the isotopes corresponding to these peaks. This information is then used to obtain the contributions of isotopes having overlapping peaks. An example of this procedure will be illustrated in one of this chapter's scenarios. Beta and alpha counting follow a similar methodology.

A positive air sample indicates that a release of material has occurred. Once radioactive material is no longer contained in its engineered structure, it can be inhaled, ingested, or absorbed through the skin. In the next section, we present a technique for a quick assessment of the radiological consequence of the released material. More detailed assessments using ICRP-30 or −60 methodology will be required, but the quick assessment will serve as a guide to the immediate recovery steps.

INTAKE OF RADIONUCLIDES

If a radioactive release from a glove box, fume hood, or other enclosure occurs, the health physicist has a number of tasks to accomplish. These include investigating the cause of the release as well as assessing its radiological consequences. In assessing the radiological hazards, the quantity of material taken into the body must be determined.

The intake can occur following accident or routine conditions and may be represented by a relationship that assumes that all potential airborne radionuclides have volatilized or evaporated and are uniformly distributed in the room or laboratory. With these assumptions, the intake is given by

$$I_i = (\text{BR})tA_i/V \qquad (3.3)$$

where

I_i = intake of radionuclide i (μCi)
A_i = activity of isotope released into the room (μCi)

BR = breathing rate (liters/min)
t = time the worker resides in the room after the activity is uniformly dispersed (min)
V = volume of the laboratory (liters)

Once a deposition of radionuclides occurs, an assessment of dose is often required. Assuming that the organ of interest may be treated sufficiently within a single compartment approximation, the initial dose rate is given by

$$\dot{D}_0 = 2.13 PE(QF)/m \qquad (3.4)$$

where

$\dot{D}_0$ = initial daily dose rate (rad/hr)
P = initial activity in the organ (μCi)
E = energy deposited in the organ (MeV)
m = organ mass (g)
QF = quality factor for the emitted radiation

OTHER RESEARCH ACTIVITIES

In addition to the research activities utilizing the isotopes of Table 3.1, additional university research activities use radioisotopes or produce ionizing radiation. A sampling of these activities include agricultural and environmental research, research reactor operations, nuclear and particle physics accelerator activity, materials research via x-ray diffraction techniques, and fusion energy research. A brief summary of the health physics considerations from these activities will be provided in the remainder of this chapter.

AGRICULTURAL/ENVIRONMENTAL RESEARCH

Agricultural and environmental research activities encompass the areas of soil fertility, irrigation, crop production, insect and pest control, plant breeding and genetics, and agrochemical usage. This work includes research on the application of isotope techniques to minimize the use of nitrogen fertilizers and studies of the application of radiation-based techniques to control the population of insect pests.

These research studies include techniques for monitoring and assessing environmental pollution, including analysis of radioactive and nonradioactive pollutants, monitoring of pesticide residues in food and in the environment, and climatic studies through the isotopic content of precipitation. Research studies also include techniques for reducing the environmental impact of chemical-induced productivity increases, radiosterilization of insects, studies of nitrogen fixation in crops and trees, use of tracers to optimize fertilizer use, and nuclear techniques to improve water management practices.

The health physicist must control the isotopes used in the studies. With tracer isotopes, the control of internal exposures and contamination control will be a primary concern. Sterilization of insects may involve use of very-high-activity gamma ray sources, and direct dose will be the major concern. Material analysis using accelerator activation techniques or proton-induced x-ray emission requires the application of sound accelerator health physics practices.

RESEARCH REACTORS

Research reactors have many problems in common with power reactors, but at a smaller scale because of the limited reactor power. Releases of radioactive material and maintenance problems are not as severe, but the neutron and gamma fluences merit an appropriate level of concern.

University reactors are used in a variety of activities, including the production of irradiated materials and special isotopes. The samples pass through high-flux regions and can become highly activated. Production of these materials challenge the health physicist. He or she must balance sound health physics and dose control efforts with the needs of the researcher.

Material radiation damage studies involve the bombardment of materials with large neutron or gamma ray fluences. This research may involve studies of basic material properties and may be related to power reactor nuclear safety. A current nuclear safety topic involves the embrittlement of reactor vessels after years of being exposed to large radiation fluences from the nuclear fuel core. These studies require that the health physicist control high activity sources, and remote handling techniques are often employed.

PARTICLE ACCELERATORS

University accelerators include a variety of research tools. These include accelerators utilized in low-energy nuclear physics research involving terminal voltages of up to 20 MeV that include ions of most nuclear cores. Intermediate-energy machines, with energies up to a few thousand MeV, produce pion and muon beams with an added complexity associated with their dosimetry. The high-energy machines, with maximum energies on the order of several thousand GeV, create a variety of new particles and associated radiations. Specific problems will be addressed in the scenarios and in Chapter 7.

Accelerators are also utilized in medical research applications. These activities are addressed in Chapter 2.

MATERIALS RESEARCH VIA X-RAY DIFFRACTION TECHNIQUES

The output from a research x-ray diffraction source is normally a thin, high-exposure-rate beam of energy on the order of 25 keV or less. The exposure rate in the primary beam may be as large as 10^5 R/hr. Therefore, very high local exposure rates are possible, and appropriate health physics controls are warranted.

The diffracted beam is also of concern, with exposure rates on the order of 100 R/hr possible. Beam leakage through improperly aligned shielding, exposure from fluorescent radiation from the material under evaluation, and exposure to x-rays from high-voltage components and power supplies are additional exposure concerns.

X-ray diffraction devices should be periodically monitored to ensure that the maximum operating voltage and other operating parameters remain unchanged. The shielding configuration, beam shutters, beam stops, scattered radiation, diffracted beam exposure rate, and fluorescence exposure rates should be verified to fall within the expected ranges. Changes in the experimental configuration or materials should be surveyed before research activity begins.

Personnel dosimetry includes both whole-body and finger rings or wristband dosimetry. Proper training to emphasize the radiation hazard is essential.

From a health physics perspective, regular monitoring of this equipment is required. The machine should have a highly visible warning indicator that signals when the beam is on and when the shutter is open. The device should be enclosed to prevent inadvertent access and be designed with beam interlocks to shut the unit off if the enclosure is breached. The strategic use of signs and warning labels should be utilized to remind the researcher of the inherent hazards of this device.

FUSION ENERGY RESEARCH

Fusion energy research, using inertial confinement or laser techniques, is concerned with the confinement and fusion of hot plasmas of light nuclei. The fusion process produces a variety of radiation types which must be controlled by the health physicist.

Potential fusion reactions under active investigation include the following:

$$D + D \xrightarrow{50\%} T(1.01 \text{ MeV}) + p(3.02 \text{ MeV})$$
$$\xrightarrow{50\%} \text{He-3}(0.82 \text{ MeV}) + n(2.45 \text{ MeV})$$
$$D + T \longrightarrow \text{He-4}(3.50 \text{ MeV}) + n(14.1 \text{ MeV})$$
$$D + \text{He-3} \longrightarrow \text{He-4}(3.60 \text{ MeV}) + p(14.7 \text{ MeV})$$
$$T + T \longrightarrow \text{He-4} + 2n + 11.3 \text{ MeV}$$

$$\text{He-3} + \text{T} \xrightarrow{51\%} \text{He-4} + p + n + 12.1 \text{ MeV}$$

$$\xrightarrow{43\%} \text{He-4}(4.8 \text{ MeV}) + \text{D}(9.5 \text{ MeV})$$

$$\xrightarrow{6\%} \text{He-5}(2.4 \text{ MeV}) + p(11.9 \text{ MeV})$$

$$p + \text{Li-6} \longrightarrow \text{He-4}(1.7 \text{ MeV}) + \text{He-3}(2.3 \text{ MeV})$$

$$p + \text{Li-7} \xrightarrow{20\%} \text{He-4} + \text{He-4} + 17.3 \text{ MeV}$$

$$\xrightarrow{80\%} \text{Be-7} + n - 1.6 \text{ MeV}$$

$$\text{D} + \text{Li-6} \longrightarrow \text{He-4} + \text{He-4} + 22.4 \text{ MeV}$$

$$p + \text{Be-11} \longrightarrow \text{He-4} + \text{He-4} + \text{He-4} + 8.7 \text{ MeV}$$

$$n + \text{Li-6} \longrightarrow \text{He-4}(2.1 \text{ MeV}) + \text{T}(2.7 \text{ MeV})$$

where D is deuterium and T is tritium. For binary events the particle energy is provided in parentheses. A negative yield or Q-value indicates that the reaction is endothermic.

The wide variety of potential reaction products and energies suggests considerable challenge for the health physicist responsible for worker radiation protection at a fusion energy research facility. For example, fusion neutrons will present an external radiation hazard. These neutrons will require shielding, and particular attention must be paid to leakage pathways which will vary with the type of reactor design, fusion reaction under consideration, and reactor operating characteristics.

Activation products will be produced by the high neutron fluence impinging on the structural components of the reactor, including the reactor or plasma containment vessel. The specific activation products will depend upon the structural material utilized in the design. Anticipated structural materials include stainless steel, niobium, and ceramic materials such as Al_2O_3. Activation products will include isotopes of Al, Na, Fe, Co, Ni, Mn, Nb, and Zr which decay by beta, positron, and electron capture with associated gamma emission.

As with fission reactors, activation products present a radiation hazard during maintenance activities. Major components require periodic replacement due to the high-energy neutron bombardment. These structural waste materials will require remote handling and processing to minimize worker exposures.

If nuclear fusion becomes a commercial electrical power source, the initial plants will likely use the D + T reaction. The D–T reaction will require that commercial fusion plants retain large tritium fuel inventories.

These tritium inventories will present an internal uptake challenge. Tritium in either molecular form or as tritiated water will diffuse through the structural materials at high operating temperatures. Tritium leakage from the reactor vessel's coolant, through seals, valves, and piping, will require health physics controls. Some tritium will also diffuse into the steam system and be released to the environment. A portion of the tritium will reside in routine work areas where it presents a skin absorption and inhalation hazard. The tritium will

78 UNIVERSITY HEALTH PHYSICS

appear as surface contamination which can be resuspended into the air or directly contaminate personnel.

Maintenance of activated structural components present an external as well as internal radiation hazard. The health physics concerns will be similar to those encountered in a commercial fission reactor. In particular, maintenance activities generate particles of a respirable size as a result of cutting, grinding, and repair activities. Additional discussion of health physics concerns associated with maintenance activities is provided in Chapter 5.

Additional radiation hazards are presented by the unique scenarios of a fusion plant accident. Initial designs propose to use liquid metal coolants and heat exchange systems. In a severe accident the liquid metal coolants contacting air, water, or steam may lead to an explosive reaction that produces hydrogen gas. Such an event could lead to a loss of structural integrity with the subsequent transport and deposition of activation products and tritium to offsite locations. These releases will differ significantly from those of a fission reactor which involve primarily noble gas and radioiodine concerns. The final safety analysis report for the commercial fusion plant will address these and other fusion accident scenarios.

SCENARIOS

Scenario 41

You are given the assignment of determining the amount of activity found on a charcoal filter used in the stack monitor at a research reactor. The detector you choose to use is a Ge(Li) detector.

Detector calibration, sample counting, and isotopic information are provided as follows:

Detector Calibration: Charcoal Filter Geometry

Gamma Energy (keV)	Calibration Source (gammas/sec) Decay Corrected	Ge(Li) Detector Net Counts (1000-sec count)
60	2000	83762
88	1840	96001
121	1590	79242
344	1490	31257
768	730	8730
963	820	7400
1408	1170	8215

Sample Count: Charcoal Filter Geometry

Peak Energy (keV)	Ge(Li) Detector (net counts in a 10-min count)
121.5	7266
136.2	8034
265.0	4467
279.2	2279
401.1	629

Possible Isotopes in the Filter

Isotope	Half-life (day)	Gamma Energy (keV)	Abundance (%)[a]
Co-57	271.8	122.1	85.9
		136.5	10.3
Se-75	119.8	121.1	17.3
		136.0	59.0
		264.7	59.2
		279.5	25.2
		400.7	11.6
Hg-203	46.7	279.2	81.5

[a] Number of gammas emitted per disintegration × 100.

3.1. Calculate the counting efficiency at each gamma photon energy for the charcoal filter geometry.

3.2. Calculate/graph the efficiency curve.

3.3. From the efficiency calculation, estimate the counting efficiency for a gamma photon having an energy of 2 MeV.

3.4. At medium energies (100–300 keV), the best explanation for the decrease in counting efficiency is:

   a. Photoelectric absorption decreases linearly when graphed on a log–log scale.

   b. Compton scattering decreases linearly when graphed on a log–log scale.

   c. Pair production does not become a significant interaction in Ge(Li) until gamma energies are about 5 MeV.

   d. Energy absorption from Compton scattering decreases with increasing energy.

   e. The K-edge for Ge(Li) occurs at 4 MeV.

3.5. The best explanation for the decrease in counting efficiency at low gamma energies (<80 keV) is:
   a. Attenuation of the gammas by the detector dead layer and housing becomes significant.
   b. These gammas do not interact with the Ge(Li).
   c. Pair production is insignificant at these low energies.
   d. The probability of interaction decreases at energies below the K-edge of Ge(Li).
   e. The signal given off by the Ge(Li) detector is no longer proportional to the energy absorbed in the crystal.

3.6. Identify the calculate and activity (μCi) of all nuclides contained in the charcoal filter sample.

Scenario 42

A laboratory contains hoods to provide graduate students safe areas to use radiochemicals. Unknown to laboratory personnel, the motor serving the exhaust fan for the hoods is turned off for repair for 1 hour. During this time, one graduate student completes a dual labeling experiment using 10 mCi of P-32 (as orthophosphate) and 10 mCi of tritium (as tritium oxide) in a hood. Another graduate student completes a procedure with 5 mCi of I-125 (as sodium iodide solution). As soon as the hood condition is recognized, the Radiation Safety Officer (RSO) is called by the graduate students.

Data

$$\text{Laboratory dimensions} = 20 \text{ ft} \times 20 \text{ ft} \times 10 \text{ ft}$$
$$1 \text{ Cubic foot} = 28.3 \text{ liters}$$
$$\text{Breathing rate} = 20 \text{ liters/min}$$

3.7. As the RSO, you want to sample the air of the laboratory. What is the airborne potential for these three isotopes?

3.8. Describe how you would perform air sampling for potential airborne radionuclides.

3.9. Should air sampling for these radionuclides be isokinetic or anisokinetic?

3.10. Assume a maximum credible case: All potentially airborne radionuclides have volatilized or evaporated and are uniformly distributed in the room. The air exchange rate is low. Calculate the maximum uptake the graduate students could have received.

3.11. How much radioactivity could be in the critical organs?

3.12. As the RSO, you perform urinalyses on the two students for appropriate sampling periods. Explain what analytical techniques would be used to assess radioactivity in the urine.

3.13. If sodium borotritide was used instead of tritium oxide, how could you assess the airborne potential?

3.14. Describe measures that should be instituted to prevent recurrence of this event.

3.15. List conditions that would affect the amount of radioactivity in the students.

Scenario 43

You are the RSO at a large university. A technician in the Physics Department reports that a radioactive source may have ruptured. The source is labeled as 100 mCi of Am-241. Preliminary wipes taken on the source, in the technician's work area, and on the technician's clothing show extensive alpha contamination. The technician believes that the source broke when a sharp object fell on it the previous day. This morning he carried the source to the machine shop in the basement and to two offices in the building. The basement and technician's work areas have linoleum floors, whereas the hallways and offices are carpeted.

Data

Lung mass = 1000 g

Am-241 particle clearance rate: $T_{1/2}$ = 120 days (ICRP-2)

Deposition energy of the alphas and recoil atoms = 5.57 MeV per decay

3.16. Given the following methods available to you, which would be best to identify all radioactive contamination from this incident in the building?
   a. Take wipes in all suspected areas, and count on a shielded gas-flow proportional counter/sample changer.
   b. Use a pancake probe GM meter to survey all suspected areas.
   c. Use a ZnS-coated/photomultiplier-tube-based portable alpha probe to survey all suspected areas.
   d. Use a 1.5-in. × 1-in. NaI crystal probe to survey all suspected areas.
   e. Use a 5-in. × 0.06-in. NaI crystal probe to survey all suspected areas.

3.17. If the source had instead been Cs-137, which means would be best to identify all contamination in the building?
   a. Take wipes in all suspected areas, and count on a shielded gas-flow proportional counter/sample changer.
   b. Use a portable pressurized ion chamber to survey all suspected areas.

c. Use a ZnS-coated/photomultiplier-tube-based portable alpha probe to survey all suspected areas.

d. Use a 1.5-in. × 1-in. NaI crystal probe to survey all suspected areas.

e. Use a 5-in. × 0.06-in. NaI crystal probe to survey all suspected areas.

3.18. Which would be the most sensitive means of detecting an intake of Am-241 by the technician who handled the source, assuming the measurement is made within the first 48 hr following the exposure?

a. Count nasal swabs on a shielded gas-flow proportional counter to estimate the activity inhaled.

b. Whole-body counting in a shielded facility via a 3-in. × 3-in. NaI detector/MCA.

c. Whole-body counting in a shielded facility via a coaxial germanium detector/MCA.

d. Analysis of 24-hr urine sample collected on day 2 by liquid scintillation counting.

e. Analysis of 24-hr fecal sample collected on day 2 by germanium detector/MCA.

3.19. If the source had instead been Cs-137, which would be the most reliable means of quantifying the technician's uptake, assuming the measurement is made within the first 48 hr following exposure?

a. Count nasal swabs on a shielded gas-flow proportional counter to estimate the activity inhaled.

b. Whole-body counting in a shielded facility via a 3-in. × 3-in. NaI detector/MCA.

c. Analysis of activity exhaled in the breath.

d. Analysis of 24-hr urine sample collected on day 2 by liquid scintillation counting.

e. Analysis of 24-hr fecal sample collected on day 2 by germanium detector/MCA.

3.20. Assuming that the technician had a deposition of 1 mCi Am-241 in the lung, calculate the initial daily dose rate to the lung. Assume that the lung behaves as a single compartment.

3.21. If a 1-mCi uptake of Am-241 to the lung occurred due to a chronic dose over a 100-day period versus an acute, 1-day dose, how would this affect the 50-year committed dose equivalent?

a. Dose by chronic exposure would be greater by a factor of ln(100).

b. Dose by chronic exposure would be greater by a factor of log(100).

c. Dose by both exposure periods would be virtually the same.

d. Dose by acute exposure would be greater by a factor of log(100).

e. Dose by acute exposure would be greater by a factor of ln(100).

3.22. ICRP-26 recommends the use of a weighting factor (W_T) to express the proportion of stochastic risk from an irradiated tissue (T) to the total risk when a body is uniformly irradiated. For the lung, this factor has a value of:

   a. 0.03
   b. 0.12
   c. 0.25
   d. 0.30
   e. 0.50

3.23. Am-241 is used in a variety of applications because of its desirable nuclear properties. The most common mode of production of Am-241 is:

   a. from naturally occurring americium in the environment.
   b. from alpha decay of Cm-245.
   c. from spontaneous fission of Cf-252.
   d. from the beta decay of Pu-241.
   e. from the Np-237(alpha, n) reaction.

Scenario 44

You have been employed as the RSO at a university for 1 month. Late on Friday afternoon (when else?) you are contacted by the Student Health Center concerning a 20-year-old female student. The SHC performed an abdominal x-ray examination on the student 5 days earlier. The student has just now reported that she was 2 weeks pregnant at the time of the examination. You must make a quick assessment of the dose to the embryo based on the data you obtained during a survey of the x-ray unit 1 week earlier.

Data from your Survey

Unit set in manual mode with 30 mA-sec and 80 kVp.
Ionization chamber readings were taken at a distance of 30 cm above the table top.

Readings from the ion chamber:

0 mm of Al added between the chamber and the tube = 500 mR
2.5 mm of Al added between the chamber and the tube = 280 mR
4.5 mm of Al added between the chamber and the tube = 180 mR
kVp on the control panel of the unit: 80 kVp

Measured kVp: 80 kVp

Patient and x-ray exam information: One anterior–posterior (AP) and one posterior–anterior (PA) film were taken with a retake needed on each film. The source image distance (SID) used was 40 inches. Technique factors = 35 mA-sec and 80 kVp. The x-ray field was collimated to a 14-in. × 17-in. size at 40 inches. The film (image receptor) was located 2 in. below the table top in the tray. The patient abdominal thickness is 20 cm.

Abdominal Embryo Data

Abdominal-Organ Dose (mrad) for 1-R Entrance Skin Exposure (Free-in-Air)

Beam Quality HVL (mm Al)	=	Dose (mrad/R)[a]					
		1.5	2.0	2.5	3.0	3.5	4.0
Ovaries	AP	97	149	203	258	313	367
	PA	60	100	146	198	255	317
	LAT	18	33	50	70	93	118
Embryo	AP	133	199	265	330	392	451
(uterus)	PA	56	90	130	174	222	273
	LAT	13	23	37	53	71	91

[a]SID 102 cm (40 inches); Film size = field size: 35.6 cm × 43.2 cm (14 inches × 17 inches).

The geometry for a x-ray situation is illustrated in Fig. 3.1.

3.24. Find the absorbed dose to the embryo based on your survey data and the given tabular data. Clearly state any assumptions used in your determination.

3.25. Identify which of the following best reflects radiation exposure *in utero* and its effect on severe mental retardation of the child. (*Note:* Your numeric answer to question 3.24 is not considered in this question.)

   a. The first 2 weeks following fertilization are the most critical with respect to radiation exposure causing severe mental retardation.

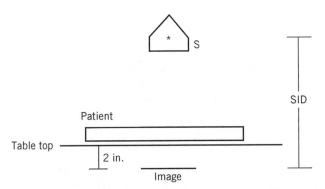

Fig. 3.1 Geometry for the unanticipated fetal x-ray situation. The labels S and SID represent the x-ray source and the source image distance, respectively.

b. Prior to the eighth week following fertilization, there is apparently little risk of severe mental retardation.
c. Lowest risk of forebrain damage based on A-bomb data occurred between the 8th and 15th week following fertilization.
d. A-bomb data show that there is no correlation between time after fertilization and exposure of the embryo or fetus resulting in severe mental retardation of the child.
e. A-bomb data indicate that the highest risk of forebrain damage occurred after the 15th week following fertilization.

3.26. The NCRP recommendations for pregnant radiation workers are to limit the dose equivalent to:
a. 100 mrem/month to a pregnant woman over the entire pregnancy.
b. 100 mrem/month to the fetus/embryo over the entire pregnancy.
c. 300 mrem/quarter to a pregnant woman.
d. 450 mrem to a pregnant woman over the entire pregnancy.
e. 500 mrem to the fetus/embryo over the entire pregnancy.

Scenario 45

During a bioassay counting session at your university, you discover that a researcher has an uptake of I-125. The following data should be considered:

Mass of thyroid = 30 g

Physical half-life of I-125 = 60 days

Biological half-life of I-125 in thyroid = 130 days

Mean energy of K x-rays = 28 keV (139.8%)

Mean energy of gamma rays = 35 keV (6.7%)

Equilibrium dose constants = 0.0884 and 0.0434 g-rad/μCi-hr

for x- and gamma rays, respectively

Absorbed fraction for thyroid (@0.03 MeV) = 0.7

w_T(thyroid) = 0.03

3.27. Thyroid bioassay results show 2.4 μCi of I-125 currently in the thyroid of a researcher on the day of your thyroid bioassay counting. Investigation determines that this uptake occurred during a single iodination procedure 14 days earlier. Consider the uptake as a single exposure and find the thyroid dose to time = infinity based on the initial uptake which occurred 14 days earlier.

3.28. Find the effective dose equivalent to a researcher with the following annual exposures:

$$\text{Whole-body film badge} = 1.25 \text{ rem}$$
$$\text{Internal dose to the thyroid} = 25 \text{ rem}$$

3.29. Using the annual internal dose equivalent guidance of ICRP-26, determine if a thyroid overexposure (for the dose equivalent reported in 3.28) has occurred. Justify your answer by giving the appropriate limit.

Scenario 46

A member of the University Chemistry Department has received a grant from the Nuclear Regulatory Commission to investigate the movement of radioactive material in the vicinity of a low-level waste burial facility. The object of the study is to obtain data on the quantities of radionuclides, particularly the I-129 in low-level radioactive waste from a typical light-water reactor facility. Data are to be obtained from several BWR and PWR reactors. The study involves the collection of a sample by passing a volume of reactor coolant through an ion-exchange column.

Data

Column dimensions
- Inside diameter: 1.0 in.
- Active length: 6.0 in.
- Wall thickness: 0.0625 in.
- Column material: Plexiglass (density = 1.0 g/cm^3)
- Resin collection efficiency: 0.95
- Resin density: 1.0 g/cm^3
- Cation/anion ratio: 1.0
- Sample flow rate: 100 ml/min
- Sample volume for this activity: 5000 liters
- Normal sampling volume: 1 liter

Reactor coolant
- Radionuclide: Co-60
- Concentration: 1.92×10^{-2} μCi/ml
- Specific gamma constant: 13.2 (R-cm^2)/(mCi-hr)
- Average beta energy: 0.096 MeV

Assume that activated corrosion products behave as cations. I-129 LLD for a normal reactor coolant sample is 3.0×10^{-10} μCi/ml.

Radiolytic Hydrogen Generation Rates:

$$G_H = 0.6 \text{ molecules } H_2/100 \text{ eV (anion bed)}$$

$$G_H = 0.13 \text{ molecules } H_2/100 \text{ eV (cation bed)}$$

1 rad = 6.242×10^7 MeV/g

3.30. Calculate the exposure rate in mR/hr at a distance of 18 in. from the center of the unshielded ion-exchange column at the end of the sample collection period. Clearly list all assumptions.

3.31. At the end of the sample collection period, the investigator performing this study notifies the university radiation safety officer that there will be a delay of approximately 6 months to 1 year in processing the sample that was just collected. State and determine two potential safety concerns that the RSO might have about storage of this sample over an extended period. Calculations must support your answer.

3.32. State one administrative and two operational radiological controls that should be imposed on this activity.

3.33. Calculate the new LLD obtained for I-129 as a result of this study. Assume that the radiochemical yield and counting efficiency are the same as that used for routine analysis.

Scenario 47

You are a university health physicist who has started a small firm manufacturing medical isotope generators. One of your products is an Mo-99/Tc-99m generator packaged within a 5-cm-thick lead shield within a $60 \times 60 \times 60$ cm^3 carton, so that the activity is in the center of the carton. Each package bears a label with the following information:

Radioactive III
Contents Mo-99
Activity 1.5 Ci
Transportation index = 3.0

The half-life of Mo-99 is 67 hr, and the half-life of Tc-99m is 6.0 hours.

3.34. What is the maximum dose equivalent rate at 1 meter from the surface of one of these packages?

3.35. Calculate the maximum dose equivalent rate at the surface of one of these packages?

3.36. Is the package legal for shipment on passenger aircraft in the United States? Why?

3.37. Four such packages comprise a shipment in the cargo hold of a commercial airliner. What is the transport index (TI) of the shipment?

3.38. A small package containing TLDs (being shipped for processing) is centered among the four packages containing radioactive materials. The TLD package is positioned 50 cm from the surface of each of these packages and left in place for 6 hours. The TLD package is separated from the generator packages by cartons of styrofoam pellets. What dose equivalents do the TLDs receive?

3.39. What provisions do commercial dosimetry vendors make to compensate for the transient dose described above and obtain valid dosimetry results?

3.40. If the generator package is returned to the manufacturer for disposal and recycling exactly 2 weeks after it was initially labeled, what will the TI be for the return package?

3.41. An Mo-99/Tc-99m generator package (not the one specified previously) has a surface dose equivalent rate of 15 mrem/hr and a TI of 2.0. What label is required and why?

Scenario 48

During a routine laboratory inspection you discover that a 10-Ci polonium-210 source being used by a group of experimenters has been badly leaking for about 2 weeks and that contamination has spread widely. This scenario addresses the actions that you would take in evaluating and correcting the uncontrolled spread of radioactive material.

3.42. List the first four actions you would take to begin evaluating and correcting this situation.

3.43. Based upon your past experience, identify the following pertinent parameters with respect to the possible radiation hazards from this source: primary mode of decay, decay energy, half-life, and hazard type.

3.44. During the investigation, you request measurements for airborne, removable, and fixed contamination. For each of these measurements, list the preferred equipment and locations, as well as precautions and other concerns you would have in evaluating the information.

3.45. Following the evaluation, list the areas you would decontaminate first. Why?

3.46. List items that should be included in the incident investigation report for advising the University President of the follow-up evaluations and corrective actions taken.

Scenario 49

You are the RSO at Excited State University where investigators use a variety of radionuclides. The major radiochemicals of concern are inorganic P-32, NaI-125 to label biological compounds, and tritiated water (HTO). All are used in millicurie quantities.

3.47. Characterize the relative hazards of these three radiochemicals in terms of skin dose potential, bioassay requirement, eye hazard, personnel dosimetry requirement, and air sampling requirement.

3.48. Discuss how you would shield 10 mCi P-32 in terms of shielding material to protect the torso when working with it on a bench top for about 3 hr per week. What material and thickness would be warranted?

3.49. If the activity of P-32 being handled is increased to 1000 mCi, how does the shielding requirements change?

3.50. Describe how P-32 radioactive waste should be processed to minimize your department's budget.

3.51. A researcher asks you for advice on whether he should take prophylactically 100 mg of potassium iodide before iodinating proteins with 2 mCi of I-125. What would you advise? Why?

3.52. You perform a wipe test in an experimental area where H-3 and I-125 are being used. If you could only afford one instrument to detect mixed H-3/I-125 contamination, what would it be? Why?

REFERENCES

ANSI N.13.1-1969, *Guide to Sampling Airborne Radioactive Materials in Nuclear Facilities*, American National Standards Institute, Washington, DC (1982).

J. J. Bevelacqua, Microscopic Calculations for the He-4 Continuum, *Canadian Journal of Physics*, **58,** 306 (1980).

J. J. Bevelacqua and R. J. Philpott, Microscopic Calculations in the He-4 Continuum (I). General Approach, *Nucl. Phys.*, **A275,** 301 (1977).

H. Blatz, *Radiation Hygiene Handbook*, McGraw–Hill, New York (1959).

A. Brodsky, *CRC Handbook of Management of Radiation Protection Programs*, CRC Press, Boca Raton, FL (1986).

HASL-300, *Health and Safety Laboratory Procedures Manual*, US ERDA, New York (1981).

IAEA Safety Series No. 35, *Safe Operation of Research Reactors and Critical Assemblies*, IAEA, Vienna, Austria (1984).

ICRP Publication No. 36, *Protection Against Ionizing Radiation in the Teaching of Science*, Pergamon Press, New York (1983).

NBS Handbook 111, *Radiation Safety for X-Ray Diffraction and Fluoresence Analysis Equipment*, National Bureau of Standards, American National Standards N43.2-1971 (1972).

NBS Handbook 114, *General Safety Standard for Installations Using Non-Medical X-Ray and Sealed Gamma-Ray Sources, Energies up to 10 MeV*, National Bureau of Standards, American National Standards N543-1974 (1975).

NBS Handbook 123, *Radiological Safety Standard for the Design of Radiographic and Fluoroscopic Industrial X-Ray Equipment*, National Bureau of Standards, American National Standards N537-1976 (1977).

NCRP Commentary No. 7, *Misadministration of Radioactive Material in Medicine-Scientific Background*, NCRP Publications, Bethesda, MD (1991).

NCRP Report No. 8, *Control and Removal of Radioactive Contamination in Laboratories*, NCRP Publications, Bethesda, MD (1951).

NCRP Report No. 32, *Radiation Protection in Educational Institutions*, NCRP Publications, Bethesda, MD (1966).

NCRP Report No. 111, *Developing Radiation Emergency Plans for Academic, Medical, and Industrial Facilities*, NCRP Publications, Bethesda, MD (1991).

J. Shapiro, *Radiation Protection—A Guide for Scientists and Physicians*, Harvard University Press (1981).

US Nuclear Regulatory Commission, *Code of Federal Regulations*, Title 10, Energy, Washington, DC (1994).

US Nuclear Regulatory Commission, Regulatory Guide 7.3, *Procedures for Picking Up and Receiving Packages of Radioactive Material*, USNRC, Washington, DC (1975).

US Nuclear Regulatory Commission, Regulatory Guide 8.13, *Instruction Concerning Prenatal Radiation Exposure*, Rev. 1, USNRC, Washington, DC (1975).

US Nuclear Regulatory Commission, Regulatory Guide 8.20, *Application of Bioassay for I-125 and I-131*, Rev. 1, USNRC, Washington, DC (1979).

US Nuclear Regulatory Commission, Regulatory Guide 8.23, *Radiation Safety Surveys at Medical Institutions*, USNRC, Washington, DC (1979).

4

FUEL CYCLE HEALTH PHYSICS

The nuclear fuel cycle includes the mining and processing of ore, the enrichment of that ore to reactor grade material, its use in a power reactor, and subsequent reprocessing. To date, the ore has been uranium- or thorium-based, but other materials may be utilized depending upon the economics of the specific chemical and physical processes of the fuel cycle. Waste disposal and weapons fabrication are also included within the fuel cycle arena.

Environmental issues, including the selection criteria for waste burial sites and radon issues associated with uranium mining, will be addressed in Chapter 6, which specifically addresses environmental issues. The health physics aspects associated with power reactors are discussed in Chapter 5.

RADIATION IN FUEL CYCLE FACILITIES

Beta radiation fields are usually the dominant external radiation hazard in facilities requiring work with unshielded forms of uranium. Table 4.1 summarizes the major uranium and decay product emissions. The majority of the U-238 beta dose is derived from the 2.29-MeV beta emitted by Pa-234m daughter radiation.

Gamma radiation from uranium is not normally the limiting radiological hazard. However, low-level (<5 mrem/hr) gamma fields can exist where large quantities of material are stored. Such fields create ALARA problems, particularly where large numbers of workers can be exposed.

Neutron radiation is also emitted from enriched uranium fluoride compounds. Large storage containers of UF_6, for low enriched (<5%) and highly

Table 4.1 Major Uranium and Decay Product Emissions

Nuclide	Alpha Energy (MeV)	Maximum Beta Energy (MeV)	Gamma Energy (MeV)
U-238	4.15 (25%) 4.20 (75%)	—	—
Th-234	—	0.103 (21%) 0.193 (79%)	0.063 (3.5%) 0.093 (4%)
Pa-234m	—	2.29 (98%)	0.765 (0.3%) 1.001 (0.6%)
U-235	4.37 (18%) 4.40 (57%) 4.58 (8%)	—	0.144 (11%) 0.185 (54%) 0.204 (5%)
Th-231	—	0.140 (45%) 0.220 (15%) 0.305 (40%)	0.026 (2%) 0.084 (10%)
U-234	4.72 (28%) 4.77 (72%)	—	0.053 (0.2%)

enriched (>97%) material, lead to neutron radiation levels of 0.2 mrem/hr and 4 mrem/hr, respectively. At high enrichments, these neutron emissions will play a dominant role in the whole-body radiological hazards of uranium fuel cycle facilities and can be up to a factor of 2 larger than the gamma fields.

OCCUPATIONAL EXPOSURE

In 1993, the U.S. Department of Energy (U.S. DOE) amended its regulations to set new occupational radiation dose limits for its employees and contractor personnel. Although the amendments to 10CFR835 have specific criteria unique to the U.S. DOE, radiation workers are normally limited to no greater than 5 rem (0.05 Sv)/year, which is the same limit as workers whose employers are regulated by the U.S. Nuclear Regulatory Commission (U.S. NRC).

The new DOE regulations also allow for the 5-rem limit to be exceeded by an additional 5 rem for workers involved in planned special exposures (PSEs). A PSE would not be classified as an emergency exposure, but would only occur under "approved, well-justified, well-controlled, highly infrequent, and unusual conditions." A worker would be limited to a total of 25 rem from planned special exposures during his or her career. The U.S. DOE rule is very similar to the 1993 10CFR20 revision promulgated by the U.S. NRC.

NUCLEAR FUEL CYCLE

The nuclear fuel cycle describes the path that is followed by nuclear reactor fuel in its successive stages from mining of the uranium or thorium ores to the final disposal of the radioactive wastes derived from the reprocessing of the spent reactor fuel. In its ideal form, the spent fuel removed from a reactor is reprocessed and the U-235 and plutonium are recovered for subsequent reuse as new reactor fuel.

Uranium Fuel Cycle

Natural uranium consists of three primary isotopes; U-238, U-235, and U-234, whose natural abundances are 99.2739, 0.7204, and 0.0057, respectively. The decay products of these uranium isotopes consist of long decay chains that decay by both alpha and beta radiation. These chains are outlined in more detail in the Chapter 6.

Uranium and its decay products are predominantly an internal radiation hazard, and their radiological hazard can be assessed by the methods noted in Appendix IV. Although uranium and its decay products are predominantly an internal radiation hazard, their beta decays produce a significant external hazard as noted in Table 4.2.

Historically, the radiological concerns of uranium have been overshadowed by chemical toxicity concerns. As a heavy metal, uranium is chemically toxic to the kidneys. Although the radiological hazards are becoming an increasing concern, the health physicist must still ensure that the chemical toxicity is properly addressed.

The theoretical cycle for uranium fuels is illustrated in Fig. 4.1. The theoretical uranium cycle assumes that the spent reactor fuel is reprocessed in order to recover uranium and plutonium for subsequent recycling as reactor fuel.

Table 4.2 Beta Surface Dose Rates from Equilibrium Thicknesses of Uranium Metal and Compounds

Source	Surface Dose Rate (mrad/hr)[a]
Natural uranium metal slab	233
UO_2	207
U_3O_8	203
UF_4	179

[a]Beta surface dose rate in air through a 7-mg/cm^2 polystyrene filter.
Source: EGG-2530.

94 FUEL CYCLE HEALTH PHYSICS

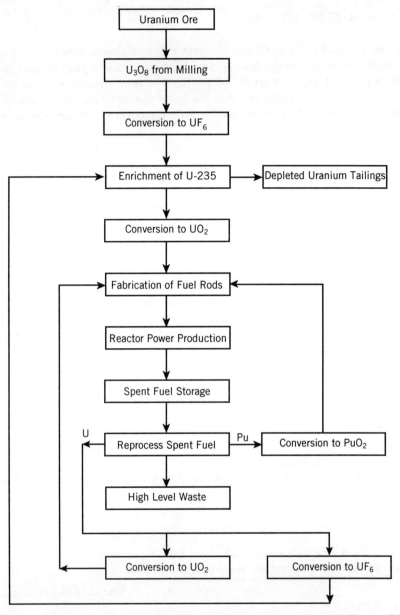

Fig. 4.1 Theoretical nuclear fuel cycle for uranium. [Adapted from the Energy Deskbook (DOE/IR/05114-1).]

URANIUM ORE AND CHEMICAL PROCESSING

The average uranium content of U.S. ores, expressed as the oxide U_3O_8, is about 0.2 weight percent; that is, 1000 kg of ore will contain the uranium equivalent of 2 kg of U_3O_8. After processing, the raw ore is converted to U_3O_8, also known as *yellowcake*. The U_3O_8 is reduced to UO_2 (brown oxide) by utilizing hydrogen gas:

$$U_3O_8 + 2H_2 \rightarrow 3UO_2 + 2H_2O \tag{4.1}$$

Uranium tetrafluoride (UF_4), also known as *green salt*, is formed by heating uranium dioxide in hydrogen fluoride gas:

$$UO_2 + 4HF \rightarrow UF_4 + 2H_2O \tag{4.2}$$

The tetrafluoride is subsequently converted into uranium hexafluoride by the use of fluorine gas:

$$UF_4 + F_2 \rightarrow UF_6 \tag{4.3}$$

The mining and chemical processing steps primarily present an internal radiation hazard from ingestion or inhalation of the material. At low enrichments, chemical toxicity will be a more limiting concern than the radiological hazard.

Enrichment

The next step in the uranium fuel cycle is the enrichment of the U-235 content from its nominal value of 0.72% by weight in natural uranium. The traditional methods for enriching uranium are the gaseous diffusion and gas centrifuge technologies. An emerging enrichment technology involves the use of selective laser photoionization of atomic uranium vapor.

GASEOUS DIFFUSION

Isotopic separation by the diffusion process is accomplished by diffusing uranium hexafluoride through a porous membrane. The different molecular weights of U-235F_6 and U-238F_6 and their resulting difference in molecular velocities are used as the basis for separating U-235 from U-238. In a mixture of U-235F_6 and U-238F_6, the average speed of the lighter U-235F_6 molecules is greater than that of the heavier U-238F_6 molecules. When the mixture contacts a porous barrier, the lighter U-235F_6 molecules strike the barrier and pass through it more frequently than the heavier U-238F_6 molecules. Because the velocity difference is small, the enrichment through each gaseous diffusion, chamber or stage is small. Consequently, thousands of stages are required to increase the assay from 0.7% U-235 to the desired enrichment.

A gaseous diffusion stage consists of a motor, compressor, and converter which contains the porous barrier or membrane and a cooler. The uranium hexafluoride is introduced as a gas and is made to flow through the inside of the barrier tube. A portion of the gas, about half, diffuses through the barrier and is fed to the next higher (increased U-235 enrichment) stage. The remaining gas that did not diffuse through the tube is fed to the next lower enrichment stage. The diffused stream is slightly enriched in U-235, and the gas remaining in the tube is slightly depleted in U-235.

A few of the stages of a diffusion cascade are illustrated in Fig. 4.2. The stages above the location of feed entry are the enriching section, and the U-235 concentration exceeds that of the nominal feed concentration. In the stripping section, below the feed point, the concentration of U-235 is less than the nominal feed concentration. The enrichment increases (decreases) the further the stage is upstream (downstream) of the feed point.

As noted in the figure, the feed for each stage in the cascade is a mixture of the enriched material from the stage immediately below and the depleted material from the stage immediately above. The enriched material from a stage will be that material that has preferentially diffused through the barrier, and the depleted material is preferentially the undiffused material. The cascade operates continuously with feed material supplied (at the right side of Fig. 4.2), enriched product drawn off the top, and depleted tails drawn off the bottom.

The number of stages in a cascade will be a function of a number of variables, including the isotopic concentration of the feed material, the desired product and tails concentrations, and the efficiency of the diffusion barrier material. For a typical application of natural uranium feed material, reactor-grade fuel product of 2.5–4.0% U-235 enrichment, and a tails assay of 0.2%, about 2000 stages will be required.

The number of stages could be altered if the product or tails assay were altered. For example, the number of stages would be reduced if the U-235 content of the tails material were increased. However, this change would reject a larger amount of U-235 that would be eliminated as tails material.

The primary hazard in a gaseous diffusion plant is from an acute exposure from a release of uranium hexafluoride from the process equipment. Chronic exposures may arise from routine maintenance or processing operations. The radiological hazard varies with the U-235 enrichment.

The uranium feed materials for the enrichment process may include small quantities of neptunium and plutonium. The radiological controls based upon the uranium hazards will usually be adequate to control the presence of these transuranic contaminants. However, these transuranics can represent a significant internal radiation hazard because their specific activities and Annual Limits on Intake (ALIs) are generally more limiting than those of the uranium isotopes.

For low enrichments, chemical toxicity remains the controlling hazard. At

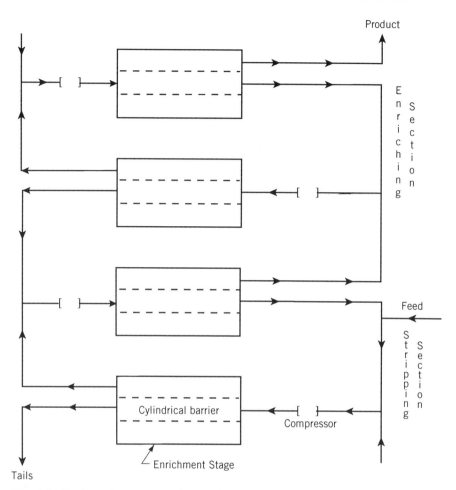

Fig. 4.2 Portion of a gaseous diffusion cascade. Material flowing radially through the cylindrical barrier will be enriched while material flowing axially along the barrier will be depleted in comparison to material entering the enrichment stage. [Adapted from the Energy Deskbook (DOE/IR/05114-1).]

higher enrichments, radiation effects become the primary concern. Criticality must also be considered at higher enrichment stages primarily at stages near the top of the cascade.

Most of the chemical compounds encountered in a gaseous diffusion plant, including uranium hexafluoride and uranyl fluoride, are class D compounds. Interactions of these materials with the process equipment and the environment can produce Class W compounds during normal and abnormal operations.

In a gaseous diffusion plant, or other facility utilizing uranium hexafluoride, the probability of a criticality is minimized by controlling the process parameters to prevent the solidification of the uranium hexafluoride. The integrity of the process stages is also maintained to prevent the inleakage of water or moist air. Radiation monitors located in key locations provide early detection of an accumulation of solidified uranium hexafluoride. For plant components containing uranium solutions or storing uranium compounds, various criticality controls are applied. These controls include geometry and batch control, limitations on the uranium concentrations and enrichment, and administrative or procedural controls.

The primary personnel hazard from a criticality event is to personnel in the immediate vicinity of the event. Timely evacuation of personnel is an effective radiation control measure. Criticality alarms will not prevent an inadvertent criticality, but they will enhance the evacuation of personnel from the immediate area of the criticality.

GAS CENTRIFUGE

Gas centrifuge technology utilizes uranium hexafluoride as its working fluid. Consequently, the health physics considerations are similar to gaseous diffusion.

A centrifuge is defined as a device for whirling an object, in a circular trajectory, with a high velocity. The force imposed on an object whirling with a velocity v is mv^2/r, where m is the object mass, v is its instantaneous velocity, and r is the radius of its circular path. For a given centrifuge design, the heavier objects will be subjected to a larger force and will tend to be moved to a larger radial distance than lighter objects. This difference in trajectories permits the heavier and lighter objects to be separated, which is the basis for the use of a centrifuge for the enrichment of uranium isotopes.

The centrifuge device is a cylinder that rotates about its long axis. Its enrichment capacity increases with the length of the device, with the radius of the device, and with an increase in its speed. Limits in material properties restrict the available values of these parameters. The actual design of a centrifuge will depend upon the enrichment desired, the technology level of the group developing the device, and the desired end-use of the enriched material.

The working fluid in a gaseous centrifuge is composed of primarily $U\text{-}235F_6$ and $U\text{-}238F_6$. Consequently, when normal uranium hexafluoride is centrifuged, material drawn off from the interior region will be somewhat enriched in the lighter U-235 isotope.

Figure 4.3 illustrates material flow in a gaseous centrifuge. The uranium hexafluoride feed material is introduced at or near the axis of the device. The circulation of the gas is illustrated by arrows. Enriched $U\text{-}235F_6$ product is drawn off at the top of the figure, whereas depleted tails material is removed at the bottom of the figure. The product withdrawal scoop will be located at a smaller radius than the tails withdrawal location.

In order to obtain the desired U-235 enrichment, the gas centrifuge process is operated in a cascade of numerous stages similar to that utilized in the gaseous diffusion technology. However, the degree of enrichment is greater for the

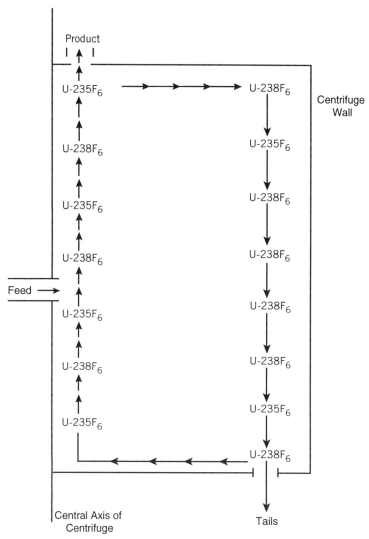

Fig. 4.3 Gaseous centrifuge isotope separation technology. The product (tails) will be slightly enriched (depleted) with respect to the feed enrichment. [Adapted from the Energy Deskbook (DOE/IR/05114-1).]

centrifuge technology. Therefore, fewer stages are needed to obtain the same enrichment with the centrifuge technology.

LASER ISOTOPE SEPARATION

Laser methods for isotope separation are an emerging technology because they are projected to be more economical than either gaseous diffusion or gaseous centrifuge technologies. Electrical energy utilization is also expected to be significantly less than existing technologies. Because of significantly higher enrichment factors, laser technology could recover the residual U-235 residing in the tailings from either diffusion or centrifuge plants.

Laser isotope separation techniques rely on the property that different isotopic species, in either an atomic or molecular form, exhibit small differences in their atomic and molecular spectra. That is, equivalent transitions from one energy level to another require a different energy to induce the transition. Thus, selective excitation is possible, and this property is a significant factor in separating isotopes.

In order to utilize the selective excitation property, an excitation source that can be selectively tuned to the desired excitation energy is required. Lasers offer an useful tool for this selective excitation.

A laser is a source of radiation that can be designed to operate at a specified frequency and intensity. Therefore, it is possible to preferentially excite one isotopic species via a precisely tuned laser and leave other isotopic species in their ground states.

Two general laser techniques are under evaluation for the enrichment of U-235. One technique involves the use of uranium vapor, and it is based on the selective photoionization of atomic uranium atoms. A second method of laser enrichment is based on the photodisintegration of U-235 hexafluoride molecules. The molecular laser isotope separation (MLIS) and atomic vapor laser isotope separation (AVLIS) technologies will be briefly addressed.

MLIS

In the molecular process an infrared laser is utilized to preferentially excite the U-235F_6 vibrational states until the U-235 hexafluoride molecule dissociates:

$$U\text{-}235F_6 + h\nu \rightarrow U\text{-}235F_5 + F \quad \text{(dissociation)} \quad (4.4)$$

$$U\text{-}238F_6 + h\nu \rightarrow U\text{-}238F_6 + h\nu \quad \text{(no reaction)} \quad (4.5)$$

The excitation process is based upon the inherent assumption that the UF_6 molecule is its ground state. Thus, it may be necessary to cool the molecules via flow through an expansion nozzle in order to ensure that all molecules reside in their vibrational ground states.

The UF_6 dissociation may be enhanced with other lasers types. For example, an ultraviolet laser could be utilized to cause electronic excitation of the vibrationally excited molecule. The electronically excited state may then dissociate immediately. Once formed, the U-235F_5 molecule precipitates as a solid, leaving the unaffected U-238F_6 molecule in a gaseous form. Preferential collection of the U-235F_5 solid is the basis for enrichment.

The health physics concerns associated with uranium hexafluoride vapor also apply to the molecular separation process. The use of laser components with high-voltage power supplies introduce x-ray hazards that need to be addressed. Other concerns associated with laser technology will be addressed in a scenario.

AVLIS

In the United States, laser enrichment methods will probably utilize uranium vapor, instead of uranium hexafluoride, as the working fluid. The use of metallic uranium will impact the fuel cycle chemical processing both prior to and post enrichment. The extent of these changes will be governed by the manner in which this technology is implemented on a production scale.

The laser technology that is currently under development by the United States Department of Energy is the atomic vapor laser isotope separation (AVLIS) process. In the AVLIS process, the uranium metal is fed into a large vacuum vessel where it is melted and then vaporized by the impingement of an electron beam. The uranium vapor is illuminated by laser radiation which is tuned to selectively ionize only the U-235 atoms. Collection of the U-235 ions is accomplished by electromagnetic alteration of the ion's trajectory via an electromagnetic field. The un-ionized U-238 atoms pass through the collection region and are separately collected on a tails collector.

Laser enrichment presents an internal as well as external radiation hazard. Internal exposure is due to the alpha decay of U-235 and U-238 and their daughter products. In contrast to UF_6-based enrichment technologies that use class D compounds, the AVLIS technology will produce primarily class W material (UO_2, UO_4, and U_3O_8) because the uranium metal will oxidize during maintenance activities. Class D UO_3 may also be produced during the wide variety of maintenance required in an AVLIS facility.

An external hazard is presented from photons and x-rays generated from the electron impingement on the uranium metal, from the various high-voltage equipment utilized in the laser and electron beam components, and by the possibility of an inadvertent criticality event following the enrichment process. As noted in Table 4.2, uranium metal and its compounds also present a beta radiation hazard. Examples of these hazards will be addressed in one of the scenarios.

SPENT POWER REACTOR FUEL

Following its use in power production, uranium-based nuclear fuel is stored in an on-site spent fuel pit. In 1977, the United States decided to postpone the reprocessing of nuclear fuel from light-water reactors. Moreover, there are no commercial plutonium breeder reactors in operation, thereby precluding the need for the fuel reprocessing step. The current U.S. fuel cycle is therefore incomplete, and it fails to operate in the most efficient or environmentally sound manner.

Thorium Fuel Cycle

A thorium-based fuel cycle is summarized in Fig. 4.4. The thorium cycle requires the fuel reprocessing operation because there is no natural source of

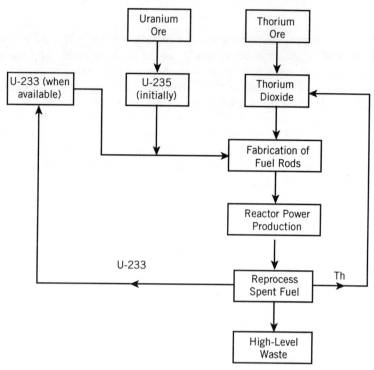

Fig. 4.4 Theoretical nuclear fuel cycle for thorium. [Adapted from the Energy Deskbook (DOE/IR/05114-1).]

U-233. In the early stages of use of the thorium cycle, some U-235 would be used to ensure an adequate fissile content of the fuel. Thorium represents an alternative to uranium, but it has not been as widely utilized as the uranium fuel cycle. However, the health physics concerns of the uranium and thorium fuel cycles will be similar.

RADIOACTIVE WASTE

One of the consequences of power production in a nuclear reactor or the end product of the utilization of medical radioisotopes is the generation of radioactive waste or effluents that may be released from the facility. A summary of these waste materials and effluent isotopes typically associated with the uranium fuel cycle is presented in Table 4.3. This listing is not necessarily complete, but it is indicative of the isotopes that will be encountered in the various fuel cycle activities. The isolation of radioactive waste to prevent access by people and potential releases to the environment are key health physics concerns.

Radioactive wastes are loosely characterized as either high-level or low-level. Low-level wastes include contaminated articles of disposable protective

Table 4.3 Radioactive Waste and Effluent Isotopes from the Fuel Cycle

Operation	Waste Form		
	Gaseous	Liquid	Solid
Mining and milling	Rn-222 Po-218 Bi-214 Po-214	—	U Ra-226 Th-230 Pb-210
Refining	—	U-238 Th-234 Pa-234 Ra-226	—
Fuel fabrication	—	—	U Pu Th
Reactor operation	A-41 Kr-87, Kr-89 Xe-135, Xe-138	H-3	Co-58, Co-60 Fe-59 Cr-51
Chemical reprocessing	H-3 Kr-85 Xe-133 I-129, I-131	Fission products dissolved in acid solutions	Pu Cm Am Fission products

clothing, spent ion-exchange resins, trash, animal carcasses, or other items commonly used in power reactor, medical, research, or industrial environments. Low-level wastes are typically compacted in steel drums and buried at a federally licensed burial site.

Another broad waste category is transuranic (TRU) wastes. TRU wastes contain appreciable quantities of elements heavier than uranium (plutonium, americium, and curium). These wastes are produced in the nuclear weapons program and in the fabrication of fast reactor fuel derived from recycled light-water reactor spent fuel.

High-Level Wastes

High-level wastes include spent nuclear fuel and the wastes associated with the reprocessing of spent nuclear fuel. Both of these waste streams contain fission products with their associated beta–gamma activity. As part of the reprocessing, the spent fuel is dissolved and the fissile material is removed. The remaining waste is a liquid containing high levels of fission fragments. Many of these fission products have half-lives of 30 years or more, which suggests that these wastes remain a hazard for hundreds of years.

CRITICALITY

Criticality is defined as the process of achieving a self-sustaining chain reaction. Although criticality is anticipated within the nuclear fuel core of a power reactor, its occurrence in fuel storage areas or reprocessing equipment is neither anticipated nor desired. The energy produced by the unanticipated criticality may be sufficient to terminate the criticality event by destroying the process equipment or by rearranging material in its immediate vicinity. Although the event may be short-lived, 10^{16} or more fission events can occur.

An unanticipated criticality is one of the most serious radiation hazards that a health physicist must confront. It may occur during the processing or handling of enriched uranium or plutonium, and its consequences for personnel and equipment are very severe. An accidental criticality generally requires the violation of a facility's physical and administrative controls.

Critically safety may be affected by the following factors:

a. Quantity and type of fissile material
b. Geometry or physical arrangement of the fissile material
c. Enrichment of the fissile isotope
d. Presence of a neutron moderator, reflector, or absorber materials

These factors will be addressed in more detail below.

Critical Mass

The minimum mass of material which sustains a nuclear chain reaction for a given set of conditions is called the *critical mass*. The critical mass will depend on the fissile isotope, the isotope's enrichment, its geometry, and the presence and type of moderator and reflector material. Examples of critical masses for selected fissile isotopes for moderated and unmoderated conditions are summarized in Table 4.4.

Fissile material may also be dissolved in a solvent. For solutions, the key parameter is the density or concentration defined in terms of mass per unit volume. For example, a uniform U-235 solution of less than 10.8 g U-235/liter will be subcritical at any volume. However, large concentrations could

Table 4.4 Minimum Critical Mass Parameters (kg)[a]

Isotope	Moderated	Unmoderated
U-235	0.82	22.8
U-233	0.59	7.5
Pu-239	0.51	5.6

[a]Moderation is by water. Water reflection is assumed for all values.
Source: U.S. AEC, *Nuclear Safety Guide*, TID 7016 (1961).

Table 4.5 Minimum Critical Concentration and Volume Parameters[a]

	Moderated Conditions	
Isotope	Concentration (aqueous) (g/liter)	Volume of Solution (liters)
U-235	12.1	6.3
U-233	11.2	3.3
Pu-239	7.8	3.5

[a]Moderation is by water. Water reflection is assumed for all values.
Source: U.S. AEC, Nuclear Safety Guide, TID 7016 (1961).

become critical under the proper conditions. Table 4.5 summarizes the minimum critical densities and volumes for various fissile isotopes.

Geometry or Shape

The leakage of neutrons from a system depends on its shape and on the properties of the moderating and reflecting material. Shapes that minimize the surface-to-volume ratio (i.e., spheres) will have a smaller critical mass than other fissile material configurations. For example, minimum U-235 critical properties for a variety of geometries are summarized in Table 4.6.

Enrichment of the Fissile Isotope

Higher enrichment values of the fissile isotope will result in smaller values of the critical mass and volume. Enrichment is required to sustain a critical reaction with a reasonable size and volume of fissile material. Table 4.7 summarizes experiments used to illustrate the impact of enriched U-235 on the critical volume of an equilateral cylinder in water-moderated lattices. The lattices are formed using enriched metal rods whose spacing is optimized to provide the minimum critical volume. As expected, higher enrichments lead to smaller critical volumes.

Table 4.6 Minimum Critical Parameters for U-235[a]

Geometric Shape	Critical Value
Mass of sphere	820 g
Diameter of infinite cylinder	5.4 in.
Thickness of infinite slab	1.7 in.

[a]Moderation is by water. Water reflection is assumed for all values.
Source: U.S. AEC, Nuclear Safety Guide, TID 7016 (1961).

Table 4.7 Lattices of Slightly Enriched Uranium Metal Rods in Water

U-235 Enrichment (%)	Rod Diameter (cm)	Average U-235 Density (g/cm^3)	Critical Volume of Equilateral Cylinder at Optimum Lattice Spacing (liters)
1.027	0.98	0.055	524
	1.52	0.06	430
	1.90	0.065	393
1.143	0.98	0.055	274
	1.52	0.065	238
1.299	0.98	0.06	175
	1.52	0.075	155
2.0	1.52	0.095	58.2
	2.35	0.12	56.6
3.063	0.445	0.09	32.0
	0.762	0.105	29.8
	1.52	0.15	30.1
	2.35	0.175	35.1

Source: LA-10860-MS (1987), p. 46.

Moderation and Reflection

A moderator is a material which thermalizes, or slows down, fast neutrons. The most effective moderators include materials containing hydrogen. The hydrogen concentration is usually expressed as the ratio of the number of hydrogen atoms to the number of fissile atoms. This ratio varies from zero for pure fissile material metal or dry unhydrated salt to several thousands for dilute aqueous solutions. The critical mass may vary from a few tens of kilograms for small ratios to a few hundred grams for an optimum moderated configuration. The presence of a moderator allows a smaller mass to become critical.

A reflected system is an assembly of materials containing a fissile material that is wholly or partly surrounded by another material having a neutron-scattering cross section that is larger than that of air. Therefore, the presence of a reflector causes a fraction of the neutrons escaping from the fissile material to be reflected back into the material rather than escape from the system. Good reflector materials have large neutron-scattering cross sections and small neutron-absorption cross sections. Water, concrete, polyethylene, graphite, and stainless steel are good reflector materials.

Neutron Absorbers or Poison Material

Neutron absorbers or poisons are nonfissionable materials which absorb neutrons and reduce the number of neutron-induced fissions. These materials are characterized by large neutron-capture cross sections and include material such as cadmium, hafnium, and boron. Boron is used as a soluble poison in commercial pressurized water reactors. Cadmium is a key component in power reactor control rods.

Consequences of a Criticality Event

In addition to direct radiation exposure, a criticality results in numerous fission events that produce a variety of gaseous and particulate fission fragments. These materials may be released into the environment. The following sections of this chapter will address methods used to characterize a radioactive release that are independent of their initiating mechanism.

DISPERSION OF RADIOACTIVE GAS FROM A CONTINUOUS SOURCE

Estimates of the dispersion of gases into the atmosphere are based on mathematical models that consider the meteorological characteristics of the atmosphere. One of the most commonly used models for estimating the ground-level concentration of a gaseous effluent is the Gaussian plume model. In this model, the plume is assumed to travel in a straight-line trajectory and the contaminant is assumed to be normally distributed around the central axis of the plume. The downwind dispersion characteristics depend upon the atmospheric stability, wind speed, and release height according to the Pasquill–Gifford equation:

$$X(x, y) = \frac{Q}{\pi a_y a_z u} \exp[-(1/2)(y^2/a_y^2 + H^2/a_z^2)] \qquad (4.6)$$

where

$X(x, y)$ = ground-level concentration in Bq(Ci) per cubic meter at the location (x, y)
x = downwind distance on the plume centerline (m)
y = cross-wind distance (m)
Q = release rate Bq(Ci)/sec
a_y = horizontal standard deviation of the contaminant concentration in the plume (m)
a_z = vertical standard deviation of the contaminant concentration in the plume (m)
u = mean wind speed (m/sec) at the plume centerline
H = effective release height (m)

Equation (4.6) assumes total reflection of the gas by the ground.

If the released material has a significant exit velocity or it exits at a high temperature, then it rises to an elevation that is greater than the physical release height. The effective release height is given by

$$H = h + d(v/u)^{1.4} (1 + \Delta T/T) \quad (4.7)$$

where

h = actual release elevation or stack height (m)
d = stack exit diameter at the release point (m)
v = release velocity of the gas (m/sec)
T = absolute temperature of the released gas (°K)

Equation (4.7) requires that ΔT be defined

$$\Delta T = T - T_0 \quad (4.8)$$

where T_0 is the absolute ambient temperature (°K).

The dispersion of the plume as it moves downwind is determined by the atmospheric stability, wind speed, and downwind distance. Pasquill proposed atmospheric stability categories to facilitate calculations of ground-level concentrations. These stability classes, denoted as A, B, C, D, E, F, and sometimes G, are described in Table 4.8.

Class A represents the least stable atmospheric conditions, and it results in a broad plume with the lowest ground-level concentration. The most narrow plumes are associated with stable atmospheric conditions (F and G) and yield the largest ground-level concentrations. A qualitative assessment of the Pasquill atmospheric stability classes in terms of observable meteorological conditions is summarized in Table 4.9.

Table 4.8 Description of Stability Classes

Stability Class	Description
A	Extremely unstable conditions
B	Moderately unstable conditions
C	Slightly unstable conditions
D	Neutral conditions[a]
E	Slightly stable conditions
F	Moderately stable conditions
G	Extremely stable conditions

[a] Neutral conditions are applicable to heavy overcast conditions during both day and night.

Source: ORO-545.

Table 4.9 Qualitative Description of Atmospheric Stability Classes

Surface wind speed (m/sec)	Daytime Insolation			Cloud Cover[a]	
	Strong	Moderate	Slight	< 1/2	≥ 1/2
<2	A	A–B	B	F	F–G
2	A–B	B	C	E	F
4	B	B–C	C	D	E
6	C	C–D	D	D	D
>6	C	D	D	D	D

[a]The degree of cloudiness is defined as that fraction of the sky above the local horizon which is covered by clouds.
Source: ORO-545.

DISPERSION OF RADIOACTIVE PARTICULATES FROM A CONTINUOUS SOURCE

The dispersion of a continuous particulate release is similar to a gaseous release, but the ground-level concentration is smaller than the gas release because some particulates may be retained in the ground. For a particulate release the dispersion is given by the approximation

$$X_P(x, y) = \frac{Q}{2\pi a_y a_z u} \exp[-(1/2)(y^2/a_y^2 + H_P^2/a_z^2)] \quad (4.9)$$

For particulates (*P*), the depletion of activity in the plume is due to gravitational settling, impaction on the surface or surface structures, precipitation settling, humidity, particulate solubility, and particle size. Therefore, calculations of the deposition of particulates is more difficult than for gaseous releases. Gravitational settling is taken into account through the effective height (H_P):

$$H_P = H - xv_t/u \quad (4.10)$$

where v_t is the terminal settling velocity.

Another common parameter is the ground deposition (Bq/m²-sec) of particulates (*w*) at the point (*x*, *y*) which is obtained from the product of the ground-level particulate concentration $X_P(x, y)$ and the deposition velocity v_g:

$$w(x, y) = X_P(x, y)v_g \quad (4.11)$$

Experimental values of the deposition velocity range from about 0.001 to 0.10 m/sec, with an average of about 0.01 m/sec.

Table 4.10 Health Physics Impact of Uranium Fuel Cycle Material Characteristics

Characteristic	Material	Radiological Hazard	
		External	Internal
Enrichment	Depleted, natural, and slightly enriched uranium	Minimal penetrating (gamma) radiation.	Inhalation results in chemical toxicity.
		Moderate beta hazard from bare uranium.	Inhalation intake leads to exposure to bone surfaces, lungs, and kidneys.
	Moderate to highly enriched uranium	Penetrating radiation increases with enrichment.	Higher enrichments imply higher doses per unit mass.
		Moderate beta hazard from bare uranium.	For chronic exposures, doses become more significant and chemical toxicity less significant as enrichment increases.
		Criticality hazards increase with enrichment.	
Chemical form	Class D: UF_6, UO_3, UO_2F_2, and $UO_2(NO_3)_2$	Chemical form does not significantly affect hazard.	For acute exposures to any enrichment, chemical toxicity is more limiting.
		Neutron levels increase in homogeneous fluorine compounds.	For chronic exposures, chemical toxicity is more limiting up to 15% enrichment.
			Beyond 15% enrichment, the nonstochastic limit for bone surfaces is limiting.

Table 4.10 (*Continued*)

Characteristic	Material	Radiological Hazard	
		External	Internal
	Class W: U_3O_8, UF_4, UO_4, and UO_2	Chemical form does not significantly affect hazard, except for fluorine mixtures and compounds.	For acute exposures, chemical toxicity is limiting up to 39% enrichment. Beyond 39%, the effective dose equivalent becomes limiting.
			For chronic exposures, chemical toxicity is more limiting up to 1.3% enrichment. Beyond 1.3%, the effective dose equivalent becomes limiting.
	Class Y: UC_2, UAl_x, UZr, and UO_2 (high-fired)	Same as Class W.	Chronic and acute inhalation are limited by the effective dose equivalent due primarily to lung dose.
Physical Form	Fixed	Beta dose rates are slightly higher from metal than from compounds.	Do not pose an internal hazard.
		Penetrating dose rates are not significantly affected by the physical form.	
	Loose	Same as fixed.	Dispersible hazard as noted under enrichment and chemical form.

Source: EGG-2530.

FUEL CYCLE FACILITIES

Fuel cycle facilities include the associated infrastructure needed to support the processes in Fig. 4.1. Fuel cycle facility design and operation will require health physics controls to protect the worker from both internal and external radiation hazards. Table 4.10 summarizes the hazards associated with the uranium fuel cycle in terms of the material's enrichment, chemical form, and physical form.

Detection of Fuel Cycle Facility Activity

Routine swipe and air samples and direct radiation surveys are used to quantify the uranium and daughter activity within the facility. These surveys consist of direct measurement of gamma radiation along with the alpha and beta counting techniques.

Alpha particles can be counted using ionization, proportional, scintillation, or solid-state detectors. Care must be taken to account for the self-shielding of the alpha particles by the filter medium or by dust loading. The energy spectrum of the collected alpha particles can be used to identify the collected alpha emitter. Typical spectroscopy applications include the use of semiconductor detectors and membrane filters or other surface collecting filters with low dust loading.

A typical problem in counting alpha samples involves distinguishing between natural short-lived alpha activity and the alphas derived from the fuel cycle facility. The naturally occurring radionuclides–radon and thoron and their decay products–will be present in widely varying concentrations. These radionuclides will normally be at higher concentrations than the uranium isotopes in the fuel cycle facility. Moreover, radon, thoron, and their daughters will interfere with the sample analysis unless they are given sufficient time to decay after sample collection.

Radon daughters, usually more abundant than the thoron daughters, decay with an effective half-life of about 30 minutes, whereas the effective thoron daughter half-life is about 10.6 hours. The presence of natural products can be determined by counting the sample several times at intervals of a few hours.

Beta particles are counted using thin-window Geiger-Mueller, ionization, proportional, or solid-state detectors. The wide range of potential beta energies requires care in calibrating the detector. Beta counting is less dependent upon self-absorption in the filter medium and dust loading. Beta spectroscopy can be obtained from tissue-equivalent plastic detectors.

Gamma emitters can be identified by using NaI and GeLi detectors. Direct dose rates may be determined from a variety of hand-held survey instruments.

SCENARIOS

Scenario 50

You are the lead health physicist at a plutonium processing facility and are investigating an exposure incident. A worker has reported hearing a hissing sound coming from a nearby glove box. About 1 min after hearing the hiss, the continuous air monitor (CAM) in the room alarmed. The CAM gamma count rate meter was noted by the worker to have pegged high. The worker then walked over to the hissing glove box and noticed that an extensive rupture of process vessels within the box had occurred, that material was spread throughout the internals of the box, and that one of the gloves had ruptured. He estimates that 20 sec elapsed between the CAM alarm and the time he exited the room. The worker was not wearing a respirator.

The on-duty health physics technician determined that the worker was contaminated (hair, face, and clothing), and a positive indication of alpha contamination was found on a nasal smear.

The following information should be considered:

Isotope	Activity Fractions in the Mixture
Pu-238	1.2×10^{-2}
Pu-239	1.4×10^{-1}
Pu-240	3.1×10^{-2}
Pu-241	8.1×10^{-1}
Am-241	1.5×10^{-3}

The CAM data suggest that the concentration is 2.0×10^{-5} μCi/ml (alpha + beta)

Particle size = 1-μm activity median aerodynamic diameter

Breathing rate of worker = 20 liters/min

Inhalation Dose, First Year, 1.0 μm AMAD (Sv/Bq intake)

Nuclide	Lung	Liver	Bone Surface	Red Marrow
Pu-238(Y)	7.0×10^{-5}	9.7×10^{-7}	3.5×10^{-6}	3.0×10^{-7}
Pu-239(Y)	6.7×10^{-5}	9.2×10^{-7}	3.5×10^{-6}	2.7×10^{-7}
Pu-240(Y)	6.7×10^{-5}	9.2×10^{-7}	3.5×10^{-6}	2.7×10^{-7}
Pu-241(Y)	7.0×10^{-9}	1.1×10^{-9}	3.8×10^{-9}	2.7×10^{-10}
Am-241(W)	1.8×10^{-5}	1.5×10^{-6}	5.4×10^{-5}	4.3×10^{-6}

Weighting Factors

Tissue	W_T
Gonads	0.25
Breast	0.15
Red bone marrow	0.12
Lung	0.12
Thyroid	0.03
Bone surfaces	0.03
Remainder (five highest other organs)	0.30

4.1. Calculate an estimate of the annual (first year) dose equivalent to (1) the lung, (2) red bone marrow, (3) bone surfaces, and (4) the liver of the worker. Assume that the CAM data represents the average concentration to which the worker was exposed during the incident prior to leaving the room.

4.2. Calculate the annual effective dose equivalent (first year) assuming that the organs identified in (question) 4.1 above are the only important target organs for internal plutonium exposure and that any external gamma exposure during this incident was negligible.

4.3. What are the relevant ICRP-26/30 stochastic and nonstochastic exposure limits appropriate for this circumstance? Was either limit exceeded?

4.4. What additional follow-up would you recommend to assess more accurately the dose to this worker? What clinical symptoms are expected for the doses calculated in (question) 4.1.

Scenario 51

As a result of process flow problems at a depleted uranium facility, two maintenance workers are assigned a task to cut into an overhead process line and replace the pipe section suspected of being clogged. The maintenance workers note that the radiation work permit (RWP) for the job specifies that the utility services department was to relieve all line pressure prior to starting the job, and they assume that this was done. Worker "A" is standing on a ladder to make the cut while Worker "B" is holding the base of the ladder at ground level. Both workers are wearing coveralls, gloves, booties, and lapel-type personnel air samplers as required by the RWP. Worker "A" is also wearing a half-face respirator. Worker "B" has no respiratory protection.

As Worker "A" initially cuts into the line, a heavy mist begins to be evolved from it, indicating that line internal pressure had not been shut off as assumed. Worker "A" nonetheless continued the cutting operation, while Worker "B" went to look around the process module to try and find the shut off valves. Failing to do so, Worker B then rejoined Worker "A" at the work location and they decided to call health physics via the telephone located within the module. The workers later indicated that approximately 4 minutes passed between the time the release first occurred and the time they called health physics.

Immediately upon arriving, health physics technicians removed the workers from the work area and confirmed contamination of the workers via facial surveys and positive nasal swipes. It took approximately 6 min for health physics to arrive following the phone call from the workers. As the Senior Health Physicist, you are asked to perform an initial dose assessment for this incident. The following data should be utilized in your assessment:

Flow rate of the lapel air samplers = 2 liter/min
Breathing rate (both workers) = 20 liters/min
Gross alpha count on filter:
  Worker A = 30,000 cpm
  Worker B = 20,000 cpm
Alpha counting efficiency = 0.3
Respiratory protection factor (half-face) = 10
Average particle size for the mist evolved from the process line = 1 μm
The mist or aerosol is enriched uranium as highly insoluble metallic uranium/UO_2
Uranium isotopic composition (alpha activity fractions):
  U-234   0.10
  U-235   0.02
  U-238   0.88
(*Note:* For dosimetric purposes, daughter activity can be ignored.)
Weighting factor (W_T) = 0.12 for the lung
For a 1-μm particle size, the pulmonary deposition factor is 0.25.

Inhalation lung dose equivalent conversion factors (rem/μCi for 1 μm AMAD particles) are listed below:

For the First Year

Nuclide	D	W	Y
U-234	1.1	59	230
U-235	1.1	54	210
U-238	1.0	52	200

For 50 Years

Nuclide	D	W	Y
U-234	1.2	59	1100
U-235	1.1	55	1000
U-238	1.0	53	980

116   FUEL CYCLE HEALTH PHYSICS

The mathematical model used to describe clearance from the respiratory system is given below. The values for the removal half-times (T) and compartmental fractions (F) are tabulated below for each of the three classes of retained material. The values given for D_{NP}, D_{TB}, and D_P (left column) are the regional depositions for an aerosol with an AMAD of 1 μm. Figure 4.5 identifies the various clearance pathways from compartments a–j in the four respiratory regions, NP, TB, P, and L.

		Class					
		D		W		Y	
Region	Compartment	T(day)	F	T(day)	F	T(day)	F
NP(D_{NP} = 0.30)	a	0.01	0.5	0.01	0.1	0.01	0.01
	b	0.01	0.5	0.40	0.9	0.40	0.99
TB(D_{TB} = 0.08)	c	0.01	0.95	0.01	0.5	0.01	0.01
	d	0.2	0.05	0.2	0.5	0.2	0.99
P(D_P = 0.25)	e	0.5	0.8	50	0.15	500	0.05
	f	n.a.	n.a.	1.0	0.4	1.0	0.4
	g	n.a.	n.a.	50	0.4	500	0.4
	h	0.5	0.2	50	0.05	500	0.15
L	i	0.5	1.0	50	1.0	1000	0.9
	j	n.a.	n.a.	n.a.	n.a.	∞	0.1

[a]n.a., not applicable.

4.5. Calculate the total uranium intake (gross alpha activity) for each worker.

4.6. Estimate the committed dose equivalent to the lung for the long-term retention component resultant from the deep pulmonary deposition for each worker. Consider as trivial (ignore) the dose from the short-term elimination component (biological half-life) ≤ 1 day), the dose to other regions of the respiratory tract, and any external exposure. Assume the maximum deposition occurs for the particle size considered in the problem.

4.7. For a uranium inhalation incident such as this, what percent of the committed dose equivalent to the lung is represented by the first year's lung dose?

4.8. Evaluate if applicable ICRP dose limit recommendations have been exceeded for either worker.

4.9. What follow-up actions would you recommend to more completely assess the long-term pulmonary dose to these workers?

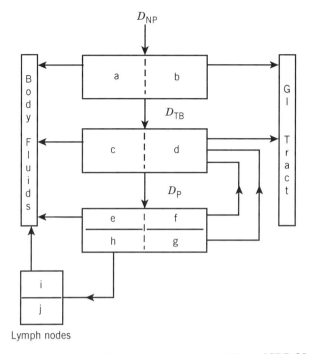

Fig. 4.5 Inhalation model for the uranium uptake. (From ICRP-30, Part 1.)

Scenario 52

You have been selected as the Health Physics Manager for the atomic vapor laser isotope separation (AVLIS) facility at Moose River, Idaho. This facility represents the U.S. DOE's most advanced technology for the enrichment of uranium. The AVLIS technology is based upon the selective photoexcitation and subsequent U-235 ionization of the uranium vapor by laser irradiation. The photo-ions formed by the interaction of the laser beam and the uranium vapor are extracted from the vapor stream by the combined action of magnetic and electric fields. Enriched and depleted streams are collected on separate surfaces. The process unit is a large vacuum chamber into which solid uranium feed and laser light are admitted. The internal surfaces of the process unit, particularly those that contact uranium, will be removed periodically from the vacuum chamber for cleaning and refurbishment. The uranium vapor is generated by the impingement of an electron beam upon the surface of the solid uranium source.

Use your experience with components similar to those described above to answer the following questions. Clearly state all assumptions about the process equipment and their corresponding parameters.

118 FUEL CYCLE HEALTH PHYSICS

Data

The lasers utilize 30,000-V power supplies.

The AVLIS technology has the potential to enrich uranium to 90%.

The electron beam and associated components utilize 50,000-volt power supplies.

Cleaning of the vacuum chamber internals will be performed manually.

Internal components will be contaminated with uranium dust.

The electromagnetic field strength is high enough to warrant consideration.

The intensity of the laser radiation warrants personnel protection.

Assume that each fission from a criticality would produce eight prompt gammas with an average energy of 1 MeV and three prompt neutrons with an average energy of 2 MeV.

Ignore delayed radiation from fission products.

Assume that the average energy of the neutrons and gamma rays are representative of the entire spectrum.

The density of polyethylene is 1.4 g/cm^3.

The gamma fluence rate = 5.5×10^5 gammas/cm^2-sec to give 1 R/hr.

The mass attenuation coefficient for polyethylene is 0.0727 cm^2/g.

The dose attenuation factor for 2.5-MeV neutrons passing through 4 inches of polyethylene = 0.3.

The flux to dose conversion factor for 2.5-MeV neutrons: 2.5 mrad/hr = 20 neutrons/cm^2-sec.

4.10. List ionizing radiation hazards and their sources, nonionizing radiation hazards and their sources, and other physical hazards and their sources that can occur in this facility.

4.11. List engineering design features which could be included in the AVLIS facility to reduce the ionizing radiation hazards.

4.12. List engineering design features which could be included in the AVLIS facility to reduce the nonionizing radiation hazards.

4.13. Assume that the facility has been in operation for 6 months. In the process of removing enriched material from the unit, a technician violates standard operating procedures and stacks enriched uranium into a critical geometry. Following additional procedure violations, a criticality results in which 1×10^{16} fissions occur. What is the approximate dose to the technician (in rad) who is standing behind a 4-inch-thick polyethylene shield located 20 feet from the source?

Scenario 53

You have been selected as the health physics manager at the U.S. DOE's Frostbite Falls Fuel Reprocessing Facility (FFFRF), which consists of two large processing buildings. Building A includes a tank farm used to process highly enriched uranium. During a batch-processing operation in Building A, a technician violates standard operating procedures, which leads to a critical geometry in a small tank. A criticality results in a burst of energy in which 1.0×10^{16} fissions occur. The plant manager is standing behind a 12-in. polyethylene shield, and he is 10 ft from the center of the tank.

FFFRF Building B contains a plutonium processing tank but contains no shield walls. It is unoccupied and operated by remote control.

Data for FFFRF Building A

The density of the polyethylene shield is 1.4 g/cm^3.

Each fission event produces three neutrons and eight gammas.

The neutron spectrum for the criticality is represented by a dose conversion factor of 2.5 mrem/hr per 20 neutrons/cm^2-sec.

The mean neutron and gamma energies are 2.5 MeV and 1.0 MeV, respectively.

The neutron dose attenuation factor for 2.5-MeV neutrons through 12 in. of polyethylene is 0.005.

The mass attenuation coefficient for polyethylene for the fission gamma spectrum is 0.0727 cm^2/g.

The gamma spectrum is represented by an exposure rate conversion factor of 5.5×10^5 gammas/cm^2-sec per 1 R/hr.

4.14. What is the neutron dose equivalent in rem received by the plant manager during the Building A criticality?

4.15. What is the gamma dose equivalent in rem received by the plant manager during the Building A criticality?

4.16. The Building B criticality monitor is a gamma response instrument with an alarm set point of 500 mR/hr. If the detector responds to 1/2500 of the actual gamma exposure rate during a short transient in Building B, what is the maximum distance over which the device will be effective in signaling an unshielded, 1-msec criticality of 1.0×10^{16} fissions? Neglect air absorption. Assume that 1.0×10^{15} fissions yields a gamma exposure of 2.5 R at 6.0 feet.

Scenario 54

At nuclear facilities, noncontaminated and very-low-level radioactive materials present unique regulatory, economic, and radiological health concerns. The following questions involve establishing a reasonable and practical approach of dealing with these materials.

Miscellaneous containers of waste oil from within the radiation control area of an operating nuclear facility have been consolidated into a single 500-gallon tank.

4.17. Give two major characteristics of this raw (unprocessed) waste that make collecting a representative sample particularly difficult.

4.18. Describe two acceptable methods for obtaining representative samples from the 500-gallon tank.

4.19. Give two general categories of analyses that should be performed on the waste oil prior to the determination of the most appropriate disposal method.

As a nuclear facility health physicist, you have been asked to establish the survey protocol for the release of clean trash (paper, glass, and plastic) from within the radiation control area. Assume that alpha contamination is not a problem.

4.20. Give two general survey techniques which trash should undergo.

4.21. State the regulatory accepted limits for release as nonradioactive waste.

As a nuclear facility health physicist, you have been assigned the responsibility of developing an application for alternative disposal of slightly contaminated lagoon sludge. You have decided that because of the large volume (10,000 ft^3) and low specific activity (1.0×10^{-8} μCi/cm^3), land farming on facility controlled property is the appropriate disposal scenario.

4.22. List four main radiation exposure pathways which must be evaluated in the application.

4.23. What level of exposure to the maximum exposed individual has been found acceptable for alternative disposal methods? You may use Title 10 of the Code of Federal Regulations (10 CFR 20.302) as a guide.

4.24. List six general environmental categories which must be evaluated for impact on waste stability and transport when submitting a request for approval of alternative disposal methods (10 CFR 20.302).

4.25. List six other types of information which must be included in the application.

Scenario 55

The following data are to be utilized to prepare an incident report concerning a 27-year-old male radiation worker at a Department of Energy production reactor:

Period of Year (Quarter)	External Deep Dose Equivalent (rem)	Intake-Inhalation Class D (μCi)
1	0.7	Cs-137: 80.0
		I-131: 19.5
2	1.2	None
3	0.8	None
4	0.3	None

Nuclide	Inhalation ALI-ICRP-30 (μCi)
Cs-137(D)	200
I-131(D)	50 nonstochastic thyroid
	(200 stochastic)

4.26. What is the dose equivalent to the thyroid from the I-131 intake?

4.27. What is the committed effective dose equivalent from the internally deposited radionuclides? State all assumptions.

4.28. Have any of the occupational dose equivalent limit recommendations in NCRP Report No. 91 (1987) been exceeded? State which ones and justify your responses.

Scenario 56

You are the senior health physicist at a facility that is involved in the treatment and solidification of high-level radioactive waste that was generated from the reprocessing of spent reactor fuel. You receive a telephone call at home late one evening from the shift supervisor at the plant. He informs you that an hour ago it was discovered that a process line had ruptured, spilling high-level waste sludge into a processing area.

Upon your arrival at the plant (30 min later), you are informed that a worker had entered the process area when the leak was first discovered. He turned off the processing pump, which terminated the spill, and participated in the cleanup of the spill event. You instruct the shift supervisor to immediately remove the worker from the contaminated area and to have him report to the decontamination and survey station.

Upon removing the worker's anticontamination clothing (booties, coveralls, and gloves) and lapel air sampler, it was determined that the worker had facial contamination and a positive nasal swab. An initial survey of the air sample was performed, and it showed considerable beta/gamma activity. Prior to permitting the cleanup activity to continue, you must assess the radiological aspects of the spill event and subsequent contamination. Using the data provided, answer the following questions.

Data

Breathing rate of worker: 20 liters/min

The worker was wearing a functional lapel-type air sampler throughout the period of time he was in the contaminated area. You had the filter removed and analyzed immediately after you arrived on the scene. Assume all activity deposited on the filter resulted from resuspension of the spilled high-level waste.

Spill dimensions: Circular spill with a diameter of 5 m.

Spill volume: 100 liters

Major radionuclide components of the high-level waste

Isotope	Waste Concentration (Ci/L)	Gamma Constant $\frac{\text{R-m}^2}{\text{Ci-hr}}$	DCF[a] (1 year)	DCF[a] (50 year)	MPC$_a$[b] (μCi/ml)	DAC[c] (μCi/ml)
Cs-137	0.2	0.33	2.9×10^{-2}	3.2×10^{-2}	1×10^{-8} (I)	7×10^{-8} (D, W, Y)
Sr-90	0.01	—	0.32	1.3	1×10^{-9} (S)	2×10^{-9} (Y)
Pu-238	0.001	0.0	30.0	310	2×10^{-12} (S)	3×10^{-12} (W)

[a]Dose conversion factor expressed as effective dose equivalent, rem per μCi inhaled, 1-μm particle size.
[b]10 CFR 20, Appendix B, Column 1.
[c]DOE Order 5480.11, Table 1, Derived Air Concentration (DAC) for Controlling Radiation Exposures to Workers at DOE Facilities.

Air Sampling Data:

> Flow rate of lapel sampler = 2 liters/min
>
> Sampling time = 90 min

Total "long-lived" activity on lapel air sampler filter (corrected for radon progeny decay):

Gross alpha = 5×10^3 dpm

Gross beta = 5×10^3 dpm (excludes Cs-137 and Y-90 contribution)

Gross gamma = 1×10^5 dpm (i.e., photons/min)

4.29. Determine the external dose equivalent from gamma rays to the worker. Assume that the worker was located at the center of the spill during the entire 90-min exposure period and that the point of interest is 1 m above the spill. Assume that 1 R = 1 rem for the external dose equivalent calculation.

4.30. Calculate the effective dose equivalent to the worker from radionuclide intakes that would be received during the first year immediately following the incident.

4.31. Calculate the committed effective dose equivalent to the worker from radionuclide intakes.

4.32. Name a currently implemented, occupational regulatory dose equivalent limit, and state whether it has been exceeded as a result of this incident. Explain your answer and state the applicable regulatory limit (i.e., 10 CFR 20, DOE Orders, etc.). Ignore considerations of any other occupational exposures independent of the incident.

4.33. Assume that you are the first person who is contacted by the shift supervisor. List four instructions that you should provide to the shift supervisor.

4.34. For Cs-137, the 1-year and 50-year dose conversion factors given above are nearly equal, while for Pu-238 they differ by more than a factor of 10. Explain the characteristics of these radionuclides that lead to this difference.

REFERENCES

N. Adams and N. L. Spoor, Kidney and Bone Retention Functions in the Human Metabolism of Uranium, *Physics in Medicine and Biology*, **19**, 460 (1974).

R. E. Alexander, *Applications of Bioassay for Uranium*, WASH-1251, US Atomic Energy Commission (1974).

ANSI N13.3-1969, *Dosimetry for Criticality Accidents*, ANSI, New York (1970).

ANSI N13.14-1983, *American National Standard for Dosimetry: Internal Dosimetry Programs for Tritium Exposure—Minimum Requirements*, ANSI, New York (1983).

ANSI/ANS-8.3-1986, *Criticality Accident Alarm System*, American Nuclear Society, LaGrange Park, IL (1986).

DOE/EV/1830-T5, R. L. Kathern and J. M. Selby, *A Guide to Reducing Radiation Exposure and As-Low-As-Reasonably-Achievable (ALARA)*, US Department of Energy, Washington, DC (1980).

EGG-2530, *Health Physics Manual of Good Practices for Uranium Facilities*, U.S. DOE (1988).

M. Eisenbud, *Environmental Radioactivity*, Academic Press, New York (1987).

EPA/520/1-75-001, *Manual of Protective Action Guides and Protective Actions for Nuclear Incidents*, US Environmental Protection Agency, Washington, DC (1980).

S. Glasstone, *Energy Deskbook*, DOE/IR/05114-1, US Department of Energy, National Technical Information Center, Springfield, VA (1982).

ICRP Publication No. 46, *Radiation Protection Principles for the Disposal of Solid Radioactive Waste*, Pergamon Press, New York (1986).

ICRP Publication No. 47, *Radiation Protection of Workers in Mines*, Pergamon Press, New York (1986).

ICRP Publication No. 48, *The Metabolism of Plutonium and Related Elements*, Pergamon Press, New York (1986).

R. A. Knief, *Nuclear Criticality Safety, Theory and Practice*, American Nuclear Society, LaGrange Park, IL (1985).

LA-10860-MS, *Critical Dimensions of Systems Containing U-235, Pu-239, and U-233*, 1986 revision, Los Alamos National Laboratory (1977).

R. L. Murray, *Understanding Radioactive Wastes*, 2nd edition, Battelle Press, Columbus, OH (1983).

NCRP Report No. 110, *Some Aspects of Strontium Radiobiology*, NCRP Publications, Bethesda, MD (1991).

NCRP Report No. 111, *Developing Radiation Emergency Plans for Academic, Medical, or Industrial Facilities*, NCRP Publications, Bethesda, MD (1991).

ORO-545, *Graphs for Estimating Atmospheric Dispersion*, Oak Ridge National Laboratory (1960).

F. Pasquill, *Atmospheric Diffusion*, 2nd edition, Halsted Press, New York (1974).

T. H. Pigford and K. P. Ang, The Plutonium Fuel Cycles, *Health Physics*, **29,** 451 (1975).

M. S. Singh, Production and Shielding of X-Rays from Electron Beam Vapor Sources, *Proceedings of the 20th Mid-year Topical Meeting of the Health Physics Society*, Reno, Nevada (1987).

D. Slade, editor, *Meteorology and Atomic Energy*, Technical Information Division, US Atomic Energy Commission, Washington, DC (1968).

TID 7016, Rev. 1, *Nuclear Safety Guide*, U.S. AEC (1961).

U.S. DOE Order 5530.1, *Response to Accidents and Significant Incidents Involving Nuclear Weapons*, Washington, DC (1983).

U.S. NRC Regulatory Guide 3.35, *Assumptions Used for Evaluating the Potential Radiological Consequences of Accidental Nuclear Criticality in a Plutonium Processing Plant*, U.S. NRC, Washington, DC (1979).

U.S. NRC Regulatory Guide 8.11, *Applications of Bioassay for Uranium*, U.S. NRC, Washington, DC (1974).

5

POWER REACTOR HEALTH PHYSICS

The field of power reactor health physics is concerned with protecting the health and safety of the public, including the plant's workers, from a wide variety of radiation environments associated with the use of the fission process to produce electricity. These environments include external radiation sources (gamma, beta, and neutron) as well as internal sources of beta and alpha radiation. The situation is often complicated by the occurrence of mixed radiation fields consisting of combinations of the previously noted radiation types.

OVERVIEW

All power reactors have a fuel core in which fissions occur, and most of the fission energy appears as heat. The nuclear fuel core contains fissile material (i.e., U-233, U-235, or Pu-239) and fertile material (U-238 or Th-232). Most commercial reactors use uranium fuel enriched to about 2–5% in U-235. The fuel consists of uranium dioxide pellets contained within zirconium rods that are arranged in a fuel bundle. The core is composed of 100–200 or more fuel bundles or assemblies.

The fuel core is cooled by water which also serves to moderate the fission neutrons to enhance reactor performance. The coolant that directly cools the core, the primary coolant, contains radioactive material which is of concern to the health physicist.

The primary reactor coolant contains a variety of dissolved and suspended radionuclides that present an external as well as internal hazard. The primary system consists of a reactor vessel, containing the reactor core, cooling water piping, heat exchanger equipment (including a pressurized water reactor's steam generators), and a multitude of support pumps, valves, and system control components.

The various components containing primary fluid may be approximated by point, line, disk, or slab sources. For example, primary piping behaves like a line source, and hot particles are approximated by point sources. The dose rates above demineralizer beds and surface contamination are often approximated by the use of either a thin disk or slab source.

The specific isotopes that comprise the reactor sources include both fission and activation sources. Their relative importance will depend upon the type of activities performed and their timing relative to reactor shutdown.

Commercial power reactors are of two basic design types: pressurized-water reactors (PWRs) and boiling-water reactors (BWRs). In PWRs, the pressure within the reactor vessel, containing the core and primary coolant, is high enough to prevent the water from boiling. After flowing through the core, the primary coolant is heated, pumped through a steam generator (heat exchanger), and then returned to the core region. Heat is transferred from the primary coolant to the lower pressure feedwater surrounding the steam generator tubes. These tubes form a boundary between the radioactive primary coolant and the clean feedwater or secondary coolant. Because the feedwater is at a lower pressure, it boils and produces steam that is used to drive the turbine generator to produce electricity. The exhaust steam from the turbine is condensed and returned as feedwater to the steam generator.

The second type of commercial power reactor is the BWR. The pressure of the primary coolant in a BWR is lower than that in a PWR and the primary water boils. Steam is produced directly by the fission heat, and there is no separate steam generator. The BWR steam is radioactive, which adds to the concern of the health physicist.

HEALTH PHYSICS HAZARDS

The power reactor health physicist must deal with a wide variety of health physics issues. Issues, such as internal and external exposure control, are not unique to the power reactor field, but their application is unique to the reactor environment. In order to address the power reactor health physics concerns, we will present a sampling of exposure concerns associated with a commercial reactor.

A summary of contamination and direct dose radiation hazards from typical power reactor activities are summarized in Table 5.1. Examples of these hazards include:

1. The buildup of activity on a demineralizer bed or filter
2. Activation of reactor components
3. Fuel element cladding failures

Table 5.1 Power Reactor Health Physics Concerns

Activity	Concern Internal	Concern External[a]	Hazard
Primary component maintenance during an outage	Yes	Yes	Activation products and fission products, depending on the fuel integrity. Hot-particle controls are warranted.
Primary component maintenance during power operation	Yes	Yes	Activation and fission products, hot particles, neutrons, and N-16.
Steam generator eddy current surveillance and repair (outage)	Yes	Yes	Activation and fission products and hot particles.
Spent-fuel-pool fuel movements	Yes	Yes	Hot particles and criticality.
Containment at power inspections	Yes	Yes	Noble gas, skin exposure, tritium, iodine, neutrons, and N-16.
Radioactive waste processing	No[b]	Yes	Activation and fission products and hot particles.

[a]Direct radiation exposure.
[b]Unless personnel error or procedure violations occur.

4. Reactor coolant system leakage
5. Hot-particle skin dose

These hazards are typical of the challenges that are part of a power reactor environment. They are, however, not a complete listing of the challenges faced by nuclear power plant health physics personnel.

Buildup of Filter or Demineralizer Activity

A common power reactor problem involves the calculation of the buildup of activity on a filter element or demineralizer bed. Demineralizers remove radionuclides by an ion-exchange mechanism. The buildup of activity (A) on a filter or in a demineralizer bed can be determined by knowing the system properties and the isotopes present. Important parameters that impact the buildup of the activity include the concentration of the isotope entering the demineralizer, the system flow rates, and the time the demineralizer is operating. The demineralizer activity can be defined in terms of these parameters through the following relationship:

$$A = (C_i)(F)(e)[1 - \exp(-\lambda t)]/(\lambda) \tag{5.1}$$

where

A = activity of the isotope at time t
C_i = influent activity (μCi/ml)
F = flow rate (ml/min)
e = efficiency of the filter/demineralizer for removal of the isotopes of interest
λ = decay constant
t = time demineralizer/filter is on line

This activity can be used in dose rate estimates. Estimates of streams involving multiple isotopes require the application of Eq. (5.1) for each nuclide present in the influent stream.

Activation of Reactor Components

Another source of activity is the direct irradiation of reactor components and the activation of corrosion products. Corrosion or wear material, dissolved or suspended in the coolant, passes through the core region. Within the core region, it is exposed to the core's neutron fluence. Activation occurs by a variety of neutron-induced reactions, and its magnitude depends upon the neutron fluence impinging upon the material in the core region.

The activity derived from the irradiation is given by

$$A = N\sigma\phi[1 - \exp(-\lambda t_{\text{irrad}})] \exp(-\lambda t_{\text{decay}}) \qquad (5.2)$$

where

N = number of atoms of a given isotope that are activated
σ = cross section for the reaction induced by the flux ϕ (barns/atom)
ϕ = fluence rate or flux (neutrons/cm^2-sec) inducing the activation reaction
λ = decay constant of the activated material
t_{irrad} = time the sample was irradiated or exposed to the core flux
t_{decay} = decay time or time the sample was removed from the reactor's core region or activating flux

Once the activity of a source is known, its dose rate impact can be determined from a knowledge of its basic geometry properties. Common geometries include the point, line, disk, and slab sources.

For example, the dose rate at a distance r from a small particle source can be obtained from the point source approximation. The point source approximation is applicable whenever the distance from the source is at least three times the largest source dimension.

A second useful approximation encountered in a power reactor environment is the line source approximation. The line source equation is often useful when assessing the dose from sample lines or piping carrying primary coolant or other radioactive fluids.

The third useful relationship for estimating the dose rate from typical power reactor components is the thin disk source approximation. A disk source provides a reasonable approximation to the dose rate from a radioactive spill or from a demineralizer bed.

Slab sources can be useful in approximating the dose rates from contaminated soil or contaminated pools. Dose rates from a spent fuel pool whose coolant activity is known or from contaminated concrete floors or walls can be addressed with reasonable accuracy with a slab source approximation.

Cladding Failures

A nuclear reactor contains a number of barriers designed to prevent fission products from escaping from the reactor core to the environment. These barriers include the fuel element cladding, the reactor coolant system and included piping, and the containment building. A breach of any of these barriers warrants serious attention in order to prevent the release of radioactivity to the environment.

A cladding failure immediately releases fission products contained within the fuel element and increases the primary coolant activity. The gaseous activity that is released to the primary coolant is released to containment via leakage paths and directly to off gas systems. These gaseous fission products are an early indication that a fuel cladding failure has occurred. BWRs normally detect fuel failure by detection of fission gases in the off-gas system. However, PWRs normally monitor the primary coolant line or letdown filter lines for these fission products or monitor the containment atmosphere for released xenon and krypton and their daughter products. The analysis of primary coolant samples by gamma spectroscopy is a routine follow-up action in either type of reactor.

Reactor Coolant System Leakage

Leakage from the primary coolant system is an undesirable but inevitable problem at a power reactor. Leaking values, pump seals, value packing, or instrument lines provide pathways for small leaks that contaminate the vicinity of the leak. This contamination must be controlled in order to limit station external and internal exposures. In addition to leaks directly from the primary system, PWR health physicists must address leakage from the primary to secondary systems.

Leakage from the primary system steam generator tubes to the secondary system presents a health physics concern because additional plant areas may become contaminated. Because the secondary components are considered as clean systems, the presence of contamination can have a negative impact on facility operation and expand the areas requiring stringent radiological controls at the facility. In addition to direct leakage, a pathway to the secondary system permits the release of fission gases to the clean side of the facility.

The secondary side gaseous activity can be calculated by assuming an activity balance. Assuming that all gaseous activity released from the primary system, via the leaking steam generator tubes, is released without holdup to the atmosphere through the condenser air ejector, the secondary activity may be determined from the activity balance relationship

$$A_p^i L k_1 = A_s^i \dot{F} k_2 \tag{5.3}$$

where

A_s^i = the air ejector activity for isotope i (μCi/cm^3)
A_p^i = primary coolant activity of isotope i (μCi/cm^3)
L = primary to secondary leak rate (gal/min)
$\dot{F}$ = air ejector flow rate (ft^3/min)
k_1 = conversion factor (3785 cm^3/gal)
k_2 = conversion factor (28,317 cm^3/ft^3)

Activity from fission or activation products can lead to significant buildup of surface contamination from the primary to secondary leakage. If the leak is not repaired in a timely manner, an area (A) of the plant will be contaminated to a level S at the end of a year's time. The secondary plant surface contamination is given by

$$S = A_s K L / A \tag{5.4}$$

where

S = surface contamination level (dpm/100 cm^2-year)
A_s = secondary system activity (μCi/cm^3)
$K = \dfrac{(3785 \text{ cm}^3/\text{gal})(365 \text{ days/year})(2.2 \times 10^6 \text{ dpm-}\mu\text{Ci})}{929 \text{ cm}^2/\text{ft}^2 (100)}$
$\times$ (60 min/hr)(24 hr/day)
A = area of secondary system exposed to the secondary system activity (ft^2)

Contamination buildup over a shorter period of time may be obtained by adjusting the conversion factor K.

PWR secondary coolant contamination has a number of negative health physics aspects. The leakage activity tends to concentrate in components such as the main steam isolation valves and high pressure turbine piping such that locally higher surface contamination areas and local hot spots result. Secondary ion-exchange resins and filters will become contaminated, which adds to the unit's contamination problems and increases the volume of radioactive waste generated. Steam generators cleanup systems will also become contaminated, and the reassessment of the plant's effluent releases may be required.

Hot-Particle Skin Dose

Particulate matter is produced by a variety of power reactor activities. Normal maintenance of pumps, valves, and piping create small particles in the process of cutting, grinding, and welding. The operation of valves and pumps leads to the wearing of active surfaces, and this wear often produces small particulate material. Cladding failures or erosion of control rod surfaces contribute additional matter to the reactor coolant system. This material is often too small to be removed by the reactor coolant filters. Therefore it passes through the core and is activated in the core's neutron fluence. The result of this activation is the creation of highly activated, microscopic material that is called a "hot particle."

Hot particles are very small highly radioactive particles with high specific activities. The particles may be composed of activation products and possibly fission fragments depending upon the core's fuel clad integrity. Particles may contain either single isotopes or a large number of radioisotopes. Skin exposure is the primary radiation hazard from these particles. Beta radiation is the dominant contribution to the skin dose.

Table 5.2 summarizes the types of hot particles that are found in power reactor environments and classifies them into three broad categories: fission products, activation products, and zirconium/niobium particles. The product hot particles consist of fission nuclides that have escaped from the fuel cladding. Activation products represent corrosion or wear products that have been subsequently activated by the core's neutron fluence. The zirconium/niobium particles are derived from cladding material wear.

Table 5.2 summarizes typical particulate compositions and are classified according to their dominant constituent isotopes. The reader should note that each particle may contain isotopes from any of the three classifications.

A common problem encountered in a power reactor involves the assessment of skin exposures from a hot particle. The hot particle may reside directly on the skin, on protective clothing, or on personal clothing. Table 5.3 provides calculated beta and gamma dose rates to the skin from the particles of Table 5.2. Dose rates are provided for a variety of particle locations. The skin doses are provided for unit activity and assume an exposure area of 1 cm^2 at a depth of 7 mg/cm^2.

The use of the data of Table 5.3 would require a knowledge of the total particulate activity. Obtaining this information requires counting in a calibrated geometry which may take an hour or more. In order to obtain a quick estimate of the dose to the exposed worker, it would be desirable to be able to convert the survey instrument reading used to detect the particle into an exposure rate. Table 5.4. provides factors that are useful in obtaining quick estimates of the hot-particle dose rate from open-window ion-chamber readings.

Table 5.2 Representative Radionuclide Distribution in Hot Particles at a PWR for Three Types of Particles

Nuclide	Fission Product	Activation Product	Niobium/Zirconium
Na-22	2.7×10^{-3}	—	—
Sc-47	—	—	1.3×10^{-2}
Cr-51	—	4.8×10^{-1}	8.8×10^{-3}
Mn-54	—	4.0×10^{-3}	1.5×10^{-3}
Co-57	—	1.0×10^{-3}	—
Co-58	8.5×10^{-4}	4.3×10^{-1}	2.0×10^{-3}
Fe-59	—	5.0×10^{-3}	4.1×10^{-4}
Co-60	5.6×10^{-4}	6.9×10^{-2}	1.3×10^{-3}
Nb-95	—	1.0×10^{-3}	5.2×10^{-1}
Zr-95	—	—	4.0×10^{-1}
Ru-106	1.2×10^{-1}	—	—
Rh-106	1.2×10^{-1}	—	—
Sn-113	—	2.0×10^{-3}	1.4×10^{-2}
In-113	—	2.0×10^{-3}	1.4×10^{-2}
Sb-125	9.2×10^{-3}	—	1.5×10^{-2}
Te-125	9.2×10^{-3}	—	1.5×10^{-2}
Cs-134	5.8×10^{-2}	—	—
Cs-137	2.9×10^{-1}	—	—
Ce-144	1.9×10^{-1}	—	—
Pr-144	1.9×10^{-1}	—	—
Hf-181	—	—	2.8×10^{-3}
Am-241	2.6×10^{-3}	—	—

Source: W. W. Doolittle et al. (1992).

Follow-up action by the health physicist will require a more refined dose assessment. These assessments are required from both regulatory and legal perspectives and may be performed by a variety of methods.

The skin dose from a hot particle residing on the skin is given by

$$D = \sum_i A_i (DF_i) t / S \quad (5.5)$$

where

D = hot-particle skin dose (rad)
A_i = particle activity (μCi) for radionuclide i
DF_i = dose factor for radionuclide i (rad-cm^2/μCi-hr)
t = residence time on the skin (hr)
S = area over which the dose is averaged (cm^2)

Table 5.3 Skin Exposure Rate per μCi for Various Protective Clothing Configurations

	Density Thickness (g/cm^2)	Skin Dose Rate for Particle Type (mrad/hr to 1 cm^2)						
		Fission Product		Activation Product		Niobium/Zirconium		
PCC[a]		Beta	Gamma	Beta	Gamma	Beta	Gamma	
Bare skin	0.0070	5.2×10^3	1.7×10^2	7.8×10^2	3.2×10^2	2.4×10^3	4.0×10^2	
One pair of PCs	0.0278	2.2×10^3	—[b]	1.3×10^2	—	3.6×10^2	—	
Two pair of PCs	0.0530	1.4×10^3	—	2.4×10^1	—	5.6×10^1	—	
One pair of rubber gloves	0.0567	1.3×10^3	4.0×10^1	1.9×10^1	7.3×10^1	4.3×10^1	9.2×10^1	
Two pairs of rubber gloves	0.1028	8.9×10^2	—	1.3	—	3.5	—	
Three pairs of rubber gloves	0.1487	7.0×10^2	2.0×10^1	9.9×10^{-2}	3.7×10^1	8.7×10^{-1}	4.6×10^1	
Boots-I	1.6822	2.0	—	1.2×10^{-9}	—	1.0×10^{-10}	—	
Boots-II	1.9050	8.9×10^{-1}	5.2×10^{-1}	0.0	9.7×10^{-1}	0.0	1.2	

Source: W. W. Doolittle et al. (1992).
[a]Protective clothing configuration.
[b]Not calculated.

Table 5.4 Ratio of Skin Dose Rate to Ion-Chamber Survey Instrument Reading for Various Protective Clothing Configurations (RSO-50)

Protective Clothing Configuration	Ratio of Skin Dose Rate to Observed Instrument Reading for Particle Type		
	Fission Product	Activation Product	Niobium/Zirconium
Bare skin	2.2×10^2	3.0×10^2	1.3×10^2
One pair of PCs	9.2×10^1	7.4×10^1	2.5×10^1
Two pairs of PCs	5.8×10^1	3.1×10^1	7.8
One pair of rubber gloves	5.4×10^1	2.6×10^1	6.4
Two pairs of rubber gloves	3.7×10^1	1.6×10^1	3.4
Three pairs of rubber gloves	2.9×10^1	1.0×10^1	2.2
Boots—I	1.0×10^{-1}	3.1×10^{-1}	3.8×10^{-1}
Boots—II	5.6×10^{-2}	2.8×10^{-1}	3.3×10^{-1}

Source: W. W. Doolittle et al. (1992).

Following NCRP 106, the skin dose is generally averaged over 1 cm^2 and evaluated at a distance of 7 mg/cm^2 at the basal cell layer depth. Because the dose from a point source falls off rapidly as 1 over r-squared, the dose from a hot particle is highly localized.

HEALTH PHYSICS PROGRAM ELEMENTS

In addition to addressing radiation hazards, the health physics program at a power reactor must consider a variety of other factors. The radiation protection program at a power reactor includes the goal of maintaining radiation exposures as low as reasonably achievable (ALARA), the assessment of plant effluent releases, assessment of internal and external exposures, radiation surveys, personnel monitoring, radiological environmental surveillance, radioactive waste management, and control of work activities. These program elements that are unique to a power reactor environment will be addressed in more detail below.

ALARA

New regulatory requirements and professional ethics suggest the need to minimize radiation exposures of plant workers. An effective ALARA program requires management commitment and cooperation between the various facility work groups, particularly health physics, maintenance, and operations. Examples of this cooperation include:

1. Development of pre-job ALARA briefings and post-job ALARA critiques for high-dose activities. These activities include reactor vessel head re-

moval and undressing, steam generator surveillance and repair activities, reactor coolant pump seal replacement, reactor vessel head installation, and reactor cavity work.
2. Development of outage pre-planning and post-outage critiques from a radiation exposure perspective. Personnel involved with the work activities must take an active part in these activities.
3. Establishing exposure goals for each outage activity and major nonoutage evolutions.
4. Development of a system for tracking exposure by work request to identify improvement areas.

In addition to these cooperative activities, the radiation protection group can significantly reduce worker's exposures by minimizing the use of protective clothing. The utilization of respiratory protection should focus on minimizing the total worker radiation exposure. Because the total dose is a sum of both internal and external exposure, respirators should not be used to reduce only the internal dose.

Contamination control is another key to an effective ALARA program. For example, minimizing contamination levels in the refueling cavity minimizes direct dose and reduces the protective clothing requirements. The use of remote technology, including cameras, state-of-the-art radiation monitors, and wireless headsets, is also effective in minimizing exposures.

Finally, the support of plant management is required for a successful ALARA program. Plant management support will be required in the following types of areas:

1. Budgetary support to upgrade health physics equipment
2. Improving preventive maintenance of plant equipment
3. Training personnel in ALARA techniques
4. Training and utilizing dedicated crews for specific high-dose tasks
5. Providing sufficient training for contractor outage personnel
6. Chemical decontamination of highly contaminated primary system components
7. Reduction of filter sizes to remove suspended particulates

Effluents

Although off gas systems are designed to trap most gaseous effluents, quantities of Kr-85, H-3, and C-14 and smaller amounts of Xe, I, and Br isotopes may also be released. The Kr, Xe, and I isotopes are generated in the fission process. Their release is facilitated by breaks in the fuel clad. Tritium arises from the neutron activation of the primary coolant, from neutron capture in B-10 [B10$(n, 2\alpha)$H-3], and from tertiary fission. C-14 is produced from the N-14(n, p)C-14 reaction.

Table 5.5 Typical Power Reactor Activation Products

Radionuclide	Neutron Energy Region	Source	Reaction
H-3	Thermal	Lithium hydroxide[a]	$^{6}\text{Li}(n, \alpha)^{3}\text{H}$
	Thermal	Boric acid[a]	$^{10}\text{B}(n, 2\alpha)^{3}\text{H}$
	Fast	Primary coolant	$^{2}\text{H}(n, \gamma)^{3}\text{H}$
Mn-54	Fast	Stainless steel[b]	$^{54}\text{Fe}(n, p)^{54}\text{Mn}$
Fe-59	Thermal	Stainless steel[b]	$^{58}\text{Fe}(n, \gamma)^{59}\text{Fe}$
Co-58	Thermal	Stainless steel[b] and stellite	$^{57}\text{Co}(n, \gamma)^{58}\text{Co}$
	Fast	Stainless steel[b] and stellite	$^{58}\text{Ni}(n, p)^{58}\text{Co}$
Co-60	Thermal	Stainless steel[b] and stellite	$^{59}\text{Co}(n, \gamma)^{60}\text{Co}$
Zr-95	Thermal	Zirconium fuel cladding	$^{94}\text{Zr}(n, \gamma)^{95}\text{Zr}$

[a] Primary coolant chemistry control.
[b] Corrosion or wear products and structural material.

Liquid effluents include fission product and activation products as well as tritium. Liquid-waste cleanup systems, including filtration and demineralization, remove much of these radionuclides which are then processed for burial as low-specific-activity waste.

Fission product radionuclides generated from binary fission include Kr-85, Kr-87, Kr-88, Xe-133, Xe-135, Xe-137, I-131, Cs-137, Ba-137, Ce-141, Ce-144, Ru/Rh-103, Ru/Rh-106, and Sr/Y-90. Activation products are produced by neutron capture by materials in the vicinity of the nuclear core. These include (a) chemical control agents dissolved in the primary coolant (Li-6 and B-10), (b) Stainless steel or stellite corrosion or wear products resulting from system maintenance, (c) primary coolant system piping and the reactor vessel, and (d) the core structural material. Examples of these activation products are summarized in Table 5.5.

RADIOACTIVE WASTE

Power reactor waste occurs in solid, liquid, and gaseous forms. Solid waste includes spent reactor fuel, ion-exchange resins, filters, evaporator bottoms, articles of protective clothing, glassware, tools, and materials used in the control of contamination (paper, plastic, and absorbant materials). Liquid waste includes the primary reactor coolant, cooling water in the spent-fuel pool, contaminated solvents and pump oil, spent chemical reagents, and scintillation fluids. Sources of gaseous radioactive waste are fission product gases and gases due to neutron activation of the primary coolant.

Table 5.6 Radioactive Waste Processing Methods

Solid	Liquid	Gas
Compaction of low-specific-activity material into boxes or drums.	Concentrate by evaporation. The remaining material is usually solidified or stabilized.	Dilution with large air volumes, monitoring, and stack discharge.
On-site storage to allow decay prior to burial.	Dilution of liquid followed by monitoring and discharge to large bodies of water or rivers.	Hold up gases in tanks to permit decay prior to release.
Solidification of resin and evaporator bottoms.	Ion exchange to remove dissolved radioactivity.	Off-gas treatment (activated charcoal beds, silica, gel, and alumina). These media are then treated as solid waste.
Burial of filters in high-integrity containers with concrete overpacks.	Filtration	
Demineralization via ion-exchange resins as an alternative to evaporation and solidification. Resins are then dewatered and placed in high-integrity containers with concrete overpacks.	Holdup lines or tanks to permit decay of short-lived isotopes.	

Processing of these wastes is summarized in Table 5.6. With rising burial costs, volume reduction and source term control are becoming important considerations in the economics of plant operation.

OUTAGES

Outages occur at commercial power reactors for scheduled refueling and maintenance and for unscheduled maintenance. Scheduled outages occur at a frequency of between 12 and 24 months depending upon the core enrichment and the philosophy of the operating utility.

In both PWRs and BWRs, the primary sources of radiation exposure during scheduled outages are activation products derived from corrosion and wear products. Co-58 and Co-60 are significant contributors to the radiation exposure. PWR outage doses are dominated by steam generator work (eddy current testing of generator tubes, tube plugging, and the installation/removal of nozzle dams), reactor coolant pump maintenance, refueling cavity work, and reactor vessel head installation/removal and associated activities. Other outage activities include valve and pump repair and maintenance, fuel transfer system maintenance, plant modifications, control rod drive repairs, in-core instrument replacement, and radioactive waste processing and packaging.

BWR outage doses are spread over similar activities, excluding steam generator work. In addition, BWR exposures include steam turbine and condenser work.

RADIOLOGICAL CONSIDERATIONS DURING REACTOR ACCIDENTS

Reactor accidents may take a variety of forms, but the most radiologically significant events will involve core damage that could lead to the potential of radioactive releases to plant areas and to the environment. Other events (namely, failure of waste-gas decay tanks or spent-fuel element breaches) are less severe, but more likely, scenarios. Reactor accidents are classified into broad categories. These accidents vary in their severity and are included in the design basis of a commercial nuclear power reactor. The accident categories include:

1. *Loss of Coolant Accidents (LOCAs).* In a LOCA, the reactor cooling water is reduced or lost and the nuclear fuel begins to heat up. The event may be caused by a piping rupture, seal failure, instrument line failure, and/or leaks in piping, valves, or components. If the LOCA is severe, the fuel will eventually melt. Fuel cladding degradation may occur even without fuel melting. Breaches in the clad will release fission radionuclides into the reactor coolant. Subsequent breaches in the RCS or containment building will offer a release path to the environment.

2. *Steam Generator Tube Ruptures (SGTRs).* In a PWR, the steam generator tubes form a barrier between the primary and secondary coolants. If a tube rupture or leak occurs, a pathway is created that mixes the primary (radioactive) and secondary (nonradioactive) fluids. As a minimum the secondary (clean) part of the plant will be become contaminated and its radiation levels will show a significant increase. Failures of atmospheric or steam generator safety values or of other secondary system piping, valves, or components provide a direct release pathway to the environment.

3. *Fuel-Handling Accidents (FHAs).* The nuclear fuel residing in the spent-fuel pool contains fission products that have been decaying since the fuel assembly was removed from the core. These fuel elements are periodically moved within the pool, and accidents during these evolutions can damage the clad and lead to a release of radionuclides into the radiologically controlled plant areas.

4. *Waste Gas Decay Tank Ruptures (WGDTRs).* Gas decay tanks store fission gases and permit their decay prior to release to the environment. Failures of the tank structure, valves, or associated components will release fission gases into the plant.

The extent to which these accidents lead to radiological consequences depends largely on the condition of the reactor core. If the fuel cladding remains intact, the releases will be characterized by the steady-state activity of the primary coolant. The radiological hazards increase proportionally with the degree to which the fuel cladding degrades and releases fission products into the primary coolant.

In each of these events, radioactive gas can be released to plant areas and then to the environment. The gamma dose rate from nuclide K in the radioactive gas cloud is often assessed using the semi-infinite cloud model:

$$\dot{D}_K^\gamma = k\overline{E}_K^\gamma X_K(r, t) \tag{5.6}$$

where

$\dot{D}_K^\gamma$ = dose rate (rad/sec) from nuclide K in the cloud

k = conversion factor for a semi-infinite cloud

$$= \frac{1 \text{ rad}}{100 \text{ erg/g}} \frac{1.6 \times 10^{-6} \text{ erg}}{\text{MeV}} \times \frac{3.7 \times 10^{10} \text{ dis/sec}}{\text{Ci}}$$

$$\times \frac{\text{m}^3}{1293 \text{ g}} \times 0.5$$

$$= 0.23 \frac{\text{rad}}{\text{sec}} \frac{\text{dis}}{\text{MeV}} \frac{\text{m}^3}{\text{Ci}}$$

$\overline{E}_K^\gamma$ = average gamma energy per disintegration for nuclide K (MeV/dis)

$X_K(r, t)$ = air concentration of nuclide K (Ci/m^3) at a distance r from the release point and at time t

It is often more convenient to measure the source term at the plant rather than at the receptor location. This is particularly true in the early stages of an accident when field measurements are not available. The air concentration can be related to the source term:

$$X_K(r, t) = Q_K \frac{X}{Q} \tag{5.7}$$

where

Q_K = source term or release rate for nuclide K (Ci/sec)
X/Q = atmospheric dispersion parameter (sec/m^3)

With this relationship, the dose rate equation becomes

$$\dot{D}^\gamma_K = k\bar{E}^\gamma_K Q_K \frac{X}{Q}$$

$$= Q_K \, \text{DRCF}_K \frac{X}{Q} \qquad (5.8)$$

where the dose rate conversion factor (DRCF_K) is

$$\text{DRCF}_K = k\bar{E}^\gamma_K \qquad (5.9)$$

If more than a single isotope is released, the total dose rate is the sum of the dose from the individual radionuclides:

$$\dot{D}^\gamma = \sum_K \dot{D}^\gamma_K = \sum_K Q_K \, \text{DRCF}_K \frac{X}{Q} \qquad (5.10)$$

The semi-infinite cloud model assumes that the release rate is constant and that the atmospheric conditions, as described by the dispersion parameter, are also constant. Accident events are not likely to meet either of these conditions for extended periods of time. Another assumption is that the plume dimensions are large compared with the distance the gamma rays travel in air. This assumption is not valid close to the source, but it is more easily achieved further from the source.

Additional model shortcomings include exclusion of the air attenuation of gamma rays. The model also does not account for radiation buildup factors caused by the Compton scattering of the gamma ray photons.

These shortcomings are easily overcome with computer codes that include these factors and perform the requisite numerical integration. However, knowledge of the semi-infinite cloud models are invaluable for quick assessments during the initial stages of accident conditions.

This general formulation also applies to thyroid, bone, and other organ doses. The dose conversion factors will change to reflect the type of dose calculation being performed.

MITIGATION OF ACCIDENT CONSEQUENCES

The health physicist must maintain worker and public radiation exposures ALARA during an accident. This includes critical reviews of plant repair activities to terminate the radiological release. The use of plant safety systems should also be considered. These safety systems include the use of sodium hydroxide spray within containment to reduce the iodine source term, the use of emergency core cooling systems (ECCS) to cool the core, and the use of filters to reduce the particulate and iodine source terms.

Table 5.7 Reactor Accident Mitigation

Accident Type	Release Type	Mitigation	Termination
LOCA	Iodine Noble gas Particulate	NaOH spray Filtration ECCS	In-plant repairs Reestablish core cooling
SGTR	Iodine Noble gas	Filtration Release via condenser ECCS	Cool and depressurize the RCS In-plant repairs
FHA (< 1-year-old fuel)	Iodine Noble gas	Filtration	Fuel assembly depressurizes
FHA (> 1-year-old fuel)	Noble gas	Filtration	Fuel assembly depressurizes
WGDTR	Noble gas Iodine (if fuel defects)	Filtration	Tank depressurization In-plant repairs

Table 5.7 summarizes the various power reactor accident types, the types of radiological releases that could occur, plant systems that could be used to mitigate the release, and methods that could be utilized to mitigate the release.

SCENARIOS

Scenario 57

A pressurized water reactor demineralizer is loaded with resin to reduce the activity of Co-60 in the primary coolant. During the current 200-day cycle, the demineralizer processed reactor coolant. Specific operational characteristics of the demineralizer are provided below.

The following data may be useful:

The demineralizer is a vertical cylindrical vessel that is 4 ft in diameter and 12 ft high.

Its wall thickness is $\frac{3}{8}$ in., and the demineralizer is made of stainless steel. The demineralization bed is 4 ft in diameter.

Assume that the dose rate from the resin bed can be approximated by a thin disk source which is described by the equation

$$\dot{D}(h) = 3.14 \times G \times C \times \ln \frac{h^2 + R^2}{h^2}$$

C = source strength in Ci/m^2

Attenuation Coefficients (Lead)

E (MeV):	0.6	0.7	0.8	1.0	1.25	1.50	2.75
u (1/cm):	1.36	1.12	0.97	0.78	0.65	0.58	0.47

Demineralizer and Radiation Characteristics

Nuclide	Half-life	Radiation Energy (MeV)	Percent Yield	RCS Activity (μCi/cm^3)
Co-60	5.26 years	1.173(gamma)	100.0	6.0 × 10^{-4}
		1.332(gamma)	100.0	
		1.480(beta)	0.12	
		0.314(beta)	99.0	

Dose Buildup Factors for a Point Isotropic Source

		\multicolumn{7}{c}{ux^a}
Material	MeV	1	2	4	7	10	15	20
Lead	0.5	1.24	1.42	1.69	2.00	2.27	2.65	2.73
	1.0	1.37	1.69	2.26	3.02	3.74	4.81	5.86
	2.0	1.39	1.76	2.51	3.66	4.84	6.87	9.00
	3.0	1.34	1.68	2.43	2.75	5.30	8.44	12.3

$^a ux$ = mass attenuation coefficient (u/density) × shield thickness (cm) × shield density (g/cm^3).

Other Data

Flow rate through the demineralizer: 350 liters/min
Demineralizer efficiency for Co-60 removal: 100.0%
Co-60 specific gamma ray emission (gamma constant):

$$G = 1.3 \frac{R - m^2}{hr - Ci}$$

1 R = 87.7 erg/gram in air.
Assume the dose (rads) to air equals the dose (rads) to tissue.

5.1. Calculate the gamma dose equivalent rate in rem/hr at a point 1 ft above the centerline of the demineralizer bed immediately after the 200-day run. For the purposes of this problem, assume that the demineralizer contains no water above the resin bed at the end of the run and that all activity is distributed uniformly over the top surface of the demineralizer bed.

5.2. A valve located 25 ft above the demineralizer bed required repair. When isolated, the demineralizer was loaded with 80.0 Ci of Co-60.

Assume that the Co-60 is uniformly distributed over the top surface of the bed. The repair must be performed by a pipe fitter whose remaining annual dose equivalent is 300 mrem. The valve repair will take 3 hr. The only shielding available is a previously installed lead mat which is composed of 2 in. of equivalent lead. Will the installed shielding keep the worker below his annual limit? State all assumptions and calculate the total dose in rem received by the worker. Assume that the repair occurs 6 months after isolation, that there is no water above the resin bed, and that the valve resides on the demineralizer centerline. For simplicity, only consider the gamma dose contribution.

Scenario 58

As a health physicist at a power reactor, you have been assigned the lead role in evaluating hot particle contamination of station personnel. Hot particles are very small highly radioactive particles with high specific activity. These particles have recently been detected with increasing frequency at your facility. Isotopic analyses indicate that the particles are composed of fuel or neutron-activated corrosion and wear products.

On Tuesday morning, you receive a phone call about a hot particle which was removed from under a pipe-fitter's fingernail. Upon reporting for work, the pipe fitter alarmed the entrance portal contamination monitor. The contamination was detected during a follow-up frisk, and it was successfully removed on the first attempt using sticky tape. Although not visible to the naked eye, the particle was analyzed for its constituent radionuclides.

Data

An analysis of the contamination on the sticky tape led to the following results:

Radionuclide	Activity (μCi)	Dose Factor[a] (rad-cm^2/μCi-hr)
Co-60	3.9×10^{-1}	4.13
Zn-65	1.0×10^{-6} (LLD)	0.106
Zr-95	1.0×10^{-6} (LLD)	4.87
Nb-95	1.0×10^{-6} (LLD)	0.865

[a]Beta dose factors applicable for skin dose calculations averaged over 1 cm^2.

Based upon a review of the event, you determine the following:

Time of contamination: 3:45 p.m. on Monday
Time of removal: 7:30 a.m. on Tuesday (following day)
Source of contamination: Unknown, although the worker's protective clothing is strongly suggested.

Using the above data and your experience, choose the single best answer to the following questions:

5.3. In general, for a hot particle on the skin, the principal contributor to the dose to the skin in the vicinity of the particle is:
   a. The neutron radiation emitted from the particle
   b. The gamma radiation emitted from the particle
   c. The alpha radiation emitted from the particle
   d. The conversion electron emitted from the particle
   e. The beta radiation emitted from the particle

5.4. The most plausible explanation for the increasing frequency of detection of hot particles at nuclear power plants is:
   a. The increased use of more sensitive instrumentation for detecting and measuring contamination by these particles
   b. An increase in the rate of production of these particles at nuclear power plants
   c. The changes in plant chemistry which have enhanced fuel reliability but have increased the corrosion rate of other components
   d. A decrease in the average corrosion particle size as plants have aged
   e. The trend toward increasing the time between refueling outages with a subsequent decrease in preventive maintenance

5.5. Prior to 1988, explicit recommendations on limits for radiation exposure of skin by hot particles were provided by:
   a. The ICRP, but not the NCRP
   b. Both the NCRP and the ICRP
   c. The NCRP but not the ICRP
   d. Neither the NCRP nor the ICRP
   e. The ICRU, NCRP, and ICRP

5.6. Radiobiological evidence suggests that, when compared to more uniform irradiation by the same quantity of radioactive material, highly localized beta irradiation of skin, such as from a particle on the skin, is:
   a. Less likely to cause skin cancer
   b. More likely to cause skin cancer
   c. About equally likely to cause skin cancer
   d. Likely to cause an erythema within a few hours
   e. Likely to cause a small necrotic lesion on the skin of an individual after only a few hours of exposure.

5.7. For a typical beta–gamma survey meter with a 5-cm pancake probe, what approximate instrument efficiency is appropriate for a 2.0-cm-

diameter stainless steel disk of Sr-90 if the source detector distance is 1 cm?
a. 1%
b. 5%
c. 35%
d. 50%
e. 65%

5.8. The beta dose equivalent to the skin of the worker's finger is:
a. 1.6 rem
b. 8.0 rem
c. 25 rem
d. 33 rem
e. 33,000 rem

5.9. For this scenario, assume that the calculated dose for regulatory purposes, beta and gamma, was 60 rem. Which one of the following statements best describes compliance with U.S. Nuclear Regulatory Commission regulations?
a. No limit was exceeded. All tissues received less than the yearly allowable regulatory limit of 75 rem for the extremities.
b. The whole-body exposure was less than the quarterly allowable regulatory limit of 3 rem.
c. The skin of the whole-body limit of 15 rem was exceeded.
d. The dose limit of 18.75 rem for the extremities was exceeded.
e. The annual whole-body dose limit of 5 rem was exceeded.

5.10. Which one of the listed follow-up actions would not be appropriate if you calculated a 60-rem total dose for this scenario?
a. Survey of the worker's home, car, and girlfriend's home and of the local restaurant where he ate the previous evening.
b. Review, recalibrate, and evaluate the station's laundry monitoring systems.
c. Interview the worker to advise him of his rights under federal law and to answer the worker's questions regarding biological effects from his exposure.
d. Evaluate the sensitivity of portal monitors and other contamination detection instrumentation.
e. Initiate thyroid monitoring and increased urinalysis frequency.

5.11. Which one of the following detectors would provide the best sensitivity for detecting these particles at the exit station from the plant?
a. An energy compensated G-M probe
b. Air proportional detector

c. Zinc sulfide scintillation probe
   d. Intrinsic germanium crystal with single channel analyzer
   e. Ionization chamber
5.12. In 1989, NCRP 106 established the following limit for hot particles:
   a. 40 µCi-hr
   b. 75 µCi-hr
   c. 150 µCi-hr
   d. 250 µCi-hr
   e. 500 µCi-hr

Scenario 59

You are a member of the professional health physics staff at a nuclear power plant and have been asked to prepare a briefing for management on occupational radiation exposure in the nuclear power industry. The briefing is to include recent trends in collective doses, means of dose reduction and maintaining doses ALARA, and potential effects of changing regulatory requirements. Your briefing will recognize that in the United States, collective occupational exposure (person-rem per reactor per year) increased to maximum in 1980. Since then, the general trend has been down. However, when compared to other countries, the collective dose in the United States is still among the highest, indicating a substantial potential for improvement. In addition to these facts, your briefing will include points covered by the following questions:

5.13. All of the following factors, with one exception, have contributed significantly to the downward trend in annual occupational exposure per reactor in the United States since 1980. That exception is:
   a. Replacement of highly contaminated major components of the reactor system
   b. Completion of NRC-mandated safety actions (such as fire protection, seismic upgrading, etc.)
   c. Industry actions including ALARA programs
   d. New plants going into service
   e. Higher capacity factors for U.S. plants

5.14. In general, the most important source of occupational radiation exposure at nuclear power plants is:
   a. Gamma radiation from the core
   b. Long-lived fission product activity
   c. Neutron activation product activity
   d. Short-lived fission product activity
   e. Neutron radiation from the core

5.15. The radionuclide that is the main cause of shutdown radiation fields at nuclear power plants is:
   a. N-16
   b. Cs-137
   c. Co-58
   d. Co-60
   e. Xe-135

5.16. Comparative assessment of U.S. and foreign nuclear power plant occupational dose experience has indicated that reductions in out-of-core radiation fields can result in substantial reductions in occupational doses. Of the following actions, the one that is the least effective in reducing out-of-core radiation fields is:
   a. Hydrogen water chemistry
   b. Control of impurities in the reactor coolant
   c. Preconditioning of out-of-core surfaces (polishing, prefilming)
   d. Reduction or elimination of cobalt in reactor system components
   e. Chemical decontamination

5.17. Studies of radiation exposure incidents have identified six radiation protection errors that are common to most incidents of unplanned, unnecessary radiation exposures at nuclear power plants. One error that is not included among these six errors is:
   a. Inaccurate or incomplete radiation surveys
   b. Inadequately prepared radiological work permits
   c. Failure of the radiological protection technician to react to changing or unusual conditions
   d. Failure of workers to follow procedures or good radiological work practices
   e. Lack of involvement on the part of supervisors or foremen
   f. A neutral or negative attitude toward radiological protection on the part of the plant operators and overall plant management
   g. Lack of knowledge of NRC regulations by the radiation protection technician

5.18. All of the following, with one exception, have been reported to be useful in monitoring the performance of the radiation protection program at a nuclear power plant. That exception is:
   a. Number of Certified Health Physicists on the staff
   b. Collective radiation exposure
   c. Skin/clothing contamination events
   d. Solid radwaste volumes
   e. Radiological incident reports

5.19. If personnel with doses less than 100 mrem/year are excluded, the average annual exposure for nuclear reactor workers is typically:
   a. 400–800 mrem
   b. 750–1500 mrem
   c. 1200–2400 mrem
   d. 2000–3000 mrem
   e. 2700–4000 mrem

5.20. Which of the following requirements of the January 1, 1993 revision of 10 CFR Part 20 is least likely to result in a significant change in the radiation protection programs at nuclear power plants?
   a. Imposition of a skin dose limit of 50 rem/year.
   b. Evaluation of the skin dose at a depth of 7 mg/cm^2 and evaluation of the eye lens dose at a depth of 300 mg/cm^2.
   c. Imposition of an eye (lens) dose limit of 15 rem/year.
   d. Imposition of a dose limit for the embryo/fetus of 0.5 rem during pregnancy.
   e. Changes in the air concentration limits.

5.21. In which of the following situations will the new (1993) 10 CFR20 dose limit for protection of the embryo/fetus apply?
   a. A female worker is known to her employer to be pregnant.
   b. A female worker is suspected by her employer of being pregnant.
   c. A female worker has voluntarily made her pregnancy known to her employer.
   d. It would apply to all female workers of child-bearing age.
   e. It would apply to all female workers.

5.22. As a result of reevaluations of the radiation doses received by the Japanese atomic bomb survivors, the ICRP and other authoritative advisory groups may revise their recommendations on occupational dose limits. Current information indicates that a new recommended annual exposure limit is most likely to be about:
   a. 0.5 rem
   b. 1.0 rem
   c. 3.0 rem
   d. 4.0 rem
   e. 6.0 rem

Scenario 60

As the station health physicist, you are assigned the task of performing the dose rate analysis for an upcoming outage in which maintenance work is planned on a primary coolant sampling system. Prior to the maintenance work, the primary system will be decontaminated. In support of this effort, you are asked to provide exposure rate estimates for a number of source configurations. In calculating exposure rates, assume that an individual can be positioned at points P, Q, and X. Ignore any shielding provided by the individual.

Data

The scrap material was originally 100% Co-59 and weighed 10 g (question 5.23).

The Co-59 (n, γ)Co-60 cross section is 37 barns.

The Co-60 half-life is 5.27 years.

Air is the medium between the sources and points P, Q, and X.

The activity of the sources in questions 5.24, 5.25, and 5.26 are each 3.0 Ci, and the activity is uniformly distributed within each source.

Each source is Co-60 which has a specific gamma-ray emission or gamma constant (G):

$$G = 1.3 \frac{\text{R-m}^2}{\text{hr-Ci}}$$

Neglect gamma-ray self-shielding within the source regions and in containing piping.

Avogadro's number = 6.02×10^{23} atoms per gram atomic weight.

5.23. Six months ago, a small scrap of material was removed from the reactor vessel. Plant records indicate that it had been irradiated for 10 years. The material is natural cobalt and was subjected to an average thermal neutron fluence rate (flux) of 1.0×10^{10} n/cm^2-sec and a fast neutron fluence rate (flux) of 5.0×10^{10} n/cm^2-sec. What exposure rate in R/hr is expected at a point that lies 2.0 m from the material? The scrap's dimensions are 2 cm × 3 cm × 0.2 cm.

5.24. Calculate the exposure rate in R/hr at point P, 2.0 m from a spherical particle of Co-60 that has a radius of 0.5 mm. The particle's activity is 3.0 Ci. The geometry for the exposure assessment is illustrated in Fig. 5.1.

5.25. Calculate the exposure rate in R/hr at point Q. Point Q is a distance of 2.0 m from the end of a sample line containing a uniform distri-

Fig. 5.1 Point source geometry for computing the exposure rate. P is the point of interest for calculating the exposure rate, and S is the small particle source.

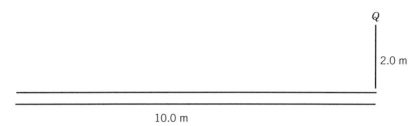

Fig. 5.2 Line source geometry for computing the exposure rate. The sample line is 10.0 m long. Point Q is located above the end of the sample line at a perpendicular distance of 2.0 m.

bution of Co-60. The sample line, illustrated in Fig. 5.2, is 10.0 m long and has a diameter of 0.5 cm. The total activity contained within the sample line is 3.0 Ci.

5.26. Calculate the exposure rate in R/hr at point X defined in Fig. 5.3. Point X is 2.0 m above the center of the spill which contains a uniform distribution of Co-60. The spill is in the shape of a thin disk (0.2 cm thick) which has a diameter of 20.0 m. The total spill activity is 3.0 Ci.

Scenario 61

You have been requested to perform an ALARA evaluation for the first spent-fuel pool cleanup campaign. The facility is 10 years old, and a variety of materials have accumulated. The irradiated hardware inventory consists of 24 control rod blades that were removed during the last refueling outage, 96 in-core neutron flux detectors removed during various refueling outages, and six Sb(Be) startup neutron sources that were removed during the first refueling outage. Other materials that were left in the pool include 18 pleated paper underwater vacuum cleaner filters that are approximately 5 years old.

The processing of irradiated hardware will require the use of several tools. For each of the following tools, list and briefly discuss one potential radiological problem and possible engineering solution.

Fig. 5.3 Disk source geometry for computing the exposure rate. The spill has a diameter of 20.0 m. Point X is 2.0 m above the spill along the central axis of the disk.

5.27. Underwater manipulator pole, consisting of one hollow 30-ft. length that can be used with a variety of small tools such as hooks and grapples.

5.28. Hydraulic cutter used to cut neutron and startup sources into approximately 6-in. lengths. The processing vendor predicts that the cutter jaws will require replacement after each 100–150 cuts.

5.29. Temporary jib crane set up to lower control rods into a crusher/shearer assembly.

Figures 5.4 and 5.5 illustrate the configuration of the installed area radiation monitor in the spent-fuel pool area. The pool is 30 ft × 80 ft and the detector is at a height of 6 ft above the floor on a wall 20 ft from the edge of the pool. The normal spent-fuel pool water level is 18 in. below the floor, and the railing is at a height of 42 in. above the floor.

5.30. What is the minimum line-of-sight distance from the edge of the pool that a source on the surface of the water could be detected by the installed area radiation monitor (ARM) without being shielded by the refueling floor.

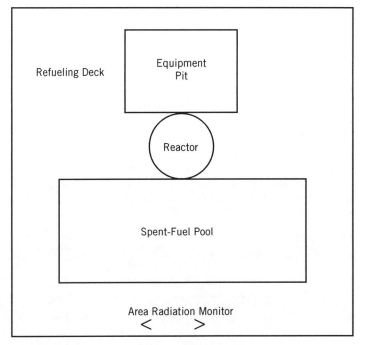

Fig. 5.4 Refueling deck floor plan illustrating the location of the reactor cavity, equipment pit, the spent-fuel pool, and the area radiation monitor.

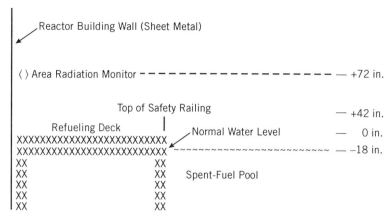

Fig. 5.5 Cross section of the refueling deck illustrating the elevation of the area radiation monitor, spent fuel pool water level, and personnel safety railing relative to the top of the refueling deck.

5.31. The alarm setpoint of the installed ARM is to be adjusted so that it will alarm if a point radiation source at the center of the pool produces an exposure rate of 1 R/hr at the pool railing. What is the appropriate ARM setpoint?

A major task in the spent-fuel pool cleanup is the radiological characterization of neutron-activated hardware. Two characterization techniques are available.

The first of these involves collecting actual specimens of the activated hardware for radiochemical analysis. This technique requires the use of a small abrasive grinding wheel, glass fiber filters, and a small pump to impinge samples on the filter media.

The second characterization technique involves neutron activation calculations and requires underwater surveys to be performed by health physics personnel. This technique requires the use of an underwater survey instrument.

5.32. For each characterization technique, state and explain one advantage and disadvantage from an overall health physics perspective.

5.33. For each characterization technique, state the major source of uncertainty. Clearly identify any assumptions that you make.

The underwater vacuum cleaner filters need to be removed from the pool for disposal. This task involves moving the vacuum cleaner filters underwater from a storage location to a disposal liner at the opposite corner of the pool.

5.34. State the principal hazard associated with movement of the filters and two major radiological consequences associated with this evolution.

5.35. For each hazard cited above, describe two appropriate radiological controls to maintain personnel doses ALARA.

Scenario 62

You are a health physicist at a nuclear utility operating a single pressurized water reactor. The utility has assigned you the responsibilities of the Radiological Control Manager (RCM) at the off-site Emergency Operations Facility (EOF) during a declared emergency in which a radioactive release to the environment is possible.

Data

Letdown radiation monitor reading (primary reactor coolant system activity):	7.3×10^2 μCi/cm^3
Blowdown radiation monitor reading (secondary system activity):	2.4 μCi/cm^3
Steam generator "A" radiation monitor reading:	42 mR/hr
Iodine partitioning factor:	0.015
Atmospheric relief valve flow rate:	1.4×10^7 cm^3/sec
Containment Pressure:	5.0 psig and increasing
Wind Speed:	15 mph
Pasquill stability class:	E
I-131 dose conversion factor:	(77.2 rem/sec)/μCi/cm^3
1 mile = 1609.36 m	

Steam Generator "A" Blowdown Sample Isotopic Results

Radionuclide	Concentration (μCi/cm^3)
I-131	6.3×10^{-1}
Xe-133	9.1×10^{-1}
Xe-135	9.3×10^{-1}
Cs-134	7.6×10^{-1}
Cs-137	1.4×10^{-1}

Pasquill Class E Atmospheric Dispersion Factors

Distance (miles)	$X u/Q$ (1/m^2)
1	1.57×10^{-2}
2	2.69×10^{-3}
5	1.56×10^{-3}
10	6.19×10^{-4}

5.36. List three of the primary responsibilities of the RCM during activation of the EOF.

5.37. List the three fission product barriers that protect the public from a release of radioactivity.

5.38. List and define the three categories of fission product barrier status used in determining off-site protective action recommendations.

5.39. Based on the data provided, what is the status of the fission product barriers?

5.40. List four factors that can affect off-site dose calculations during a declared emergency.

5.41. At this time, a release has not occurred, but you have been asked to provide an assessment of off-site doses. What is the projected thyroid dose rate 2.0 miles downwind from the facility?

5.42. Based upon your projected thyroid dose calculation and information from the plant indicating that the reactor has been stabilized and that a release of radioactivity is no longer likely to occur, describe the Protective Action Recommendation you would recommend to downwind sectors within 2.0 miles of the plant.

5.43. A small town with a population of 2500 is located 2.7 miles downwind from the plant. If an atmospheric relief value, off the main steam line, were opened for 15 min and then closed, what Protective Action Recommendations would you make?

Scenario 63

A particulate air sample is to be used in estimating Kr-88 concentrations in a PWR containment building. The filter paper is counted for a 30-min period to detect the Rb-88 beta particles. The Kr-88 source input has been continuous and the Kr-88 concentration unchanged for 2 days. Assume that the collection of the Kr-88 on the filter paper is negligible.

Given:

Kr-88 → Rb-88 + β^-

Kr-88 half-life = 2.84 hr

Rb-88 half-life = 17.7 min

Net counts in 30 min = 1.50×10^5 counts

Beta efficiency = 0.1 counts/disintegration

Transit time = 10 min (time from the air sampler being turned off to the start of the count)

Air sample time = 30 min

Sample flow rate = 30 liters/min

Filter retention = 100%

5.44. What is the Rb-88 filter activity at the beginning of the counting interval (μCi)?
5.45. What is the Rb-88 activity concentration?
5.46. What is the Kr-88 activity concentration?

REFERENCES

Code of Federal Regulations, *Standards for Protection Against Radiation*, Title 10, Part 20, US Government Printing Office, Washington, DC (1994).

W. W. Doolittle, R. S. Bredvad, and J. J. Bevelacqua, *Radiation Protection Management*, **9**, 27 (1992).

G. G. Eichholz, *Environmental Aspects of Nuclear Power*, Ann Arbor Science Publications, Ann Arbor, MI (1977).

M. M. El-Wakil, *Nuclear Power Engineering*, McGraw–Hill, New York (1962).

J. I. Fabrikant, Guest Editorial: Health Effects of the Nuclear Accident at Three Mile Island, *Health Physics*, **40**, 151 (1981).

A. R. Foster and R. L. Wright, Jr., *Basic Nuclear Engineering*, 3rd edition, Allyn and Bacon, Boston (1977).

S. Glasstone and A. Sesonske, *Nuclear Reactor Engineering*, D. Van Nostrand, New York (1963).

S. Glasstone and W. H. Walter, *Nuclear Power and Its Environmental Effects*, American Nuclear Society, La Grange Park, IL (1980).

IAEA Safety Series No. 50-SG-05, *Radiation Protection During Operation of Nuclear Power Plants*, International Atomic Energy Agency, Vienna (1983).

IAEA Technical Report Series No. 189, *Storage, Handling, and Movement of Fuel and Related Components at Nuclear Power Plants*, IAEA, Vienna (1979).

ICRP Publication No. 40, *Protection of the Public in the Event of Major Radiation Accidents: Principles for Planning*, Pergamon Press, New York (1984).

ICRP Publication No. 43, *Principles of Monitoring for the Radiation Protection of the Public*, Pergamon Press, New York (1984).

NCRP Commentary No. 1, *Krypton-85 in the Atmosphere with Specific Reference to the Public Health Significance of the Proposed Controlled Release at Three Mile Island*, Pergamon Press, Elmsford, NY (1980).

NCRP Commentary No. 4, *Guidelines for the Release of Waste Water from Nuclear Facilities with Special Reference to the Public Health Significance of the Proposed Release of Treated Waste Waters at Three Mile Island*, Pergamon Press, Elmsford, NY (1987).

NCRP Report No. 55, *Protection of the Thyroid Gland in the Event of Releases of Radioiodine*, NCRP Publications, Bethesda, MD (1977).

NCRP Report No. 92, *Public Radiation Exposure from Nuclear Power Generation in the United States*, NCRP Publications, Bethesda, MD (1988).

NCRP Report No. 106, *Limit for Exposure to "Hot Particles" on the Skin*, NCRP Publications, Bethesda, MD (1989).

USNRC Regulatory Guide 8.8, *Information Relevant to Ensuring that Occupational Radiation Exposures at Nuclear Power Stations Will Be as Low as Reasonably Achievable (ALARA)*, U.S. NRC, Washington, DC (1979).

USNRC Regulatory Guide 8.13, *Instruction Concerning Prenatal Radiation Exposure*, U.S. NRC, Washington, DC (1988).

USNRC Regulatory Guide 8.29, *Instruction Concerning Risks from Occupational Radiation Exposure*, U.S. NRC, Washington, DC (1981).

6

ENVIRONMENTAL HEALTH PHYSICS

Environmental Health Physics deals with the assessment of relatively low levels of radioactivity. Problems in environmental health physics include (a) the assessment of effluent releases from facilities utilizing or producing radioactive material and (b) radon measurements, dose assessments, and mitigation measures. An important aspect of effluent release calculations involves the characterization of the facility site prior to its commercial operation, the continuing assessment of effluent releases during its operation, and characterization of releases during its closure and decommissioning. Increasingly, these estimates are being tied to population risks.

The environmental health physicist is also concerned with the measurement and characterization of naturally occurring radioactive material (NORM). Radon and thoron are particularly important naturally occurring radioactive materials. The impact of radon in homes and public buildings, calculations of radon exposures, and the mitigation of elevated radon levels are becoming increasingly important aspects of environmental health physics.

NATURALLY OCCURRING RADIOACTIVE MATERIAL

Natural radiation and naturally occurring radioactive material in the environment provide the principal source of radiation exposure to the general public. For this reason, natural radiation is commonly used as a basis for comparison with man-made exposures received by radiation workers and members of the public.

NCRP Report No. 93, *Ionizing Radiation Exposure of the Population of the United States*, assesses population exposures from natural background, con-

Table 6.1 U-238 Series (Radon)

Nuclide	Half-Life	Dominant Decay Mode
U-238	4.51×10^9 years	Alpha
Th-234	24.1 days	Beta
Pa-234m	1.17 min	Beta
U-234	2.47×10^5 years	Alpha
Th-230	8.0×10^4 years	Alpha
Ra-226	1602 years	Alpha
Rn-222	3.823 days	Alpha
Po-218	3.05 min	Alpha
Pb-214	26.8 min	Beta
Bi-214	19.7 min	Beta
Po-214	164 μsec	Alpha
Pb-210	21 years	Beta
Bi-210	5.01 days	Beta
Po-210	138.4 days	Alpha
Pb-206	Stable	—

sumer products, and other man-made sources. Among the conclusions reached in NCRP 93, none is as significant as the importance of radiation exposure from radon and its daughters. The average annual exposure to radon in the United States is assessed at a dose equivalent of 2400 mrem to the bronchial epithelium. By means of the lung weighting factor, this tissue dose is converted to an effective dose equivalent of 200 mrem. When added to the annual effective dose equivalent of 100 mrem from other sources [contributed by cosmic (27 mrem), terrestrial (28 mrem), and internal radionuclides (42 mrem)], natural radiation from all sources yields a combined annual effective dose equivalent of 300 mrem/year.

Because radon and thoron and their daughters contribute two-thirds of the natural background radiation, it is a significant dose contributor and an important topic in environmental health physics. Tables 6.1, 6.2, and 6.3 describe the radon, thoron, and actinon natural decay series, respectively.

RADON

Radon is a gaseous radioactive element produced from the decay of radium, thorium, and uranium. Its isotopes are short-lived, with a mean half-life of about 30 min. Radon-222 is an alpha emitter, having a half-life of 3.8 days. Radon is heavier than air and is soluble in water; and although it is chemically inert, it will attach to dust or other particulates dispersed in the air. These properties play an important role in the behavior of radon and influence its impact upon the environment.

Table 6.2 Th-232 series (Thoron)

Nuclide	Half-Life	Dominant Decay Mode
Th-232	1.41×10^{10} years	Alpha
Ra-228	6.7 years	Beta
Ac-228	6.13 hr	Beta
Th-228	1.91 years	Alpha
Ra-224	3.64 days	Alpha
Rn-220	55 sec	Alpha
Po-216	0.15 sec	Alpha
Pb-212	10.64 hr	Beta
Bi-212[a]	60.6 min	a. Alpha (36%)
a. Tl-208	3.10 min	Beta
Pb-208	Stable	—
		b. Beta (64%)
b. Po-212	304 nsec	Alpha
Pb-208	Stable	—

[a]Bi-212 decays by alpha emission (36%): Bi-212 $\xrightarrow{\alpha}$ Tl-208 $\xrightarrow{\beta^-}$ Pb-208 and by beta emission (64%): Bi-212 $\xrightarrow{\beta^-}$ Po-212 $\xrightarrow{\alpha}$ Pb-208.

Naturally occurring uranium-238 and thorium-232 decay to gaseous radon-222 (radon) and radon-220 (thoron), respectively. Gaseous radon enters the atmosphere and ground water by diffusion through the soil. Radon is present in ground water, granite, pumice, clay, brick, and other construction materials such as concrete made from fly ash and industrial slag.

Radon diffuses into the soil and into a building or structure and will increase in concentration, eventually reaching equilibrium. For a given structure, the radon concentration will vary significantly with the ventilation rate. For ex-

Table 6.3 U-235 Series (Actinon)

Nuclide	Half-Life	Dominant Decay Mode
U-235	7.1×10^8 years	Alpha
Th-231	25.5 hr	Beta
Pa-231	3.25×10^4 years	Alpha
Ac-227	21.6 years	Beta
Th-227	18.2 days	Alpha
Ra-223	11.43 days	Alpha
Rn-219	4.0 sec	Alpha
Po-215	1.78 msec	Alpha
Pb-211	36.1 min	Beta
Bi-211	2.15 min	Alpha
Tl-207	4.79 min	Beta
Pb-207	Stable	—

ample, structures with low ventilation rates are more likely to exhibit elevated indoor radon air concentrations. For these conditions, increased ventilation will reduce the building's radon levels.

The radiological hazard associated with radon is the inhalation of the element and its alpha emitting daughters that have collected on dust particles suspended in the air. Radon inhalation effects have been established from studies of uranium mine workers. Exposure to radon and its decay products at elevated concentrations has resulted in several hundred excess lung cancers among uranium mine workers in the western United States. These statistics indicate that there is a direct correlation between radon exposure and the occurrence of lung cancer. However, excess cancers have only been seen in association with exposure to radon concentrations that are two or three orders of magnitude larger than those found in normal indoor environments. Extrapolating the uranium miner data to the case of the general population is complicated by factors including cigarette smoking and silica dust exposure among the miners.

Radon concentrations are often expressed in terms of picocuries per liter (pCi/liter) of air, where 1 pCi/liter = 37 Bq/m^3. The concentration of radon daughters is often expressed in working level units (WL). A working level is any combination of the short-lived radon daughters in 1 liter of air that has a potential alpha energy release of 1.3×10^5 MeV.

The concentration of radon, in equilibrium with its daughter products, that is equal to 1 WL is 100 pCi/liter. Unfortunately, the short-lived daughters are not necessarily in equilibrium with their radon parent. Conversion of radon concentrations to daughter concentrations utilizes a factor of about 0.7 for uranium mines and 0.4 for homes. Therefore, 1 pCi/liter in a home is about 0.004 WL.

Based on available data, the average radon daughter exposure for homes in the United States is estimated to be about 0.004 WL. Outdoor radon concentrations average about 0.001 WL.

The occupational limit for radon exposure in the United States is 4 WLM/year. The Environmental Protection Agency has established a 0.02-WL limit for homes. In order to assess the health effect in homes, an evaluation of the radon concentration will be required.

A common problem in radon assessments is the assessment of the number of working level months of exposure in 1 year that an individual receives from living in a home with a known annual average potential-alpha-energy working level concentration. The occupancy factor of the home will be a required input parameter. The number of working level months (NWLM) is given by

$$\text{NWLM} = \frac{C_{\text{WL}} F_{\text{OCC}}}{k} \qquad (6.1)$$

where

NWLM = number of working level months of exposure/year

F_{OCC} = occupancy factor

C_{WL} = working level concentration (WL)

k = conversion factor

$$= \frac{170 \text{ hours}}{\text{WLM/WL}} \frac{1 \text{ day}}{24 \text{ hours}} \frac{1 \text{ year}}{365 \text{ days}}$$

The radon progeny equilibrium factor for the home may be determined from careful measurements. With this value, the average potential alpha-energy air concentration in units of working level for this home can be estimated from the relationship

$$C_{WL} = C(EF)k \qquad (6.2)$$

C_{WL} = average alpha-energy air concentration (WL)
C = average radon air concentration (pCi/liter)
EF = radon progeny equilibrium factor
k = conversion factor (1 WL/100 pCi/liter)

The effective dose equivalent for a member of this household can be determined if you are given the average indoor radon concentration in WLM. To calculate the effective dose equivalent, the average number of working level months per year must be calculated:

$$\text{AWLM} = (F_{OCC})(C_{WL})k_1/k_2 \qquad (6.3)$$

where

AWLM = average number of working level months/year (WLM/year)
F_{OCC} = occupancy factor for the home
C_{WL} = average working level concentration in the home
k_1 = time conversion factor (365 days/year × 24 hr/day)
k_2 = 170 hr/month for WLM estimates

The committed dose equivalent rate to the tracheobronchial (TB) region of the lung is obtained from the AWLM:

$$D_L = (\text{DCF}_{TB})(\text{AWLM})(\text{QF}) \qquad (6.4)$$

where

D_L = committed dose equivalent rate to the TB region of the lung (rem/year)
DCF_{TB} = tracheobronchial dose conversion factor
QF = quality factor for alpha particles (rem/rad)

RADON   163

The committed effective dose equivalent (H) is derived from the ICRP Methodology:

$$H = w_L D_L \qquad (6.5)$$

where

w_L = weighting factor for the lung

Buildup of Radon from Inleakage

The mitigation of radon in a detached, single-family home with a full basement is a common radon problem. Assuming that radon is removed only by outside air infiltration, one can estimate the annual average radon source strength or entry rate into the structure (dC/dt) in units of pCi of Rn-222 per hour with the relationship

$$dC/dt = S/V - \lambda C - IC \qquad (6.6)$$

where

C = radon concentration (pCi/liter) in the air
t = time
S = radon source strength (pCi/hr) which could be determined by a suitable measurement technique such as an extended alpha-track measurement.
V = free space volume of home (liters)
λ = decay constant for radon (hr^{-1})
I = outside air infiltration rate (hr^{-1})

For steady-state conditions, $dC/dt = 0$, the radon source strength is defined by

$$S = VC(\lambda + I) \qquad (6.7)$$

This expression may be used to determine the radon evolution from a variety of steady-state situations.

Evolution of Radon from the Household Water Supply

An important radon problem is the assessment of radon air concentration as a result of the household water supply. The contribution to the air radon concentration from the water utilization is given by

$$S_w = NUC_w f_w \qquad (6.8)$$

where

S_w = radon source strength (pCi/hr) from the water
N = number of family members residing in the home
U = average water usage per member per day (liters/member-day)
C_w = radon concentration in the water (pCi/liter)
f_w = fraction of radon released from the water into the air

ENVIRONMENTAL MONITORING PROGRAMS

Environmental monitoring is an important aspect of a facility's radiation protection program. The environmental monitoring program is often mandated by the facility's licence or by Federal regulations. The purpose of environmental monitoring is to measure the radiation exposure to the general public from the facility's effluent releases.

Prior to a facility's operation, the monitoring program will determine the location of any radiation anomalies, document local radiation levels, identify local radionuclide concentrations, and document the annual meterological conditions. The preoperational program provides a baseline set of radiological conditions and parameters that can be used to assess the impact of the facility on the environment and the population living in the vicinity of the facility.

After the facility is in operation, the focus of the environmental monitoring program shifts to ensure compliance with all applicable regulations. Periodic reports will address plant releases by nuclide and pathway. Air, water, vegetation, fish, animal, plant samples, direct radiation measurements, and environmental air samples will further characterize the facility's impact upon the environment. A properly functioning environmental monitoring program will also foster a positive and credible public image.

ENVIRONMENTAL RELEASES

Effluent releases from nuclear facilities are often characterized in terms of the committed effective dose equivalent (CEDE) from various isotopes and their release pathways. This requires knowledge of the concentration of the radionuclides released via their associated pathways. The following equation outlines the conceptual methodology utilized to assess the effect of effluent releases upon downstream population groups:

$$D = \sum_i^N \sum_j^P C^{ij} I_a^{ij} D_F^{ij} \mathrm{CF}^i \tag{6.9}$$

where

D = committed effective dose equivalent (rem)
N = total number of different radionuclides released

i = index labeling the ith radionuclide
j = pathway label index
P = total number of pathways (air, water, and food stuffs)
C^{ij} = average concentration of radionuclide i in pathway j (μCi/cm^3)
I_a^{ij} = annual intake of isotope i in the environmental medium in pathway j to an individual in population group a (cm^3 or kg of material)
a = population group composed of infants, children, teenagers, and adults
DF^{ij} = dose factor for the particular nuclide and pathway (rem/μCi)
CF^i = conversion factor to ensure the CEDE is in mrem or μSv

More specific pathway equations will be provided in a subsequent section of this chapter. In particular, both gaseous and liquid effluent releases will be outlined in terms of Federal regulatory guidance. Although these equations will be more detailed than Eq. (6.9), they will still contain its essential elements.

Accumulation of Activity in Ponds and Surfaces

Gas or liquid effluents and their accumulation on flat surfaces (ponds or fields) are of concern in assessing facility releases. Assuming a constant release rate, the rate of buildup of activity on a flat surface is given by

$$D = (V)(Q)(X/Q) \qquad (6.10)$$

where

D = surface deposition rate (Bq/m^2-sec)
V = deposition rate (m/sec)
Q = source term (Bq/sec)
X/Q = relative concentration (dispersion factor) (sec/m^3)

In addition to surface deposition, the accumulation of activity in a body of water, such as a pond, must often be assessed as part of an environmental monitoring program. The steady-state concentration of a soluble isotope in a pond assuming a constant daily input rate is given by

$$C_{eq} = (r_d)(S)/(\lambda_e)(V) \qquad (6.11)$$

where

C_{eq} = steady-state concentration in the pond (Bq/m^3)
λ_e = effective disintegration constant
 = ln (2)/T_e = ln (2)(1/T_p + 1/T_b)
T_p = physical half-life of the released radionuclide
T_b = biological half-life of the radionuclide in the pond
T_e = effective half-life
r_d = daily input rate (Bq/m^2-day)

S = surface area of the pond (m²)
d = average pond depth (m)
V = pond volume (m³)
= $S \times d$

The accumulation of activity in a pond requires the assessment of the intake of radionuclides by organisms living in that environment. The concentration of a radionuclide in an organism living in the pond is given by

$$C_f = (C_{eq})(I)/(\lambda_e) \tag{6.12}$$

where

C_f = concentration of activity in the organism (Bq/kg)
C_{eq} = steady-state concentration in the pond (Bq/m³)
λ_e = effective disintegration constant (day^{-1})
I = daily intake of pond water by the organism per its unit mass (m³/kg-day)

Ultimately, the effluent results are characterized in terms of dose which can be related to an impact upon humans. Currently, a common approach is to quantify this dose in terms of risk.

The risk (r) expressed in excess cancer deaths is given by

$$r = cD \tag{6.13}$$

where c is the risk coefficient (excess cancer deaths per person-rem) and D is the dose (person-rem) received by the population at risk. The EPA risk coefficient is 2.8×10^{-4} health effects/person-rem.

REGULATORY GUIDANCE FOR EFFLUENT PATHWAYS

The Federal Government promulgates regulatory guidance for the calculation of environmental releases. Liquid, gaseous, particulate, and radioiodine pathways are considered. Herein, we will illustrate the guidance provided to facilities regulated by the Nuclear Regulatory Commission (NRC). The NRC approach is similar to other regulatory approaches and is outlined in Regulatory Guide 1.109, *Calculation of Annual Doses to Man from Routine Releases of Reactor Effluents for the Purpose of Evaluating Compliance with 10 CFR 50, Appendix I*. The dose limits of 10 CFR 50, Appendix I are summarized in Table 6.4. All limits apply to the highest off-site dose calculated for a maximum exposed individual within the model framework.

The NRC considers a population group composed of infants (0–1 year), children (1–11 years), teenagers (11–17 years), and adults (17 years and older).

Table 6.4 Summary of NRC Effluent Dose Limits (10 CFR 50, Appendix I)

Dose	Effluent Type	Limit per Unit
Total body from all pathways	Liquid	3 mrem/year
Any organ from all pathways	Liquid	10 mrem/year
Gamma dose in air	Noble gas	10 mrad/year
Beta dose in air	Noble gas	20 mrad/year
Dose to total body of an individual	Noble gas	5 mrem/year
Dose to the skin of an individual	Noble gas	15 mrem/year
Dose to any organ from all pathways	Radioiodine and particulates (including H-3 and C-14)	15 mrem/year

For the purpose of evaluating dose commitment, the maximum exposed infant is assumed to be newborn, the maximum exposed child is assumed to be 4 years old, the maximum exposed teenager is assumed to be 14 years old, and the maximum exposed adult is assumed to be 17 years old.

The "maximum" individuals are characterized as "maximum" with respect to food consumption, occupancy, and other usage of the region in the vicinity of the plant site. As such, a maximum exposure is realized from the model's assumptions.

Because the radiation dose commitment per unit intake of a given radionuclide varies as a function of age, four sets of internal dose conversion factors have been calculated. These dose factors are appropriate for the four different age groups defined above. Specifically, these dose conversion factors are based upon continuous intake over a 1-year environmental exposure period. The associated dose commitment extends over a 50-year period following the intake.

DOSES FROM LIQUID EFFLUENT PATHWAYS

The NRC requires the assessment of radiation doses from the use of potable water, aquatic food, shoreline deposits, and irrigated foods by members of the general public. The following dose assessment models are recommended by the NRC guidance in order to evaluate liquid effluent pathways for the maximum exposed individual and the population within 50 miles of the facility.

Potable Water

One of the most basic pathways is the ingestion of water. The water may contain a variety of radionuclides that are utilized or generated in the nearby nuclear facility. Water may be released from the facility through a variety of pathways such as release to wells, lakes, or streams. The dose from potable water from these pathways is given by

$$R_{apj} = k_1 \frac{U_{ap} M_p}{F} \sum_i Q_i D_{aipj} \exp(-\lambda_i t_p) \qquad (6.14)$$

where

R_{apj} = the total annual dose to organ j for individuals of age group a from nuclide i in pathway p (mrem/year)
a = label for age group
  = 1 for infants
  = 2 for children
  = 3 for teenagers
  = 4 for adults
i = label for ith nuclide of interest
j = label for organ of interest
p = label for pathway
U_{ap} = usage factor that specifies the exposure time or intake rate for a specific individual and pathway (hr/year, liters/year, or kg/year)
M_p = mixing ratio (or reciprocal of the dilution factor) at the point of exposure (or the point of withdrawal of drinking water or point of harvest of aquatic food). It is dimensionless.
F = flow rate of the liquid effluent (ft^3/sec)
Q_i = release rate of nuclide i (Ci/year)
D_{aipj} = dose factor for a specified age group, radionuclide, organ, and pathway, which can be used to calculate the radiation dose from an intake of a radionuclide (in mrem/pCi) or from exposure to a given concentration of a radionuclide in sediment, expressed as a ratio of the dose rate (mrem/hr) and the areal radionuclide concentration (pCi/m^2).
λ_i = ith radionuclide's decay constant (1/hr)
t_p = average transit time required for a radionuclide to reach the point of exposure. For internal dose, it is the total time elapsed between release of the nuclide and ingestion of food or water (hours).
k_1 = conversion factor of 1100, the factor to convert from (Ci/year)/(ft^3/sec) to pCi/liter

Aquatic Foods

Aquatic foods include biota such as fish, crabs, and mussels that are consumed by the nearby population. The pathway will include all likely routes by which the facility's radionuclides are released, transported to the water body, incorporated into the water body, sediment, plants, or other food chain elements, consumed by the fish, and then consumed by humans. The dose to humans from the ingestion of aquatic foods is given by

$$R_{apj} = k_1 \frac{U_{ap} M_p}{F} \sum_i Q_i B_{ip} D_{aipj} \exp(-\lambda_i t_p) \tag{6.15}$$

where

B_{ip} = the equilibrium bioaccumulation factor for a given nuclide and pathway expressed as the ratio of the concentration in biota (pCi/kg) to the radionuclide concentration in water (pCi/liter), in liters/kg

The other quantities were previously defined in Eq. (6.14).

Shoreline Deposits

The calculation of an individual's dose from shoreline deposits of radionuclides is complex. It involves the estimate of sediment radionuclide loading, transport, and concentration of radionuclides associated with suspended or deposited radionuclides released from the facility. The calculation of radiation dose from these shoreline deposits is given by

$$R_{apj} = k_2 \frac{U_{ap} M_p W}{F} \sum_i Q_i T_i D_{aipj} \exp(-\lambda_i t_p)$$
$$\times [1 - \exp(-\lambda_i t_b)] \tag{6.16}$$

where

W = shoreline width parameter (dimensionless)
T_i = ith radionuclide's half-life (days)
t_b = period of time for which the sediment or soil is exposed to the contaminated water (hours)
k_2 = conversion factor of 110,000, the factor to convert from (Ci/year)/(ft^3/sec) to pCi/liter and to account for the proportionality constant used in the sediment radioactivity model.

Irrigated Foods

Liquid effluents can migrate or be directly released to adjacent water bodies that may supply water to agricultural crops. This contaminated water will provide a source of radioactivity that is taken up by the plant. Once the plant is consumed by humans, its radionuclides will be available for transport within the human body. The dose from the consumption of irrigated foods from tritium and from other radionuclides will be presented in the next two sections.

Irrigated Foods (Tritium).

The dose from the consumption of irrigated foods from tritium is given by

$$R_{apj} = U_{ap}^{veg} C_v D_{apj} + U_{ap}^{animal} D_{apj} F_A (C_v Q_F + C_{Aw} Q_{Aw}) \qquad (6.17)$$

where

U_{ap}^{veg} = ingestion rate of vegetables for a specific age group and pathway (kg/year)
C_v = tritium concentration in vegetables (pCi/kg)
U_{ap}^{animal} = ingestion rate of meat and poultry for a specific age group and pathway (kg/year)
F_A = the stable element (hydrogen) transfer coefficient that relates the daily intake rate by an animal to the concentration in an edible portion of an animal product, in pCi/liter (milk) per pCi/day or pCi/kg (animal product) per pCi/day.
Q_F = consumption rate of contaminated feed or forage by an animal in kg/day net weight
C_{AW} = tritium concentration in water consumed by animals (pCi/liter)
Q_{AW} = consumption rate of contaminated water by an animal (liters/day)

Irrigated Foods (Radionuclides Other than Tritium). The dose from the consumption of irrigated foods from radionuclides other than tritium is given by

$$R_{apj} = U_{ap}^{veg} \sum_i d_i \exp(-\lambda_i t_h) D_{aipj}$$

$$\times \left[\frac{r[1 - \exp(-\lambda_{Ei} t_e)]}{Y_v \lambda_{Ei}} + \frac{f_I B_{iv}[1 - \exp(-\lambda_i t_b)]}{P\lambda_i} \right]$$

$$+ U_{ap}^{animal} \sum_i F_{iA} D_{aipj} \left(Q_F d_i \exp(-\lambda_i t_h) \right.$$

$$\times \left[\frac{r[1 - \exp(-\lambda_{Ei} t_e)]}{Y_v \lambda_{Ei}} + \frac{f_I B_{iv}[1 - \exp(-\lambda_i t_b)]}{P\lambda_i} \right]$$

$$\left. + C_{iAw} Q_{Aw} \right) \qquad (6.18)$$

where

d_i = deposition rate of a radionuclide (pCi/m²-hr)
t_h = holdup time that represents the time interval between harvest and consumption of food (hours)
r = fraction of deposited activity that is retained on crops

λ_{Ei} = effective removal rate constant for a radionuclide from crops (1/hr)
    = $\lambda_i + \lambda_w$
λ_w = removal rate constant for physical loss by weathering (1/hr)
t_e = time period that crops are exposed to contamination during the growing season (hours)
Y_v = agricultural productivity (yield) in kg (net weight)/m^2
f_I = fraction of the year's crops that are irrigated
B_{iv} = concentration factor for the uptake of a radionuclide from soil by the edible portion of crops, in pCi/kg (wet weight) per pCi/kg dry soil
P = effective "surface density" for soil, in kg (dry soil)/m^2
F_{iA} = stable element transfer coefficient that relates the daily intake rate by an animal to the concentration in an edible portion of animal product, in pCi/liter (milk) per pCi/day or pCi/kg (animal product) per pCi/day
C_{iAw} = concentration of a radionuclide in water consumed by animals in pCi/liter

These model equations contain numerous parameters that are defined in Regulatory Guide 1.109. The models are necessarily qualitative in nature, and the parameters and assumptions are chosen to ensure that the Federal effluent release limits are conservatively achieved.

DOSES FROM GASEOUS EFFLUENT PATHWAYS

The NRC effluent pathway analytical model permits the calculation of doses from exposure to noble gases discharged to the atmosphere. Separate models are given for air and tissue doses due to gamma and beta rays. With the exception of the case for a noble gas released from an elevated stack, all models assume immersion in a semi-infinite cloud.

Only a sample of the possible release pathways and dose models will be presented. This selection will be sufficient to illustrate the calculational methodology and the associated input parameters.

Annual Gamma Air Dose from Noble Gas Releases from Free-Standing Stacks Higher Than 80 Meters

The term *gamma air dose* refers to the components of the air dose associated with photons emitted during nuclear and atomic transformations. Although gamma rays and x-rays will be the dominant dose contributor, annihilation and bremsstrahlung photon radiations are also contributors to the gamma air dose. The annual gamma air dose from noble gas releases from a high stack is

$$D^\gamma(r, \theta) = \frac{k_1}{r(\Delta\theta)} \sum_n \frac{1}{u_n} \sum_s f_{ns} \sum_k u_a(E_k) E_k I(H, u, s, \sigma_z, E_k)$$
$$\times \sum_i Q_{ni}^D A_{ki} \qquad (6.19)$$

where

$D^\gamma(r, \theta)$ = annual gamma air dose at a distance r and in the sector at angle θ (mrad/year)
r = distance from release point (meters)
θ = angle of plume relative to a fixed direction. Normally, north = 0°, east = 90°, etc.
k_1 = conversion factor = 260 to obtain $D^\gamma(r, \theta)$ in mrad/year and has the units of mrad-radians-m^3-dis/sec-MeV-Ci
$\Delta\theta$ = sector width over which atmospheric conditions are averaged (radians)
n = label for the wind speed class
s = label for the stability class
k = label for the photon energy group
i = label for the radionuclide
u_n = mean wind speed of wind speed class n (m/sec)
f_{ns} = joint frequency of occurrence of stability class (s) and wind speed class (n) for sector θ (dimensionless)
$u_a(E_k)$ = air energy absorption coefficient for a specified photon energy group (1/m)
E_k = energy of the kth photon group (MeV/photon)
$I(H, u, s, \sigma_z, E_k)$ = dimensionless integration constant accounting for the distribution of radioactivity according to the meterological conditions
H = effective stack height
u = wind speed
σ_z = vertical plume standard deviation
$Q_{ni}^{\tilde{D}}$ = radionuclide release rate corrected for decay during plume transit to a specified distance under a fixed wind speed (Ci/year)
A_{ki} = photon yield for gamma-ray photons in energy group k from the decay of radionuclide i (photons/dis)

Annual Air Dose from All Noble Gas Releases

The next two sections will outline the methodology to calculate the annual air dose from noble gas releases. Specific expressions for gamma and beta doses will be provided. Plumes of gaseous effluents are considered semi-infinite in the case of ground-level noble gas releases.

Annual Gamma Air Dose from All Noble Gas Releases. The equation for annual gamma air dose is as follows:

$$D^\gamma(r, \theta) = k_2 \sum_i Q_i [X/Q]^D(r, \theta) \mathrm{DF}_i^\gamma \qquad (6.20)$$

where

$D^\gamma(r, \theta)$ = annual gamma air dose at the point (r, θ) (mrad/year) with respect to the discharge point
k_2 = conversion factor = 3.17×10^4, the number of pCi/Ci divided by the number of seconds per year
Q_i = release rate of the specified radionuclide (Ci/year)
$[X/Q]^D(r, \theta)$ = annual average gaseous dispersion factor (corrected for radioactive decay) at a given point (r, θ)
DF_i^γ = gamma air dose factor for a uniform, semi-infinite cloud of the specified radionuclide (mrad-m³/pCi-year)

Annual Beta Air Dose from All Noble Gas Releases. *Beta air dose* refers to the component of the air dose associated with particle emissions during nuclear and atomic transformations. These transformations include beta decay, positron decay, and conversion electrons. The beta air dose is given by

$$D^\beta(r, \theta) = k_2 \sum_i Q_i [X/Q]^D(r, \theta) DF_i^\beta \qquad (6.21)$$

where

$D^\beta(r, \theta)$ = annual beta air dose at the point (r, θ) (mrad/year)
DF_i^β = beta air dose factor for a uniform, semi-infinite cloud of the specified radionuclide (mrad-m³/pCi-year)

Annual Total Body Dose from Noble Gas Releases from Free-Standing Stacks More Than 80 Meters High. The calculation of the total body dose from a high stack is similar to the calculation for air dose. The whole-body dose calculation includes a factor that includes the ratio of tissue to air energy absorption coefficients and is given by

$$D^T(r, \theta) = k_3 S_F \sum_k D_k^\gamma(r, \theta) \exp[-u_a^T(E_k) t_d] \qquad (6.22)$$

where

$D^T(r, \theta)$ = annual total body dose evaluated at 5 g/cm² tissue depth at the point (r, θ) (mrem/year)
S_F = attenuation factor that accounts for the dose reduction due to the shielding provided by residential structures (dimensionless)
k_3 = conversion factor = 1.11, the average ratio of tissue to air energy absorption coefficients
$D_k^\gamma(r, \theta)$ = annual gamma air dose associated with a specified photon energy group at the point (r, θ) in mrad/year
$u_a^T(E_k)$ = tissue energy absorption coefficient (cm²/g)

t_d = product of the tissue density and depth used to determine a total body dose (g/cm^2)

Annual Skin Dose from Noble Gas Releases from Free-Standing Stacks Higher Than 80 Meters. The annual skin dose is given by

$$D^S(r, \theta) = k_3 S_F D^\gamma(r, \theta) + k_2 \sum_i Q_i [X/Q]^D(r, \theta) \text{DFS}_i \qquad (6.23)$$

where

$D^S(r, \theta)$ = annual skin dose at a depth of 7 mg/cm^2 in tissue at the point (r, θ) in mrem/year

DFS$_i$ = beta skin dose factor for a semi-infinite cloud of the specified nuclide which includes the attenuation by the outer "dead" layer of the skin in mrem-m^3/pCi-year

Annual Doses from All Other Noble Gas Releases

Equations (6.22) and (6.23) provided the annual total body and skin dose from noble gas releases from free-standing stacks higher than 80 m. The following two sections provide models for total body and skin doses for all other types [ground-level and shorter (<80 m) stacks] of noble gas releases.

Annual Total Body Dose from All Other Noble Gas Releases

The annual total body dose from noble gas releases, evaluated at a depth of 5 cm into the body, is given by

$$D^T_\infty(r, \theta) = S_F \sum_i X_i(r, \theta) \text{DFB}_i \qquad (6.24)$$

where

$D^T_\infty(r, \theta)$ = annual total body dose due to immersion in a semi-infinite cloud at point (r, θ) in mrem/year

$X_i(r, \theta)$ = annual average ground-level concentration of a specified radionuclide at the point (r, θ) in pCi/m^3.

DFB$_i$ = total body dose factor for a semi-infinite cloud of radionuclide i including the attenuation of 5 g/cm^2 of tissue in mrem-m^3/pCi-year

Annual Skin Dose from All Other Noble Gas Releases

The calculation of skin dose at 7 mg/cm^2 from immersion in a noble gas cloud is given by the sum of gamma and beta contributions

$$D^S_\infty(r, \theta) = k_3 S_F \sum_i X_i(r, \theta) \text{DF}^\gamma_i + \sum_i X_i(r, \theta) \text{DFS}_i \qquad (6.25)$$

where

$D_\infty^S(r, \theta)$ = annual skin dose due to immersion in a semi-infinite cloud at point (r, θ) in mrem/year

Doses from Radioiodines and Other Radionuclides Released to the Atmosphere

NRC limits also apply to radioiodine and other radionuclides, not including noble gases. Doses due to particulate releases, as well as carbon-14 and tritium intakes from terrestrial food chains, are also included within this category. The ground deposition and inhalation pathways are specifically considered. Equations (6.26) through (6.28) describe doses due to radioiodine, particulates, and tritium. Noble gas exposures where included in Eqs. (6.19) through (6.25), but not included in Eqs. (6.26) through (6.28).

Annual Organ Dose from External Irradiation from Radionuclides Deposited Onto the Ground Surface

The annual organ dose resulting from direct exposure to the contaminated ground plane from radioiodine, particulates, and tritium released from the facility is given by

$$D_j^G(r, \theta) = k_4 S_F \sum_i C_i^G(r, \theta) \text{DFG}_{ij} \qquad (6.26)$$

where

$D_j^G(r, \theta)$ = annual dose to organ j at point (r, θ) in mrem/year
j = labels organ of interest
i = labels radionuclide of interest
k_4 = conversion constant = 8760, the number of hours in a year
S_F = shielding factor that accounts for the dose reduction due to shielding provided by residential structures during occupancy
$C_i^G(r, \theta)$ = ground plane concentration of radionuclide i at the point (r, θ) in pCi/m^2
DFG_{ij} = open field ground plane dose conversion factor for organ j from radionuclide i, in mrem-m^2/pCi-hr

Annual Organ Dose from Inhalation of Radionuclides in Air

The annual dose associated with the inhalation of radionuclides to organ j of an individual in age group a is given by

$$D_{ja}^A(r, \theta) = R_a \sum_i x_i(r, \theta) \text{DFA}_{ija} \qquad (6.27)$$

where

$D^A_{ja}(r, \theta)$ = annual dose to organ j of an individual in age group a at location (r, θ) due to inhalation, in mrem/year
a = labels the age group
R_a = annual air intake for individuals in age group a, in m³/year
$x_i(r, \theta)$ = annual average concentration of radionuclide i in air at location (r, θ), in pCi/m³
DFA_{ija} = inhalation dose factor for radionuclide i, organ j, and age group a, in mrem/pCi

Annual Organ Dose from Ingestion of Atmospherically Released Radionuclides in Food

The annual dose to organ j of an individual in age group a that results from ingestion of atmospherically released radionuclides in produce, milk, meat, and leafy vegetables is given by

$$D^D_{ja}(r, \theta) = \sum_i \mathrm{DFI}_{ija}[U^v_a f_g C^v_i(r, \theta) + U^m_a C^m_i(r, \theta) + U^F_a C^F_i(r, \theta) + U^L_a f_l C^L_i(r, \theta)] \quad (6.28)$$

where

$D^D_{ja}(r, \theta)$ = annual dose to organ j of an individual in age group a from the ingestion of produce, milk, leafy vegetables, and meat at location (r, θ), in mrem/year
DFI_{ija} = ingestion dose factor for radionuclide i, organ j, and age group a, in mrem/pCi
U^v_a = annual intake of produce (non-leafy vegetables, fruits, and grains) for individuals of age group a, in kg/year
U^m_a = annual intake of milk for individuals of age group a, in liters/year
U^F_a = annual intake of meat for individuals of age group a, in kg/year
U^L_a = annual intake of leafy vegetables for individuals of age group a, in kg/year
f_g = fraction of the produce ingestion rate that is produced in the garden of interest
f_l = fraction of the leafy vegetable ingestion rate that is produced in the garden of interest
$C^v_i(r, \theta)$ = concentration of radionuclide i in produce at location (r, θ), in pCi/kg
$C^m_i(r, \theta)$ = concentration of radionuclide i in milk at location (r, θ), in pCi/liter
$C^L_i(r, \theta)$ = concentration of radionuclide i in leafy vegetables at location (r, θ), in pCi/kg

$C_i^F(r, \theta)$ = concentration of radionuclide i in meat at location (r, θ), in pCi/kg

PATHWAY SELECTION

The pathway models are sufficient to estimate radiation exposure for maximum individuals and the population within 50 miles of the facility. These pathways appropriately describe exposure routes routinely applicable to nuclear facilities. However, other pathways that arise either from site-specific features or from changes in facility operation should be evaluated if they contribute at least 10% of the total dose from the routine pathways previously defined.

MODEL PARAMETERS

Regulatory Guide 1.109 provides recommended values for the various model parameters. Specific values of environmental data, human data, dose factors, and miscellaneous model parameters are tabulated. The NRC licensee may take into account facility-specific exposure conditions. These conditions include actual values for agricultural productivity, dietary habits, residence times, dose attenuation by structures, measured environmental transport or bioaccumulation factors, and other model parameters determined for a specific site.

SCENARIOS

Scenario 64

You have been placed in charge of an environmental measurements section for a major health physics firm that has been providing indoor radon measurements. Your involvement in providing these services to the public requires that you understand the principles and technology supporting your services. A reporter has submitted the following list of questions for your interview next week. Based upon published reports from NCRP, EPA, and the Health Physics Society, answer the following questions.

6.1. Radon and radon progeny measurements can be categorized into three types: instantaneous (grab), integrated, and continuous. Define each type and give two examples of a method or instrument that exemplifies each type.

6.2. List three passive, integrating measurement methods or detectors used to measure radon in homes and briefly describe the principle or theory of each.

6.3. The short-lived radon (Rn-222) decay products are:
   a. Po-226, Bi-214, Po-218, and Bi-216
   b. Pb-214, Po-214, Bi-216, and Po-218
   c. Bi-214, Pb-214, Po-218, and Po-214
   d. Po-214, Ra-226, Rn-220, and Pb-214
   e. Po-216, Pb-212, Bi-212, and Po-212

6.4. Indoor unattached radon decay products are commonly measured using all of the following except:
   a. Diffusion battery
   b. Cyclone precollectors
   c. Electrostatic collectors
   d. Diffusion tubes
   e. Screen samples

6.5. The fraction of unattached radon decay products in the air depends on all the following except:
   a. Condensation nuclei concentration
   b. Particle size distribution
   c. Radon concentration
   d. Diffusion coefficient of Po-218
   e. Room surface plateout rate

6.6. How many working level months of exposure in 1 year will an individual receive living in a home with an annual average potential-alpha-energy concentration of 0.09 working levels and an occupancy factor of 0.75?

6.7. How does the tracheobronchial dose per unit radon concentration depend on (1) breathing rate, (2) equilibrium factor, and (3) unattached fraction? Sketch a graph for each.

Scenario 65

From initial emissions to final exposure to humans, numerous factors act to disperse or concentrate emitted radioisotopes. To properly assess the net impact requires familiarity with general concepts and the ability to solve specific problems.

6.8. What mechanisms reduce the concentration of airborne radioactive materials during atmospheric transport?

6.9. A person continuously ingesting radioactive material at a constant rate will eventually build up to a maximum internal dose rate depending on the effective half-life of the material. Draw a graph showing internal dose rate as a function of time in days (with one as the maximum dose on the y axis). Assume an effective half-life of 1 day, and also assume a continuous intake for six effective half-lives followed by no intake for six effective half-lives.

6.10. Reproduce the previous graph, this time assuming no cessation of intake (i.e., the curve becomes/remains flat). Label this curve 1. Evaluate and superimpose on the graph the dose rate from a second isotope which has an effective half-life of 2 days. Assume that all other parameters remain the same.

6.11. Assume that I-131 settles onto a pond at a steady-state rate. Calculate the input rate to the pond's surface using the following information:

Deposition rate = 1×10^{-2} m/sec

Source term = 1×10^{8} Bq/sec

Relative concentration $(X/Q) = 1.8 \times 10^{-7}$ sec/m^3

6.12. If the daily input rate to the pond is 0.5 Bq/m^2, calculate the maximum steady-state concentration of I-131 in the pond assuming:

Pond surface = 100 m by 10 m

The pond depth = 1 m (average)

I-131 physical half-life = 8 days

I-131 biological half-life (pond) = 15 days

6.13. Calculate the average concentration of radioactive iodine expected in fish using the following information:

Daily intake = 8×10^{-5} m^3/kg-day

Biological half-life (fish) = 21 days

Equilibrium activity in the pond = 2 Bq/m^3

6.14. In question 6.13, the assumption was that all I-131 that entered the pond is available for concentration in the fish tissue. List factors that may contribute to the inaccuracy of that model.

Scenario 66

You are the health physicist for an engineering firm that is designing a new low-level radioactive waste disposal facility which will operate for 30 years. The Compact Region which needs this facility has chosen the site and has decided against using traditional shallow land disposal technology (used at Barnwell, Beatty, and Hanford). The Compact Commission wants to use an enhanced disposal technology in hopes of lessening public opposition to the

Table 6.5 Dose Estimates

Parameter	Disposal Options		
	Below Ground	Above/Below Ground	Above Ground
Average worker dose (person-rem/year)	3	3	5
Number of occupationally exposed employees	30	30	23
Maximum surface dose rate (mrem/year) 100 years post closure	7.7×10^{-7}	4.9×10^{-7}	1
Committed dose equivalent to maximally exposed individual (mrem/year), 500 years post closure	1.5	1.3	4
Average population committed dose equivalent from 100 to 500 years post closure (person-rem/year)	10	9	30

facility. Your firm is to provide comprehensive licensing and design services for three conceptual designs to the Compact Commission.

The three designs are for (1) above-ground vaults, (2) earth-mounded bunkers with large above-ground concrete canisters for Class A waste and vaults below ground for Class B and C waste, and (3) below-ground vault disposal for all classes of waste. Vaults are to be constructed of reinforced concrete, and all waste is to be grouted in place. The concrete canisters are 6 feet in diameter and over 7 ft high (large enough to contain large high-integrity containers and liners). The waste is also to be grouted in place in the canisters.

Preliminary dose estimates for the three options are presented in Table 6.5. The State and Compact Commission agree that the site meets all the technical requirements for stability and all draft licensing criteria.

6.15. Outline a generic preoperational environmental monitoring plan that will cover the three possible site designs. Include media to be sampled, locations, sample frequencies and types, and analyses to be performed.

6.16. Identify four major exposure pathways which could be expected during the institutional care period.

6.17. Discuss any potential health effects you might expect during the post-closure phase of each facility. Use a risk coefficient of 1.0×10^{-4} excess cancer deaths/rem in your assessment and assume a 400-year period of evaluation.

6.18. The Compact Commissioner has reviewed the three designs and asks you which disposal facility you recommended. As a health physicist, which option do you recommend to the Commissioner and why? Assume that all regulatory requirements can be met by each of the three designs.

Scenario 67

XYZ Radon Abatement Company has hired you as a consultant to help mitigate a detached single-family home with full basement that has a free-space volume of 100 m^3. A 1-year alpha track measurement was performed in this house, with a result of 25 pCi/liter. SF$_6$ tracer measurements have determined that outside air infiltrates the home at an average rate of 0.2 air changes per hour.

The following data apply to this question:

Decay constant (λ) for Rn-222 = 0.00755/hr
Number of persons in the household = 5
Occupancy factor = 0.7
Daily water usage per person = 200 liters
Fraction of radon released from water usage = 0.7
Tracheobronchial dose conversion factor = 0.7 rad/WLM
Quality factor (QF) for alpha = 20
ICRP organ weighting factor for the tracheobronchial (TB) region of the lung = 0.06

6.19. Assuming that radon is removed only by outside air infiltration, estimate the annual average radon source strength or entry rate into the structure in units of pCi of Rn-222 per hour?

6.20. Radon measurements of the household water supply indicate an Rn-222 concentration of 2000 pCi per liter of water. Could the water supply alone account for the elevated radon levels in the home?

6.21. The radon progeny equilibrium factor for the home was determined to be 0.30 (unitless). Estimate the average potential alpha-energy air concentration in units of working level for this home.

6.22. Calculate the annual effective dose equivalent for a member of this household if the indoor radon concentration is 0.1 WLM.

6.23. Assuming that the radon contribution from building materials used in constructing the home is negligible and that the household water is an insignificant source, what is the likely source of the elevated indoor radon? Name and briefly describe three mitigation techniques that you might recommend to reduce the home's radon levels.

Scenario 68

As an environmental health physicist at a national laboratory, you must evaluate your facility's radioactive releases and their impact upon surrounding population groups. This evaluation is complicated by the existence of natural sources of radiation. At your facility, the following natural sources are of concern: K-40, cosmic radiation, uranium series, and thorium series.

For each health physics task or decision noted below, briefly state how natural radiation sources may affect it.

6.24. Air monitoring.
6.25. Sample counting.
6.26. *In vivo* counting.
6.27. Radiation background measurements.
6.28. Calibration of low-level instruments.
6.29. Materials for construction and shielding of low-level counting facilities.
6.30. Radiochemical analyses including materials and equipment used.

Scenario 69

You are the health physicist at a facility that utilizes a large inventory of Kr-85. A mechanical failure has occurred, and Kr-85 gas is escaping into the environment via a 10-m stack.

The release began at 11:00 p.m. and is expected to last for 4 hr. On-site and off-site sampling teams are staged and available for use. The facility also has an environmental monitoring network that surrounds the plant site.

The release is occurring on a cloudless night, and the wind is blowing at 2.0 m/sec to the east. The nearest resident in that direction lives 1000 m away.

Data

Ventilation system monitor (gas channel):
Normal value: 50 cpm
Current value: 1.0×10^7 cpm
The gas channel detector has a volume of 50 cm^3 and an efficiency of 0.3 cpm/dpm.
Ventilation system monitor (particulate channel):
Normal value: 50 cpm
Current value: 60 cpm
Ventilation system flow rate: 100,000 cfm
Sampling system flow rate: 10 cfm
$X\bar{u}/Q = 5.0 \times 10^{-4}/m^2$ @ 1000 meters

Dose factors for Kr-85 (rem m^3 Ci^{-1} sec^{-1}):
  Skin dose: 6×10^{-2}
  Whole body: 4.7×10^{-4}
1 µCi = 2.2×10^6 dpm
1 ft^3 = 28.32 liters

6.31. Calculate the total body and skin dose to the nearest resident.
6.32. The shift supervisor is concerned about the radiological effects of the offsite doses. What are the off-site impacts and the actions that should be taken by the shift supervisor?
6.33. What would you recommend relative to activating the on-site and off-site monitoring teams?
6.34. What action would you initiate relative to the environment monitoring network, if any?

Scenario 70

Enormous States Power and Light has decided that it will build an advanced boiling water reactor (BWR) on the shores of Lake Erie. You have been asked to prepare a preliminary environmental impact assessment for the advanced BWR. The following questions are part of that report:

6.35. List 10 pathways for off-site exposure.
6.36. The BWR will utilize a once-through secondary cooling water system whose effluent is directly released to the lake, along with a gaseous effluent system equipped with a 30-min delay line, but will utilize no charcoal absorbers. Which pathway will be the dominant dose contributor to the population residing within 50 miles of the facility?
6.37. Outline some of the information that you should obtain in order to perform a more detailed evaluation of the relative importance of each pathway.

REFERENCES

BEIR IV, *Health Risks of Radon and Other Internally Deposited Alpha-Emitters*, National Academy Press, Washington, DC (1988).
D. Bodansky, *Indoor Radon and Its Hazards*, University of Washington Press, Seattle, WA (1987).
C. R. Cothern and J. E. Smith, Jr., editors, *Environmental Radon*, Plenum Publishing Company, New York (1987).
G. G. Eichholz, *Environmental Aspects of Nuclear Power*, Ann Arbor Science Publications, Ann Arbor, MI (1977).
M. Eisenbud, *Environmental Radioactivity*, Academic Press, New York (1987).

R. D. Evans, Engineers' Guide to the Elementary Behavior of Radon Daughters, *Health Physics*, **17,** 229 (1969).

R. J. Gardner, *Transfer of Radioactive Materials from the Terrestrial Environment to Animals and Man*, CRC Critical Reviews in Environmental Control, Chemical Rubber Co., Cleveland (1972).

ICRP Publication No. 29, *Radionuclide Release into the Environment: Assessment of Doses to Man*, Pergamon Press, Oxford, England (1979).

ICRP Publication No. 39, *Principles for Limiting Exposure of the Public to Natural Sources of Radiation*, Pergamon Press, Oxford, England (1979).

ICRP Publication No. 50, *Lung Cancer Risk from Indoor Exposures to Radon Daughters*, Pergamon Press, Elmsford, NY (1987).

R. L. Kathren, *Radioactivity in the Environment: Sources, Distribution, and Surveillance*, Harwood Academic Publishers, New York (1984).

A. W. Klement, Jr., *Handbook of Environmental Radiation*, CRC Press, Boca Raton, FL (1982).

NCRP Report No. 50, *Environmental Radiation Measurements*, NCRP Publications, Bethesda, MD (1976).

NCRP Report No. 76, *Radiological Assessment: Predicting the Transport, Bioaccumulation, and Uptake by Man of Radionuclides Released to the Environment*, NCRP Publications, Bethesda, MD (1984).

NCRP Report No. 77, *Exposures from the Uranium Series with Emphasis on Radon and Its Daughters*, NCRP Publications, Bethesda, MD (1984).

NCRP Report No. 78, *Evaluation of Occupational and Environmental Exposures to Radon and Radon Daughters in the United States*, NCRP Publications, Bethesda, MD (1984).

NCRP Report No. 94, *Exposure of the Population in the United States, and Canada from Natural Background Radiation*, NCRP Publications, Bethesda, MD (1988).

NCRP Report No. 97, *Measurement of Radon and Radon Daughters in Air*, NCRP Publications, Bethesda, MD (1988).

NCRP Report No. 100, *Exposure of the U.S. Population from Diagnostic Medical Radiation*, NCRP Publications, Bethesda, MD (1989).

NCRP Report No. 101, *Exposure of the U.S. Population from Occupational Radiation*, NCRP Publications, Bethesda, MD (1989).

NCRP Report No. 103, *Control of Radon in Houses*, NCRP Publications, Bethesda, MD (1989).

W. C. Reinig, editor, *Environmental Surveillance in the Vicinity of Nuclear Facilities*, Charles C Thomas, Springfield, IL (1970).

J. E. Till and H. R. Meyer, editors, *Radiological Assessment: A Textbook on Environmental Dose Analysis*, NRC Report NUREG/CR-3332, Washington, DC (1983).

United Nations Scientific Committee on the Effects of Atomic Radiation (UNSCEAR), *Ionizing Radiation: Sources and Biological Effects*, United Nations, New York (1982).

U.S. Department of Energy Order 5480.1A, *Environmental Protection, Safety, and Health Protection Program for DOE Operations*, Chapters XI and XIII, U.S. Department of Energy, Washington, DC (1993).

REFERENCES

U.S. Department of Energy Order 5480.1B, *Protection, Control, and Abatement of Environmental Pollution*, Chapter XII, U.S. Department of Energy, Washington, DC (1993).

U.S. Nuclear Regulatory Commission Regulatory Guide 1.109, *Calculation of Annual Doses to Man from Routine Releases of Reactor Effluents for the Purpose of Evaluating Compliance with 10 CFR 50, Appendix I*, U.S. NRC, Washington, DC (1977).

U.S. Nuclear Regulatory Commission Regulatory Guide 4.1, *Programs for Monitoring Radioactivity in the Environs of Nuclear Power Plants*, U.S. NRC, Washington, DC (1975).

U.S. Nuclear Regulatory Commission Regulatory Guide 4.8, *Environmental Technical Specifications for Nuclear Power Plants*, U.S. NRC, Washington, DC (1975).

F. W. Whicker and V. Schultz, *Radioecology: Nuclear Energy and the Environment*, CRC Press, Boca Raton, FL (1982).

7

ACCELERATOR HEALTH PHYSICS

Particle accelerators present unique challenges to the radiation protection professional. The primary beam can produce large exposure rates in localized areas, and the beam constituents often have large linear energy transfer (LET) values. There is a wide variety of beam constituents that range from conventional particles (protons and electrons) at extreme relativistic energies to heavy ions with enormous LET values.

The secondary radiations from these facilities consist of a variety of radiation types including bremsstrahlung, neutrons, scattered particles, electrons, and other leptons, hadrons, and spallation products. These produce large dose rates over large areas of the accelerator facility.

Accelerators include (a) electron linacs used in radiation therapy, (b) cyclotrons for producing Positron Emission Tomography (PET) radionuclides, and (c) the sophisticated machines used in basic nuclear and particle physics research. Each of these accelerator types have unique properties and associated sources of radiation. Three popular and illustrative accelerator types are the proton, electron, and heavy ion machines. These machine types generally illustrate the particle and radiation characteristics encountered in most accelerator facilities.

HIGH-ENERGY INTERACTIONS

Particle accelerators are designed for a variety of purposes including basic research, production of radioisotopes, generation of bremsstrahlung for use in radiotherapy, fusion ignition, pumping of research lasers, and generation of synchrotron radiation. Each purpose requires a particular energy range and particle requiring acceleration. For a particular primary particle beam, the health physicist must understand the radiation fields produced by the beam and its associated interactions because doses can exceed the lethal range in a short period of time.

PROTON ACCELERATORS

If the accelerated particle is a proton, the physics of the interaction can be discussed in terms of energy regions including elastic scattering, inelastic scattering, and particle production. The particles produced by protons in each energy region and their associated health physics concerns are summarized in Table 7.1.

Elastic proton (p) scattering involves reactions of the type $p + X \rightarrow p + X$, where X is the target nucleus. In elastic scattering, the incident proton does not have enough energy to penetrate the target nucleus' Coulomb barrier and hence nuclear reactions do not occur.

As the energy increases, the proton will begin to penetrate the Coulomb barrier of the target. Binary reactions of the type $p + X \rightarrow b + Y$ can be generated, where b is an ejected particle or cluster and Y is the exit channel or residual nucleus. Tertiary production and multiple exit channel particles are also possible. Generally, reactions of the type $X(p, n)Y$ dominate the health physics concerns. Neutrons, having a range of energies, will need to be evaluated by the health physicist.

At proton energies beyond 100 MeV, reactions become more complex with multiple particle production becoming more common. At about 140 MeV, pion production begins. Because much of the radiation is in the beam (forward) direction, shielding must be more extensive along the beam direction. Because

Table 7.1 Proton Accelerators

Beam Energy (MeV)	Region	Radiation/Particles Produced	Health Physics Concerns
<6–8	Elastic scattering	Protons (range <1 mm in most solids and <1 m in air)	Direct exposure to the beam or scattered protons.
6–100	Inelastic scattering	Neutrons and nuclear fragments	Neutrons dominate the shielding requirements.
>100	Particle production	Pions (>140 MeV); Muons from pion decay; neutrons and protons	Most particles are produced in the beam direction. As accelerator energies increase, muons become more important at increasingly larger angles.

pions decay into muons, muon radiation must be evaluated for its impact on the shielding design.

ELECTRON ACCELERATORS

If the accelerated particle is an electron, it will also interact in a variety of ways depending upon its incident energy. Table 7.2 summarizes the health physics concerns for electron beams as a function of their incident energy. The radiation and particles produced by the interactions of the incident electron beam as a function of energy are also summarized in Table 7.2.

HEAVY-ION ACCELERATORS

The term *heavy ions* refers to nuclei heavier than hydrogen that are to be accelerated. Ion beams as heavy as uranium have been accelerated in nuclear physics applications. Beams composed of Li-7, O-16, N-16, or S-32 ions have been utilized in a wide variety of research activities.

In a heavy-ion accelerator, neutrons dominate the radiation field outside the biological shielding for beam energies above the Coulomb barrier. Unlike proton and electron machines, there does not yet exist a large body of experimental source term data to compile a simple table of health physics concerns for heavy-ion accelerators. For light ions (such as H-2, H-3, or He-4) of several hundred MeV per nucleon, the dose equivalent outside a thin shield or at forward angles may be dominated by neutrons of energy greater than 20 MeV.

In the initial phase of facility design, the accelerator health physicist must provide input into the shielding design. In order to shield a low-intensity heavy-ion beam, it may be acceptable, in the absence of a complete knowledge of the source term, to calculate the neutron shielding requirements based on the average nucleon energy in the beam. However, a high-intensity heavy-ion beam will require a much thicker shield. The neutrons of lower energy will be attenuated in the first few layers of the shield so that the high-energy neutron component will govern the required shielding thickness.

For low-energy heavy-ion beams, the health physicist must determine if the incident heavy ion has sufficient energy to penetrate the target nucleus' Coulomb barrier and to produce neutrons. For ions of mass greater than that of the proton, the energy (in MeV) below which only Coulomb interactions can occur is given by Adler's relationship:

$$E_{\text{Coul}} = \frac{Z_1 Z_2 (1 + A_1/A_2)}{A_1^{1/3} + A_2^{1/3} + 2} \qquad (7.1)$$

where Z_1 and Z_2 are the charge of the heavy-ion beam and target nucleus, respectively, and A_1 and A_2 are their respective mass numbers. Adler also

Table 7.2 Electron Accelerators

Beam Energy (MeV)	Region	Radiation/ Particles Produced	Health Physics Concerns
<6	Low energy	Ionization and bremsstrahlung (x-rays)	Primary electron beam and photons it produces must be shielded. Backscattered electrons may be important within shielded enclosures.
6–50	Giant resonance (GR)	Electrons; photons produced by electrons, excite the nucleus and produce neutrons.	Bremsstrahlung is the dominant source of radiation, but neutron radiation requires hydrogenous shielding such as concrete. As the energy increases, the bremsstrahlung is increasingly forward peaked. Giant resonance neutrons are produced isotropically. For exposures inside the shield, electrons are a major concern.
30–150	Intermediate energy	Neutrons and bremsstrahlung	The neutron production cross section is lower than that in the GR region, and bremsstrahlung still dominates the shielding considerations. For exposure inside the shield, electrons and bremsstrahlung will be the major contributors to the dose.
>150	High energy	Pions, muons, and neutrons	For shields thicker than about 120 cm, neutrons become the primary design concern. Bremsstrahlung is the major source of radiation inside the shield.

provides an approximate relationship between the charge and mass number for stable nuclei:

$$Z = 0.487 \frac{A}{1 + A^{2/3}/166} \qquad (7.2)$$

RESIDUAL RADIOACTIVITY

Residual radioactivity includes the activity induced within the target or within accelerator structures or components. It includes the activation of water, air, soil, and any oils or greases utilized in the vicinity of the beam or its scattered radiation.

Residual radioactivity may make up an important part of the radiation field inside the shield. Whenever the electron energy is greater than the binding energy of a nucleon, or when a proton can penetrate the Coulomb barrier of the nucleus, a radioactive nucleus may form. Residual radioactivity is not as great in electron machines as in proton accelerators because most of the primary electron energy goes into ionization and production of photons and electron–positron pairs. Examples of various residual radioactivity at accelerator facilities are summarized in Table 7.3.

Activation of Water

Water exposed to high-energy radiation will become radioactive. The major source of activated water will be the water used to cool beam dumps, collimators, targets, magnets, and other components requiring cooling water. Commonly produced radionuclides (half-lives in parentheses) include O-15 (2 min),

Table 7.3 Residual Activity at Accelerator Facilities

Accelerator Component or Material	Accelerator Type	Isotopes Produced
Iron magnet yokes	Proton	Mn-54
		Co-60 (for certain alloys)
Copper coils	Proton	Zn-65, Co-57, Co-58, and Co-60
Iron structures and components	Electron	Fe-55 via the (γ, n) reaction
Greases and oils	Proton or electron	H-3
Water	Proton or electron	O-15, N-13, C-11, Be-7, and H-3
Air	Proton or electron	O-15, N-13, and C-11
Soil	Proton or electron	H-3 and Na-22

N-13 (10 min), C-11 (20.4 min), H-3 (12.3 years), and Be-7 (53.6 days). These radionuclides are produced by both electron and proton machines.

O-15, N-13, and C-11 are all short-lived positron emitters. If they are released to the air, they do not present an inhalation hazard, but only an external dose hazard due to their annihilation radiation. Being positron emitters, their 0.511-MeV gammas present a potent external concern. These annihilation photons can produce high radiation levels around cooling water systems.

Be-7 is longer-lived (53.6 days) and may be removed by ion-exchange resins. In some cases the resin beds, used to trap the Be-7, become sources of exposure concern and may require shielding.

Circumstances may arise that require cooling water to be immediately sent to a heat exchanger. High-power levels or component cooling restrictions may warrant such an operational requirement. Normally, heat exchangers are shielded, with lead or iron. This design approach may present an exposure problem if N-17 (4.17 sec) is produced because it is a neutron emitter. Concrete shielding should be utilized if N-17 production becomes a design consideration.

H-3 is the only long-lived radionuclide produced in water that cannot be removed by ion-exchange resins. Therefore, tritium will continue to buildup in a water system at a rate governed by the intensity of the radiation producing it, natural decay, and water leakage which removes it. The health physicist should provide guidance regarding the tritium sampling frequency and criteria for disposal of the contaminated water.

Activation of the Soil

High-energy (> 10 MeV) particles produce both H-3 and Na-22 in soil. Tritium will become part of the ground water and can be monitored by liquid scintillation counting. The transport of any leached radionuclide from the soil can be a lengthy process. Therefore, it is important to monitor the soil around accelerator facilities. Na-22 may be detected in soil samples by gamma-ray detectors such as Ge(Li). Facility design can significantly reduce soil activation by adequately shielding the target and beam dump areas with concrete and steel.

Activation of Air

Spallation reactions from high-energy beams passing through air produce radionuclides similar to those created in water (O-15, N-13, and C-11). Where air activation is a concern, it is a good practice to delay entry into areas traversed by the primary or secondary beam for a time appropriate to the half-lives and concentrations of these radionuclides. Because these nuclides are all positron emitters, the hazard is primarily an external one due to the 511-keV annihilation photons.

BUILDUP OF RADIOACTIVE AND TOXIC GASES IN AN IRRADIATION CELL

Accelerator-induced gaseous activity will build up during facility operation. The concentration of the gaseous radioactivity will decrease after the beam current is reduced or after accelerator shutdown. The activity in the vicinity of the irradiation cell will vary after shutdown following a relationship that depends on the ventilation rate and time after beam shutdown:

$$C(t) = C(0) \exp\left[-(v/V + \lambda)t\right] \tag{7.3}$$

where

$C(t)$ = concentration of the radioactive gas at time t
$C(0)$ = concentration of the radioactive gas at time $t = 0$
v = exhaust velocity (m^3/sec)
V = cell volume (m^3)
t = time after shutdown (min)
$\lambda = 0.693/T_{1/2}$ = disintegration constant of the radioactive gas
$T_{1/2}$ = half-life of the radionuclide in the chamber

Similarly, the concentration of the toxic gas varies after shutdown according to the relationship

$$Z(t) = Z(0) \exp\left[-(v/V + 1/T)t\right] \tag{7.4}$$

where

$Z(t)$ = concentration of the toxic gas at time t
$Z(0)$ = concentration of the toxic gas at time $t = 0$
T = mean lifetime of the toxic gas (min)

Toxic gases include ozone, nitrous oxides, or sulfur oxides that arise from the air or from materials in the irradiation cell.

OTHER RADIATION SOURCES

The aforementioned sources of radiation at accelerator facilities are not the only sources of concern to the health physicist. Other sources of radiation, such as klystrons, radiofrequency (RF) equipment, high-voltage power supplies, or experimental equipment in adjacent areas, should be evaluated for their radiological hazard. These sources are more difficult to control than the primary or scattered accelerator radiation because the health physicist may not be aware of their use, the experimenter utilizing this equipment may not be aware of the hazard, or the radiation source is at least partially masked by the accelerator output.

Whenever high-voltage or RF equipment is utilized, there is a strong possibility of x-ray production. The x-ray source can exist even if there is no heated filament or other obvious source of electrons. Because the physics of these x-ray fields is not completely understood, examples from accelerator facilities will illustrate the magnitude of these radiation hazards.

In an accelerator test, high RF power was applied to a 25-cm section of a standing wave accelerator. A dose rate of 6100 rad/hr (17 mGy/sec) resulted at 140 cm from the centerline of the accelerator section when 35 MW of RF power at 120 pulses/sec was applied. The resultant x-rays produced a dose of 10 mrad/hr (28 nGy/sec) at a distance of 2 m after penetrating through 4 inches of lead and 4 ft of concrete.

On the axis of the accelerator section, the dose rate was much higher. At RF fields greater than 20 MV/m, stray electrons can be continuously accelerated. In the experiment, the accelerated beam melted a hole in a stainless steel plate at the end of the accelerator section.

At another facility, a particle beam separator containing a pair of 400-kV, 1-mA high-voltage units required maintenance due to voltage difficulties. The separator is normally interlocked to exclude personnel entry. However, the maintenance work required the interlock to be temporarily bypassed. During the maintenance troubleshooting at 50 kV, a radiation survey revealed no x-ray production.

During the repair work, a wiring error caused the separator's Cockcroft Walton high-voltage stack to operate at about 400 kV instead of the indicated 50 kV. This error was identified by the sound of sparking from the separator. A follow-up radiation survey showed dose rates of about 1500 rad/hr (4.2 mGy/sec) at about 1 ft from the separator's surface centerline.

SLAC-327 provides these and other examples of x-ray hazards which are summarized in Table 7.4.

The optimum way to control these x-ray exposures is to educate the experimenters of the potential hazards of their high-voltage and RF equipment. This can be accomplished through facility initial and periodic training. However, the best approach is for the health physicist to be cognizant of activity in the workplace that can generate a radiological hazard. He or she should frequently tour facility areas and become a useful partner in assisting the researcher in safely conducting his or her experiments.

SHIELDING

Shielding is a common design tool utilized to reduce radiation exposures to facility workers and the general public. Shield design is a complex task, frequently utilizing computer models, and this subject will only be addressed in general terms in this chapter. The shielding requirements will depend on a

ACCELERATOR HEALTH PHYSICS

Table 7.4 Unanticipated X-Ray Sources at Accelerator Facilities

Device	Power/Voltage	Radiological Hazard
RF cavity at a storage ring	200 kW	500 mrad/hr[a] @ 1 m (1.4 μGy/sec)
	350 kW	8 rad/hr[a] @ 1 m (22 μGy/sec)
Secondary emission test device	110 kV dc	160 rad/hr @ 10 cm from glass viewing port (444 μGy/sec)
Doubler RF cavity	65 kW	5 rad/hr @ 1 foot (14 μGy/sec)
Three klystrons with end caps removed	50 MW	1700–3600 rads/hr @ 8 cm from end cap (4.7 to 10 mGy/sec)
Resonating microwave waveguide driven by a klystron	17.5 MW	300 rad/hr @ 6 cm from the waveguide (833 μGy/sec)
A 20-kJ KrF ultraviolet laser	Pumped by a 1.7-MeV, 40-kA, 20-nsec pulsed electron source	300 mrad per pulse @ 15 ft (3 mGy)

[a]Dose increased as the fifth power of the RF power.
Source: SLAC-327 (1988).

number of factors, including the time radiation workers or the public are exposed to the radiation sources, accelerator radiation fields, induced activity, and miscellaneous radiation sources.

The Department of Energy recommends that shielding for new facilities limit worker exposure from leakage radiation to 1 rem/year (10 mSv/year). Moreover, the facility's boundary dose is limited to 100 mrem/year (1 mSv/year).

For most accelerators, concrete is the preferred shielding material. Concrete is often selected because it has a reasonable density, high hydrogen content, and low cost, is a good construction material, and has good structural properties. Typical concrete densities are in the range of 2.2 to 2.4 g/cm^3.

Heavy concrete can be utilized in special circumstances by using aggregate of iron, magnetite, or barite. Densities as high as 6.5 g/cm^3 have been attained. The use of heavy concrete yields an increased linear attenuation for photons, charged particles, and high-energy neutrons. However, heavy concrete normally contains less hydrogen, which leads to a decreased linear attenuation for neutrons with energies below a few MeV.

As a shielding rule of thumb, neutron energy equilibrium is achieved and then remains constant after one or two attenuation lengths of shield material. Therefore, for shields thicker than a few attenuation lengths, the dose equivalent outside concrete and iron shielding will be attenuated with attenuation lengths of 120 g/cm^2 and 145 g/cm^2, respectively.

Neutron energy loss by elastic scattering requires a hydrogenous shield to maximize the energy transfer as the neutrons slow down. At energies above 10 MeV, inelastic processes are effective in attenuating the neutrons. Iron can also be used to shield the higher-energy (>10 MeV) neutrons if it is followed by a hydrogenous material. The iron degrades the neutron spectrum by reducing its energy, while the hydrogeneous material will remove the lower-energy neutrons that result from inelastic scattering with the iron nuclei. Because there is no effective removal mechanism by lower-energy (<several hundred keV) neutron interactions in iron, an iron shield will be nearly transparent to these lower-energy neutrons.

A qualitative description of the radiation characteristics dominating the accelerator shielding design is summarized in Table 7.5. Although the beam and its scattered radiation control the shielding design, this source is not the dominant dose contributor to personnel.

This situation is analogous to the power reactor shielding design. The most intense source of radiation, the nuclear fuel core, is heavily shielded, but most dose is derived from exposure to activation sources during maintenance operations.

In an accelerator, the radioisotopes Cu-64, Co-56, Co-57, Co-58, Co-60, Fe-59, Mn-54, Mn-56, Na-22, and Na-24 produce most of the personnel dose, regardless of whether the source is a proton or neutron machine. All these activation products are induced in the accelerator's components and support systems.

Detailed measurements outside thick shields are difficult to perform. Table 7.6 presents the composition of radiation fields above thick shields at the CERN Proton Synchrotron. The CERN measurements are quite complex because the generated neutrons have energies from the eV to the GeV range.

Table 7.5 Qualitative Radiation Characteristics Outside Accelerator Shielding

Accelerator	Shielding	Dominant Shielding Consideration
High-energy proton	Very thick	Neutrons at large angles
		Muons at small angles
High-energy electron	Thin to moderately thick	Photons
	Very thick	Neutrons at large angles
		Muons at small angles

Table 7.6 Composition of Radiation Fields above Thick Shields at CERN

Radiation	Percentage of Dose Equivalent (%)	
	Above Concrete Shield Bridge	Above Target Through Earth Shield
Thermal neutrons	11–12	<1–3
Fast neutrons (0.1–20 MeV)	50–70	10–37
High-energy particles ($E > 20$ MeV)	2–25	52–89
Gamma rays and ionization from charged particles	2–19	1–13

Source: Rindi and Thomas (1973).

ACCELERATOR BEAM CONTAINMENT

The accelerator beam, like the core of a nuclear reactor, must be controlled to ensure that the facility is operated in a radiologically safe manner. Specifically, the beam must arrive at its designated target location and deposit its energy there. If the beam strikes another location or burns through the target and strikes another area, the facility design assumptions are challenged and unanticipated radiological conditions may be created. These radiological conditions may include very high radiation levels in unprotected areas. Examples of such errors at accelerator facilities are summarized in Table 7.7.

The events of Tables 7.4 and 7.7 actually occurred and produced radiation levels up to the lethal range. The health physicist must be on the alert for these types of situations, must ensure that their occurrence is detected in a timely manner, and must institute corrective actions to preclude their recurrence.

Problems of beam containment is a joint responsibility shared by the health physics, accelerator operations, and beam line design groups. The health physicist must ensure that these groups clearly communicate to avoid the types of errors noted in Tables 7.4 and 7.7.

The containment of accelerator beams is normally accomplished by utilizing a combination of mechanical and electronic interlock systems. Mechanical components include slits, collimators, magnets, beam stoppers, and beam dumps.

DOSE EQUIVALENT FROM THE ACCELERATOR TARGET

The neutron dose equivalent at a distance r away from the target along the beam centerline can be formulated if the distribution is isotropic and the target is considered to be a point source of radiation:

$$H = IKk_1 P(\text{DCF})/(4\pi r^2) \tag{7.5}$$

Table 7.7 Accelerator Beam Containment Errors

Design Condition	Error/Radiological Consequences
A lightly shielded experimental hall contains a target that will absorb about 1% of the beam power. The remaining beam energy passes through the target and is absorbed in a shielded beam dump.	Because of an alignment error, the beam strikes the target's housing. Radiation levels increase by a factor of 10–100.
An experiment is designed for the beam to strike the target and to produce secondary particles or photons. The beam is then directed by a magnet into a beam dump.	As a result of a magnet failure, the beam passes out of the design area to an area occupied by personnel.
The beam is intended to dump its energy into a beam stop after its interaction with the target.	Because of a personnel error, the beam strikes an inadequate beam stopper, burns through the stopper, and then enters an occupied area.
A new beam line is designed to include a magnet as part of its beam confinement package.	On initial startup, the magnet was incorrectly connected. The 30-W beam struck the outer shielding wall and produced a dose rate >360 rad/hr (1 mGy/sec) outside the shield wall.

Source: SLAC-327 (1988).

where

H = dose equivalent rate (rem/hr)
I = proton beam current (A)
k = charge/proton = 1.602×10^{-19} coulombs/proton
K = $1/k$ = 6.24×10^{18} protons/A-sec
k_1 = time conversion factor (3600 sec/hr)
P = neutron production rate (neutrons/proton)
DCF = dose conversion factor (rem-cm^2/n)
r = distance from the target (cm)

BEAM CURRENT

A key health physics parameter that is utilized as an input to dose assessments is the beam current. It may be determined from the more readily available parameters such as the power and accelerator voltage. The beam current in an accelerator can be determined from the relationship

$$P = IV \tag{7.6}$$

where

P = beam power (watts)
V = accelerator voltage (volts)
I = beam current (amperes)

The production rate of various particles is often related to the beam current.

PULSED RADIATION FIELDS

The term "pulse" has different meanings in discussions of accelerator radiation fields. "Pulse" may refer to either beam or detector characteristics. The time that the accelerated beam interacts with the target will be denoted as the "beam pulse." It is also referred to as "pulse," "burst," and "spill." "Detector pulse" refers to instruments when used to detect discrete events and electronically process them.

Measurements in pulsed radiation fields will be influenced by the instantaneous intensity of the field, its duration, cycle time, and characteristics of the detector and its associated electronic circuitry. The amount by which the instantaneous or peak intensity of the radiation field exceeds its time-average value depends upon the accelerator repetition rate or cycle time and the length of time the beam interacts with the target. The peak radiation intensity (I_p) during the beam pulse is related to the average radiation intensity (I) by

$$I_p = I/\text{DF} \tag{7.7}$$

where DF is the duty factor of the accelerator.

Radiation instruments that function accurately during steady-state conditions may yield inaccurate results when subjected to a pulsed radiation field. However, pulsed fields do not affect all radiation detectors in the same manner.

The combination of pulsed radiation fields and high dose rates offer challenges to the measurement of accelerator radiation fields. Ion chambers, proportional counters, and scintillation/photomultiplier detectors can be utilized to measure accelerator fields. Instruments should be tested prior to routine use to ensure that they accurately measure the actual radiation field. Comparison of instrument readings and thermoluminescent dosimetry measurements may be helpful in ensuring accurate results. Modifications, such as changes in the detector volume, may be necessary to obtain sufficiently accurate instrument readings.

SCENARIOS

Scenario 71

Pions are produced in an accelerator when protons with energies of several hundred MeV or more strike a suitable target. Pions have a rest energy of 140

MeV, about 270 times that of an electron. A pion can carry one unit of charge of either sign, or it can be neutral. Beams of negative pions, extracted from accelerators, have been used for radiation therapy at several institutions. Like all charged particles, negative pions entering the body slow down. By properly selecting the incident energy, a beam of negative pions can be made to come to rest at the site of a tumor to be treated. When it stops in matter, a negative pion is captured by a positively charged atomic nucleus. The negative pion is annihilated upon capture, releasing its rest energy of 140 MeV inside the nucleus. The highly excited nucleus literally explodes, giving off energetic neutrons, protons, and heavier fragments. The average distribution of emitted particles and energies for capture by an oxygen nucleus is noted below. Similar data describe capture by carbon and nitrogen, the other principal constituents of soft tissue in addition to hydrogen, which does not effectively capture pions.

Reaction Products and Average Energies from Capture of a Stopped Negative Pion by an Oxygen Nucleus

Emitted Particle	Average Number per Pion Capture	Average Kinetic Energy per Capture (MeV)
Neutrons	2.9	61.0
Protons	1.3	20.0
Heavier fragments	2.0	17.0
Gamma photons	~3	6.0
		104.0

7.1. What would be the chief potential advantage of using a negative-pion beam versus using a Co-60 beam for treating a tumor?
   a. More accurate assessment of dose rate at the tumor site
   b. The generation of neutrons at the tumor site
   c. Lack of dose buildup in intervening healthy tissue
   d. Delivery of a localized dose at high LET
   e. Shorter patient exposure time at higher dose rate

7.2. Which of the following has the greatest effect in causing an initially parallel beam of charged particles to "spread out" as it penetrates tissue?
   a. Multiple Coulomb scattering of the particles by atomic nuclei
   b. The generation of delta rays along the particle paths
   c. Energy-loss straggling
   d. Range straggling
   e. Collisions of the particles with atomic electrons

7.3. What types of radiation are the most important considerations in the shielding design for a negative-pion therapy installation?
   a. Neutrons, prompt photons, muons, and pions
   b. Neutrons, prompt photons, and pions
   c. Neutrons, prompt and residual photons, and muons
   d. Neutrons, prompt and residual photons
   e. Neutrons and residual photons

7.4. When the accelerator is not operating, which of the radiation types would most likely contribute the greatest dose equivalent to a technician working in the immediate area where the patients are exposed in a negative-pion radiotherapy facility?
   a. Gamma rays and beta rays
   b. Gamma rays and neutrons
   c. Neutrons, gamma rays, and beta rays
   d. Gamma rays, muons, and pions
   e. Muons, beta rays, and gamma rays

7.5. The difference between the pion rest energy (140 MeV) and the average total kinetic energy released per capture (104 MeV) shown in the table is:
   a. Carried away by undetected neutrinos
   b. Spent in overcoming nuclear binding energies
   c. Not zero, because the table gives only average values
   d. Lost by the pion when captured
   e. Emitted as bremsstrahlung during rearrangement of the atomic electrons about the nuclear fragments produced

7.6. A negative-pion beam from an accelerator will also likely contain:
   a. Negative muons and electrons as well as some neutrons
   b. Negative muons only
   c. Neutrons only
   d. Electrons and photons
   e. Electrons only

7.7. The range of a 50-MeV negative pion in a material of low atomic number can be expressed as 9.1 g/cm^2. What is the range in centimeters, in soft tissue, having a density of 0.95 g/cm^3?

7.8. From the data given in the table, estimate the average absorbed dose in a 1.0-cm-radius sphere of water surrounding the site of capture of a stopped negative pion by an oxygen nucleus. State the assumptions you use in making the estimate.

7.9. In addition to the data in the table, what other information would you need in order to make a more accurate calculation of the average absorbed dose in the last problem?

Scenario 72

You are the health physicist at a 4-MeV, 200-mA, electron linear accelerator used for experimental and testing purposes in an industrial setting.

7.10. An aluminum-walled water-filled box is used in the beam as a beam monitor and stopper. The 1.78-MeV gamma ray of the 3-min half-life Al-28 was seen on a spectrometer being used in the target area. What is the most likely explanation for this?
   a. Neutron activation of Al-27 resulting from a gamma–neutron reaction in the beam
   b. Neutron activation of Al-28 resulting from a gamma–neutron reaction with deuterium in the beam stopper
   c. Isomeric transition of Al-28, initiated by a gamma–neutron reaction with tritium present in the beam stopper
   d. Neutron activation of Al-27 resulting from a gamma–neutron reaction with deuterium in the beam stopper
   e. Electron–positron pair formation with subsequent release of a gamma ray from Al-28

7.11 The energy of the beam after the first 90° scatter of a 4-MeV bremsstrahlung beam is represented by which of the following equations
   a. $E = T_e + T_p + 2m_0c^2$
   b. $E = \dfrac{E_0 m_0 c^2 (1 - \cos\theta)}{m_0 c^2 + E_0}$
   c. $E = E_0 m_0 c^2 (1 - \cos\theta)$
   d. $E = E_0 m_0 c^2 (\sin\theta)$
   e. $E = \dfrac{E_0 m_0 c^2}{m_0 c^2 + E_0(1 - \cos\theta)}$

   where
   E = energy of the scattered beam
   E_0 = initial energy of the beam
   T_e = kinetic energy of the electron
   T_p = kinetic energy of the positron
   m_0 = initial mass of the electron
   c = speed of light
   θ = scattering angle

7.12. One experimenter at your facility directs an electron beam into a copper target. The beam has been running for 4 hr using a water-cooled magnet. The water coolant is stopped and the accelerator scrams. The experimenter wants to rush in to fix his setup. You, as the health physicist, assess the primary hazard to be:
   a. O-17
   b. N-16

c. There is no radiation hazard which would prevent the researcher from taking care of his experiment.

d. Activated dust

e. Residual scatter

7.13. An experimenter wishes to test a beryllium oxide ceramic to determine its properties under electron bombardment at high beam powers. What radiological controls would be appropriate for this application?

7.14. The accelerator experiment committee is considering the modification of the existing facility to house a 40-MeV LINAC, capable of a 2-A peak current, 0.5-μsec pulse duration, and 250 pps. List the type of information you would require to evaluate the shielding design for this upgrade.

7.15. If both the old machine and the new machine had equal dose rate outputs and lead were used as the shielding material, which machine would require less lead to shield the bremsstrahlung?

a. The 4-MeV machine because 4-MeV electrons are easier to shield.

b. The 4-MeV machine because 4-MeV gammas are easier to shield.

c. The 40-MeV machine because 40-MeV electrons are easier to shield.

d. The 40-MeV machine because 40-MeV gammas are easier to shield.

e. The requirements are essentially the same because the broad-beam tenth-value layer for the 4-MeV machine is nearly the same as that for the 40-MeV machine.

7.16. What is the qualitative relative importance of the neutron source for the 4-MeV and 40-MeV machines?

a. There is essentially no neutron production with the 4-MeV machine, but significant neutron production with the 40-MeV LINAC.

b. There is some neutron production with the 4-MeV machine and essentially none with the 40-MeV machine.

c. The neutron production with the two machines is approximately the same.

d. Both machines produce significant neutrons, with considerably more neutron production with the 4-MeV machine than with the 40-MeV machine.

e. Neutron production with both machines can be neglected.

7.17. What is the qualitative relative importance of the bremsstrahlung production as compared to the neutron production for the two machines if the shielding were constructed of concrete?

a. The bremsstrahlung production is sufficiently high for both machines to control the shielding design.

b. the bremsstrahlung production controls shielding design for the 4-MeV machine, but neutron production controls shielding design of the 40-MeV machine.

c. The bremsstrahlung production controls shielding design for the 40-MeV machine, but neutron production controls shielding design of the 4-MeV machine.

d. The neutron production is sufficiently high for both machines to control shielding designs.

e. Both neutron production and bremsstrahlung production control shielding thickness equally for both machines.

Scenario 73

A small plastics manufacturing company is planning to install a 50-MeV, 100-kW electron LINAC for the radiation processing of their product. The electron beam will be used to maximize the dose rate on the product. The facility layout has been drawn and a proposed operation procedures manual has been prepared. You have been hired as a consultant to review the facility design and manual. The following questions are among the many factors that you must consider.

7.18. Both radioactive and chemically toxic gases can result from the irradiation of air by electron or x-ray beams from this accelerator. List the products that can be expected from this accelerator. Include both radioactive and toxic gas products.

7.19. How do the atomic number of the target and the beam current affect the induced radioactivity and toxic gases produced in the irradiation chamber?

7.20. As the photon energy is raised above the photonuclear reaction threshold to about 35 MeV, the rate of production of induced radioactivity increases very rapidly with the increase in energy. From about 35 MeV up to 100 MeV there is little increase in induced activity, and that increase is only proportional to the relative increase in beam power. Explain this observation.

7.21. A "Rule of Thumb" says that "under certain conditions, if an electron accelerator is properly shielded for bremsstrahlung, it is automatically shielded sufficiently for the neutrons produced." What are those conditions?

7.22. If the beam energy is kept constant but the beam current is reduced in half, the expected induced activity production rate would:

a. be reduced to 1/2.

b. be reduced to 1/4.

c. be reduced by $\exp(-0.693)$.

d. remain the same.
e. be reduced by exp(−2.303).

Scenario 74

You have been retained to evaluate a Department of Energy accelerator facility's operating practices. The delay time before the irradiation chamber may be safely entered after shutdown of the beam depends on the decay of the induced radioactivity, the rate of destructive reaction of the toxic chemicals, and the clearance of the chamber by exhaust. The following information should be considered in your assessment:

Cell volume = 560 m^3
Exhaust velocity = 4 m^3/sec
Toxic gas mean-life = 25 min
G value for toxic gas = 6.0 molecules/100 erg
Beam path length in air = 2.0 m
Toxic gas concentration in the cell at equilibrium = 3.5 ppm

Radioactive Gas Concentration in the Cell at Equilibrium

Radionuclide	Half-life	Concentration
A	10 min	6.3×10^4 Bq/cm^3
B	2 min	2.9×10^4 Bq/cm^3

7.23. Calculate the time required after beam shutdown for the concentration of the radioactive gas to be reduced to 2 Bq/cm^3.

7.24. Calculate the time required after beam shutdown for the toxic gas concentration to be reduced to 0.1 ppm.

Scenario 75

A linear accelerator (LINAC) bombards a tritium target with a 25-μA beam of 2.5-MeV protons. This produces 1.2-MeV neutrons via the T(p, n) reaction.

Data

Production rate = 1.8×10^{-6} neutrons/proton
6.24×10^{18} protons/amp-sec
Neutron removal cross section for concrete = 0.08 cm^{-1}
Dose equivalent rate = 3.5×10^{-8} rem cm^{-2} neutron^{-1}

7.25. Calculate the neutron dose equivalent at a point 40 cm away from the target along the beam centerline. State all assumptions.

7.26. An operator is located 4.0 meters from the target and is shielded by a 50-cm-thick concrete wall. Compared to the dose equivalent rate calculated in question 7.25, the dose equivalent rate at the operator's console will be reduced by a factor of:
   a. 1.8×10^{-2}
   b. 1.8×10^{-4}
   c. 6.3×10^{-4}
   d. 1.6×10^{-5}
   e. 6.3×10^{-5}

7.27. Which of the following instruments would have good sensitivity to neutrons while providing the best discrimination against gammas?
   a. BF_3 proportional counter in a polyethylene moderator.
   b. Geiger-Müller (GM) tube at greater than 2 atmospheres in a polyethylene moderator.
   c. Silver-wrapped GM tube inserted in a polyethylene moderator.
   d. LiI(Eu) scintillator inserted in a polyethylene moderator.
   e. Cadmium-wrapped LiI(Eu) scintillator.

7.28. Which of the following statements best describes the neutron distribution as viewed in the laboratory coordinate system?
   a. Isotropic fluence rate, but energy peaked in the forward direction.
   b. Isotropic energy distribution, but fluence rate peaked in the forward direction.
   c. Both energy and fluence rate peaked in the forward direction.
   d. Energy and fluence rate are peaked at 90° to the incident proton beam.
   e. Fluence rate peaked at 90° to the incident proton beam, isotropic energy distribution.

Scenario 76

You are the health physicist at an electron LINAC processing facility that has the characteristics given below. The laboratory director is concerned that residual ozone (O_3) and oxides of nitrogen (NO_x) may delay access to the processing cell by personnel after the beam is turned off. The director requests that you provide answers to the following questions on O_3 and NO_x levels.

Data

Energy	10 MeV
Peak power	5 MW
Duty Factor	0.01
Irradiation cell volume	75 m^3
Cell exhaust rate	5 m^3/sec
Beam path in air	2 m

206 ACCELERATOR HEALTH PHYSICS

In order to respond to your boss' request, you have found an empirical expression for ozone production:

Production rate (molecules cm^{-3} sec^{-1}) = (600 eV cm^{-4} A^{-1} sec^{-1}) × GId

where

G = 10.3 molecules/100 eV for ozone

I = average beam current (A)

d = length of air path traveled by the beam in air (cm)

7.29. What is the ozone production rate in molecules cm^{-3} sec^{-1}?

7.30. For an NO_x production rate of 100 molecules cm^{-3} sec^{-1}, calculate the steady-state concentration in the irradiation cell. Assume that the mean lifetime of NO_x is 1800 sec.

7.31. Assume that the ozone concentration in the cell achieves a steady-state value of 10 ppm. Calculate the delay time after beam shutdown for it to reach the TLV of 0.1 ppm. Assume that the mean lifetime of ozone is 2000 sec.

Scenario 77

An experimental physicist calls you to report that he believes he has accidentally placed his right arm in the beam of a 6-GeV proton synchrotron for approximately 1 min. The accelerator produces 10 pulses per minute at an intensity of 1.0×10^{12} protons per pulse.

Data

Beam size = 1 cm^2.

Production cross section for C-11 for protons in oxygen = 20 mb.

C-11 half-life = 20.4 min.

Efficiency of detector including geometry effects = 10%.

Time of measurement = 1 hr after suspected exposure.

The thickness of the arm is 10 cm, and its composition is H_2O.

The NaI detector yields 400 cps when subjected to an exposure rate of 10 $\mu R/hr$ due to Ra-226 gamma rays.

7.32. You have available a 3-in. × 3-in. NaI scintillation counter. Calculate the counting rate you will observe from C-11 in activated body tissue with such a counter as a result of the exposure.

7.33. The measurement of activity induced in tissue is to be taken in the radioactive environment of the accelerator where the gamma background is about 1 mR/hr. Will the induced activity be detected?

7.34. What other actions would you initiate in investigating this incident?

REFERENCES

K. Adler, *Coulomb Interactions with Heavy Ions*, CONF-720669, Proceedings of the Heavy Ion Summer School-ORNL (1972).

ANSI 43.1, *Radiological Safety in the Design and Operation of Particle Accelerators*, American National Standards Institute, US Government Printing Office, Washington, DC (1978).

M. Barbier, *Induced Radioactivity*, North-Holland, Amsterdam (1969).

S. Baker, Soil Activation Measurements at Fermilab, in *Proceedings of the Third Environmental Protection Conference*, Volume I, ERDA-92, Energy Research and Development Administration, Washington, DC (1975).

P. J. Gollon, The Production of Radioactivity by Accelerators, *IEEE Transactions on Nuclear Science*, **NS-23(4)**, 1395 (1976).

ICRU Report 28, *Basic Aspects of High Energy Particle Interactions and Radiation Dosimetry*, ICRU Publications, Bethesda, MD (1978).

ICRU Report 34, *The Dosimetry of Pulsed Radiations*, ICRU Publications, Bethesda, MD (1982).

H. Joffre and H. Vialettes, Review of Accidental Irradiations with Accelerators and Protective Measures Against Radiation, in *Proceedings of the Symposium on Accidental Irradiation at Place of Work*, EUR-3666, EURATOM, Nice, France (1967).

J. J. Livingood, *Principles of Cyclic Particle Accelerators*, Van Nostrand, Princeton, NJ (1958).

M. S. Livingston and J. P. Blewett, *Particle Accelerators*, McGraw-Hill, New York (1962).

NCRP Report No. 38, *Protection Against Neutron Radiation*, NCRP Publications, Bethesda, MD (1971).

NCRP Report No. 49, *Structural Shielding Design and Evaluation for Medical Use of X-Rays and Gamma Rays of Energies Up to 10 MeV*, NCRP Publications, Bethesda, MD (1976).

NCRP Report No. 51, *Radiation Protection Design Guidelines for 0.1–100 MeV Particle Accelerator Facilities*, NCRP Publications, Bethesda, MD (1977).

NCRP Report No. 72, *Radiation Protection and Measurements for Low Voltage Neutron Generators*, NCRP Publications, Bethesda, MD (1983).

NCRP Report No. 79, *Neutron Contamination from Medical Electron Accelerators*, NCRP Publications, Bethesda, MD (1984).

NCRP Report No. 102, *Medical X-Ray, Electron Beam, and Gamma-Ray Protection for Energies Up to 50 MeV (Equipment Design, Performance, and Use)*, NCRP Publications, Bethesda, MD (1989).

W. R. Nelson and T. M. Jenkins, Similarities Among the Radiation Fields at Different Types of High Energy Accelerators, *IEEE Transactions on Nuclear Science*, **23**, 1351 (1976).

H. W. Patterson and R. H. Thomas, *Accelerator Health Physics*, Academic Press, New York (1973).

A. Rindi and R. H. Thomas, The Radiation Environment of High-Energy Accelerators, *Annual Review of Nuclear Physics*, **23**, 315 (1973).

SLAC-327, *Health Physics Manual of Good Practices for Accelerator Facilities*, Stanford Linear Accelerator Center, Stanford, CA (1988).

W. P. Swanson, *Radiation Safety Aspects of the Operation of Electron Linear Accelerators*, IAEA Technical Reports Series No. 188, IAEA, Vienna, Austria (1979).

R. H. Thomas and G. R. Stevenson, *Radiological Safety Aspects of the Operation of Proton Accelerators*, IAEA Technical Reports Series No. 285, IAEA, Vienna, Austria (1989).

PART III

ANSWERS AND SOLUTIONS

The solution section of this text further develops the theory and introduces additional applications of the principles developed in Parts I and II. The answers and solutions presented in Part III illustrate many of the practical difficulties encountered in real-world applications. The student is encouraged to carefully examine these solutions in order to gain the maximum benefit of this text.

SOLUTIONS FOR CHAPTER 1

Scenario 1

Question 1.1: c

The specific activity of C-14 in living organisms is relatively constant during the organism's lifetime. Following the organism's death, the C-14-specific activity decreases at a rate governed by the C-14 half-life.

Question 1.2

The desired age can be determined by solving the activity equation for t:

$$A(t) = A_0 \exp(-\lambda t)$$

$A(t)$ = activity (Curies, dpm, cpm, Bq, etc.) at time t

t = time

λ = disintegration constant (time^{-1}) = $\ln(2)/T_{1/2}$

A_0 = initial activity

The age is given by

$$t = (-1/\lambda) \ln(A/A_0)$$

$$\lambda = (0.693)/5600 \text{ years} = 1.24 \times 10^{-4}/\text{year}$$

$$A/A_0 = (1.5 \times 10^{-1} \text{ Bq/g})/(1.67 \times 10^{-1} \text{ Bq/g}) = 0.898$$

$$t = -\ln(0.898)/(1.24 \times 10^{-4}/\text{year}) = 868 \text{ years}$$

Scenario 2

Question 1.3

$$a \rightarrow b \rightarrow c$$

where a = Mo-99, b = Tc-99m, and c = Tc-99.

$$A_b = CY \frac{\lambda_b A_{a0}}{\lambda_b - \lambda_a} [\exp(-\lambda_a t) - \exp(-\lambda_b t)]$$

where

C = chemical yield = 0.95
Y = fractional yield = 0.876
$\lambda_b = (0.693)/T_b = 0.693/6.0$ hr = 1.155×10^{-1}/hr for Tc-99m
$\lambda_a = (0.693)/T_a = 0.693/67.0$ hr = 1.034×10^{-2}/hr for Mo-99

$A_b = (0.95)(0.876)[(1000 \text{ mCi})(1.155 \times 10^{-1}/\text{hr})/(1.155 \times 10^{-1} - 1.034 \times 10^{-2})/\text{hr}] \times [\exp(-1.034 \times 10^{-2} \times 48) - \exp(-1.155 \times 10^{-1} \times 48)]$
= 553 mCi

Question 1.4

The Tc-99m activity obtained by a second milking 24 hr later is given by

$$A_b = CY \frac{\lambda_b A'_{a0}}{\lambda_b - \lambda_a} [\exp(-\lambda_a t) - \exp(-\lambda_b t)] + CA_{b1} \exp(-\lambda_b t)$$

where

t = 24 hr
C = 0.95
Y = 0.876
$\lambda_b = 1.155 \times 10^{-1}$/hr
$\lambda_a = 1.034 \times 10^{-2}$/hr

The Mo-99 activity remaining after 48 hours is given by

$$A'_{a0} = A_{a0} \exp(-\lambda_a t')$$

where

t' = 48 hr (assumes generator arrives at the hospital 48 hr after its manufacture)

$A'_{a0} = 1000 \text{ mCi exp}[-1.034 \times 10^{-2}/\text{hr} \times 48 \text{ hr}] = 609 \text{ mCi}$

$A_{b1} = 5\%$ Tc-99m left over from the last milking
$= \{[553 \text{ mCi}/(0.95)] - 553 \text{ mCi}\} = 29 \text{ mCi}$

$A_b = (0.95)\{(0.876)(609 \text{ mCi})(1.155 \times 10^{-1}/\text{hr})/(1.155 \times 10^{-1}/\text{hr} - 1.034$
$\times 10^{-2}/\text{hr}) \times \{\exp[-(1.034 \times 10^{-2}/\text{hr} \times 24 \text{ hr})]$
$- \exp[-(1.155 \times 10^{-1}/\text{hr} \times 24 \text{ hr})]\}$
$+ (29 \text{ mCi})(0.95) \exp[-(1.155 \times 10^{-1}/\text{hr} \times 24 \text{ hr})]\}$
$A_b = 401 \text{ mCi}$

Scenario 3

Question 1.5: d

Because $\lambda_B \gg \lambda_A$, the parent and daughter will eventually reach the condition of transient equilibrium.

Question 1.6

$$T_{1/2}(A) = T_A = 10 \text{ hr} = 0.693/\lambda_A$$
$$T_{1/2}(B) = T_B = 1 \text{ hr} = 0.693/\lambda_B$$

The daughter will reach its maximum activity at a time t given by

$$t = \{\ln(\lambda_B/\lambda_A)\}/(\lambda_B - \lambda_A)$$
$$= \{\ln[(0.693/1 \text{ hr})/(0.693/10 \text{ hr})]\}$$
$$\times [(0.693/1 \text{ hr}) - (0.693/10 \text{ hr})]^{-1}$$
$$= 3.69 \text{ hr}$$

Scenario 4

Question 1.7: b

The tissue dose from thermal neutrons is determined by the (n, γ) reaction with hydrogen and the (n, p) reaction with nitrogen.

Question 1.8: e

Tissue dose from fast neutrons is due principally to elastic scattering with nuclei.

Question 1.9: c

The most probable process for energy deposition by a 1-MeV photon in tissue is Compton scattering.

Question 1.10: a

Inelastic scattering by atomic electrons is the principal mechanism of dose deposition by a 5-MeV alpha particle that stops in tissue.

Question 11: d

Inelastic scattering by atomic electrons is the principal mechanism of dose deposition by a 100-keV beta particle that stops in tissue.

Question 1.12: d

$$100 \text{ keV} \times \frac{1000 \text{ eV}}{\text{keV}} \times \frac{1 \text{ ion pair}}{34 \text{ eV}} = 2941 \text{ ip}$$

Question 1.13: a

$$100 \text{ keV} \times \frac{1000 \text{ eV}}{\text{keV}} \times \frac{1 \text{ ion pair}}{3.4 \text{ eV}} = 29412 \text{ ip}$$

Question 1.14: b

A nuclide that undergoes orbital electron capture emits a neutrino and the characteristic x-rays of the daughter.

Question 1.15: c

The specific gamma-ray emission rate for Cs-137 is approximately 0.33 R hr^{-1} Ci^{-1} m^2.

Question 1.16: e

The thyroid is an organ for which the ALI is determined by the limit for nonstochastic effects.

Scenario 5

M_p and M_d are expressed as mass defects (p = parent and d = daughter).

Question 1.17: c

$$\text{I-126 (K-capture)} \rightarrow \text{Te-126}$$

For K-capture:

$$M_p = M_d + b + Q$$

or

$$Q = M_p - M_d - b$$
$$= [-87.90 - (-90.05) - 0.03] \text{ MeV}$$
$$= 2.12 \text{ MeV}$$

Question 1.18: c

$$\text{I-126 (positron decay)} \rightarrow \text{Te-126}$$

for positron decay:

$$M_p = M_d + 2M_e + Q$$

or

$$Q = M_p - M_d - 2M_e$$
$$= [-87.90 - (-90.05) - 2(0.51)] \text{ MeV}$$
$$= 1.13 \text{ MeV}$$

Question 1.19: e

$$\text{I-126 (beta decay)} \rightarrow \text{Xe-126}$$

For beta decay:

$$M_p = M_d + Q$$

or

$$Q = M_p - M_d$$
$$= [-87.90 - (-89.15)] \text{ MeV}$$
$$= 1.25 \text{ MeV}$$

Question 1.20: e

For both internal and external exposures, the antineutrino will be the least significant dose contributor. Neutrino and antineutrino interaction cross sections are much smaller than those for the other particles.

Question 1.21: d

The 32-keV Te x-rays are released when the electron capture event creates a vacancy in the inner shells and electrons from the outer shells fill the vacancy.

Scenario 6

Question 1.22: d

$$\lambda = \text{disintegration constant} = \ln(2)/T_{1/2}$$
$$\overline{T} = 1/\lambda = T_{1/2}/0.693 = 64.2 \text{ hr}/0.693$$
$$= 92.6 \text{ hr}$$

Question 1.23: c

$$SA = \lambda N$$

where

N = number of atoms per gram of Y-90

$$SA = \frac{0.693}{64.2 \text{ hr}} \frac{1 \text{ hr}}{3600 \text{ sec}} \frac{6.02 \times 10^{23} \text{ atoms/mole}}{90 \text{ g/mole}} \frac{1 \text{ dis}}{\text{atom}}$$

$$\times \frac{1 \text{ Bq}}{\text{dis/sec}} \frac{1 \times 10^3 \text{ g}}{\text{kg}} = 2.01 \times 10^{19} \text{ Bq/kg}$$

Question 1.24: d

Using the serial decay relationship, we obtain

$$A(\text{Y-90}) = \frac{\lambda(\text{Y-90}) A_0(\text{Sr-90})}{\lambda(\text{Y-90}) - \lambda(\text{Sr-90})} \{\exp[-\lambda(\text{Sr-90})t] $$
$$- \exp[-\lambda(\text{Y-90})t]\}$$

At $t = 72.0$ hr, $A(\text{Y-90}) = 3.4$ mCi

$\lambda(\text{Sr-90}) = (0.693/27.7 \text{ year})(1 \text{ year}/365 \text{ days})(1 \text{ day}/24 \text{ hr})$
$= 2.86 \times 10^{-6}/\text{hr}$

$\lambda(\text{Y-90}) = (0.693/64.2 \text{ hr}) = 1.08 \times 10^{-2}/\text{hr}$

The initial activity of Sr-90 is obtained by solving the first equation for this quantity:

$$A_0(\text{Sr-90}) = \frac{A(\text{Y-90}; t = 72 \text{ hr})[\lambda(\text{Y-90}) - \lambda(\text{Sr-90})]}{\lambda(\text{Y-90})}$$

$$\times \frac{1}{\exp[-\lambda(\text{Sr-90})t] - \exp[-\lambda(\text{Y-90})t]}$$

$$= \frac{(3.4 \text{ mCi})(1.08 \times 10^{-2}/\text{hr} - 2.86 \times 10^{-6}/\text{hr})}{(1.08 \times 10^{-2}/\text{hr})}$$

$$\times \frac{1}{[\exp(-2.86 \times 10^{-6}/\text{hr} \times 72 \text{ hr}) - \exp(-1.08 \times 10^{-2}/\text{hr} \times 72 \text{ hr})]}$$

$$= 3.4 \text{ mCi}/(1.0 - 0.46) = 6.29 \text{ mCi}$$

Scenario 7

Question 1.25

$$\dot{X} = \frac{i}{Vp} \times \frac{T}{T_{\text{STP}}} \times \frac{P_{\text{STP}}}{P} k$$

$T_{\text{STP}} = 0°\text{C} = 273°\text{K}$

$P_{\text{STP}} = 760 \text{ mm Hg}$

$k = [2.58 \times 10^{-4} \text{ C/kg-R}]^{-1}$

$$\dot{X} = \frac{1 \times 10^{-9} \text{ C/sec}}{4 \text{ cm}^3 \times 1.293 \times 10^{-6} \text{ kg/cm}^3} \times \frac{283°\text{K}}{273°\text{K}} \times \frac{760 \text{ mm Hg}}{755 \text{ mm Hg}} k$$

$$= \frac{2.02 \times 10^{-4} \text{ C/kg-sec}}{2.58 \times 10^{-4} \text{ C/kg-R}} = 0.783 \text{ R/sec}$$

Question 1.26

The optimum counting time for a fixed counting interval is obtained from the relationship

$$t_{s+b}/t_b = (R_{s+b}/R_b)^{1/2}$$

where

t_{s+b} = count time for the sample

t_b = count time for the background

R_{s+b} = sample count rate = $(2400 + 300)$ cpm

R_b = background count rate = 300 cpm

SOLUTIONS FOR CHAPTER 1

$$t_{s+b} + t_b = 100 \text{ min}$$

$$t_{s+b}/t_b = t_{s+b}/(100 \text{ min} - t_{s+b})$$

$$= [(2400 \text{ cpm} + 300 \text{ cpm})/(300 \text{ cpm})]^{1/2}$$

$$t_{s+b}/(100 \text{ min} - t_{s+b}) = 3$$

$$t_{s+b} = 75 \text{ min}$$

Question 1.27

$$\text{Sample count rate} = R_{s+b} = 600 \text{ counts}/10 \text{ min} = 60 \text{ cpm}$$

$$t_{s+b} = 10 \text{ min}$$

$$\text{Background count rate } R_b = 56 \text{ cpm (40-min count)}$$

$$t_b = 40 \text{ min}$$

$$\text{Net count rate} = R_s = 60 \text{ cpm} - 56 \text{ cpm} = 4 \text{ cpm}$$

The standard deviation is

$$\sigma = \{R_{s+b}/t_{s+b} + R_b/t_b\}^{1/2}$$

$$\sigma = (60 \text{ cpm}/10 \text{ min} + 56 \text{ cpm}/40 \text{ min})^{1/2}$$

$$= (7.4)^{1/2} \text{ cpm} = 2.72 \text{ cpm}$$

At the 95% confidence interval (one-tail), the sample count rate is

$$95\% \text{ C.I.} = R_s \pm 1.65 \, \sigma$$

$$95\% \text{ C.I.} = 4 \text{ cpm} \pm 1.65 \times 2.7 \text{ cpm}$$

$$= 4 \text{ cpm} \pm 4.5 \text{ cpm}$$

Because the sample count rate at a 95% confidence interval (one-tail test) includes "zero," it should be concluded that the net sample count rate is not statistically different from the background count rate. Therefore, the sample does not contain any net activity (above background).

Scenario 8

Question 1.28

The minimum sample and background counting time, required to ensure a lower limit of detection (LLD) at the 95% confidence level less than or equal to 0.10 MPC for I-131, may be obtained as follows:

$$\text{LLD} = 4.66(1/k)(R_b/t_b)^{1/2}$$

R_b = background count rate = 50 cpm

t_b = background count time (minutes)

k = correction factor (cpm/μCi/cm³)

  = (detector efficiency)(sampling efficiency)
    × (sample volume)(2.22 × 10⁶ dpm/μCi)

  = (0.2 cpm/dpm)(0.7)(5 liters/min × 10 min)
    × (1000 cm³/liter)(2.22 × 10⁶ dpm/μCi)

  = 1.55 × 10¹⁰ cpm − cm³/μCi

LLD = 0.1 MPC = (0.1)(1.0 × 10⁻⁹ μCi/cm³)

  = 1.0 × 10⁻¹⁰ μCi/cm³

Solving the basic equation for t_b, we obtain

$t_b = (4.66/k \text{ LLD})^2 R_b$

  = (50 cpm)(4.66)²/[(1.55 × 10¹⁰ cpm − cm³/μCi)(1.0 × 10⁻¹⁰ μCi/cm³)]²

  = 452 min

Question 1.29

Methods to reduce the counting time include:

1. Improving the collection efficiency increases the number of counts collected. This increases k and therefore reduces t_b.
2. Decreasing the background levels with shielding reduces R_b and therefore directly reduces t_b.
3. Increase the counter efficiency by improving the geometry to increase the collected counts or obtain a more efficient detector.
4. Increasing the sample flow rate increases the number of counts collected, which increases k and therefore reduces t_b.
5. Increase the sampling time. t_b is inversely proportional to the sample time squared.

Scenario 9

Question 1.30

In order to be representative of isokinetic sampling conditions, the linear flow into the sampling nozzle must be equal to the linear flow in the stack. This requires that the ratio of volumetric flow to flow area be a constant:

$$\frac{\text{Volumetric flow (stack)}}{\text{Area (stack)}} = \frac{\text{Volumetric flow (nozzle)}}{\text{Area (nozzle)}}$$

$$\frac{(20 \text{ m}^3/\text{min})}{(\pi)(0.25 \text{ m})^2} = \frac{200 \text{ liters/min}(10^{-3} \text{ m}^3/\text{liter})}{(\pi) r^2}$$

where r is the internal radius of the isokinetic probe.

$$r^2 = \frac{0.2 \text{ m}^3/\text{min}(0.0625 \text{ m}^2)}{20 \text{ m}^3/\text{min}}$$

$$r = 2.5 \text{ cm}$$

Therefore, the sample nozzle has an internal diameter of 5 cm.

Question 1.31

Flow patterns within a stack or duct may be distorted near blends, interferences, or transition regions, as well as at the entrance and exit to the duct. Therefore:

1. Sampling locations should be at least 5–10 stack diameters downstream of bends/duct transitions and as far upstream from the atmospheric exit as practical.
2. For large-diameter ducting, radial and axial variations in the velocity and particle composition may occur. If these differences are significant, multiple axial and radial sampling locations may be needed for a representative composite sample.

Question 1.32

Filter type	Advantages	Disadvantages
Cellulose	Low ash content	High (burial) losses for alpha counting
	Easily dissolved or decomposed	Not suitable for high-temperature applications
	Low airflow resistance	
	Good efficiency for respirable particles (0.3–10 μm)	
	Strong and not easily damaged	
Glass fibers	High collection efficiency for respirable particles	Not easily dissolved
	Low airflow resistance	Fragile/careful handling

Filter type	Advantages	Disadvantages
Glass fibers (continued)	Can be used at higher temperatures than cellulose Low burial loss for alpha counting	
Membrane	Wide range of pore sizes (0.01–10 μm)	Fragile to handle
	Easily dissolved in many solvents	High airflow resistance
	High collection efficiency for respirable particles	Not suitable for high temperatures
	Larger pore sizes (1–10 μm) are well-suited for sampling alpha emitters because lower collection efficiency is offset by minimal burial loss.	
	Small-pore-size varieties have very high collection efficiencies and minimal alpha burial losses.	
	Low ash content	

Scenario 10

Question 1.33

The larger initial count rate is due to the combination of natural products and long-lived alpha activity (Pu-239) collected on the filter. After 48 hr, it is safe to assume that all the radon decay products have decayed away because the longest half-life in the series is only 29 min.

Question 1.34

Pu-239 represents the long-lived alpha contribution. The concentration of Pu-239 can be determined from the relationship

$$C = \frac{(R_S)(\text{FA})}{(E_d)(E_f)(\text{SA}_f)(F)(T_s)(\text{DA})}$$

where

R_S = net count rate (cpm)
FA = total filter area = 500 cm^2
E_d = detector efficiency = 0.3 cpm/dpm
E_f = filtration efficiency = 0.8

SA_f = filter paper self-absorption factor = 0.4
F = filter sampling rate = 55 ft³/min
T_s = filter sampling time = 60 min
DA = detector area = 60 cm²

$$C = \frac{(220 \text{ counts}/100 \text{ min} - 20 \text{ counts}/100 \text{ min})(500 \text{ cm}^2)}{(0.3 \text{ cpm/dpm})(0.8)(0.4)(55 \text{ ft}^3/\text{min})(60 \text{ min})}$$

$$\times \frac{1}{(0.0283 \text{ m}^3/\text{ft}^3)(60 \text{ cm}^2)} = 1.86 \text{ dpm/m}^3$$

The standard deviation is

$$\sigma^2 = \sigma_b^2 + \sigma_{s+b}^2$$
$$\sigma = (R_b/t_b + R_{s+b}/t_{s+b})^{1/2}$$
$$R_b = (20 \text{ counts}/100 \text{ min}) = 0.2 \text{ cpm}$$
$$R_{s+b} = (220 \text{ counts}/100 \text{ min}) = 2.2 \text{ cpm}$$
$$\sigma = (2.2 \text{ cpm}/100 \text{ min} + 0.2 \text{ cpm}/100 \text{ min})^{1/2}$$
$$= 0.155 \text{ cpm}$$

σ may be related to concentration:

$$\sigma = \frac{(0.155 \text{ cpm})(500 \text{ cm}^2)}{(0.3 \text{ cpm/dpm})(0.8)(0.4)(55 \text{ ft}^3/\text{min})}$$

$$\times \frac{1}{(60 \text{ min})(0.0283 \text{ m}^3/\text{ft}^3)(60 \text{ cm}^2)}$$

$$= 0.14 \text{ dpm/m}^3$$

$$C \pm \sigma = (1.86 \pm 0.14) \text{ dpm/m}^3$$

Question 1.35

$$\text{LLD} = 4.66(R_b/t_b)^{1/2}$$
$$= 4.66(0.2 \text{ cpm}/100 \text{ min})^{1/2} = 0.21 \text{ cpm}$$

This LLD may be related to concentration:

$$C_{\text{LLD}} = \frac{(0.21 \text{ cpm})(500 \text{ cm}^2)}{(0.3 \text{ cpm/dpm})(0.8)(0.4)(55 \text{ ft}^3/\text{min})}$$

$$\times \frac{1}{(60 \text{ min})(0.0283 \text{ m}^3/\text{ft}^3)(60 \text{ cm}^2)}$$

$$= 0.20 \text{ dpm/m}^3$$

Scenario 11

Question 1.36

$$\overline{T} = \text{Mean lifetime} = 1.44 T_{1/2}$$

The total number of disintegrations (N) is given in terms of the half-life and activity (A) of the isotope

$$N = 1.44 T_{1/2} A$$

The total alpha energy is the sum of the products of the total number of disintegrations and the alpha energy available for each daughter. By definition, Rn-222 is not considered in the calculation. The total alpha energy is given by

$$E_{\text{total}} = \sum_i E^i_{\text{total}} N_i$$

where the sum i is over the Po-218, Pb-214, Bi-214, and Po-214. An activity of 100 pCi/liter for radon and its daughters is assumed.

For Po-218,

$\overline{T} = 1.44(3.05 \text{ min}) = 4.4 \text{ min}$
$N = 1.44(3.05 \text{ min})(100 \times 10^{-12} \text{ Ci})(3.7 \times 10^{10} \text{ dis/sec-Ci}) \times (60 \text{ sec/min})$
$\quad = 976 \text{ dis}$
$E_{\text{alpha}} = (6.00 + 7.68) \text{ MeV/dis} = 13.68 \text{ MeV/dis}$
$E_{\text{total}} = 13.68 \text{ MeV/dis} \times 976 \text{ dis} = 1.34 \times 10^4 \text{ MeV}$

For Pb-214,

$\overline{T} = 1.44(26.8 \text{ min}) = 38.6 \text{ min}$
$N = 1.44(26.8 \text{ min})(100 \times 10^{-12} \text{ Ci})(3.7 \times 10^{10} \text{ dis/sec-Ci}) \times (60 \text{ sec/min})$
$\quad = 8567 \text{ dis}$
$E_{\text{alpha}} = 7.68 \text{ MeV/dis}$
$E_{\text{total}} = 7.68 \text{ MeV/dis} \times 8567 \text{ dis} = 6.58 \times 10^4 \text{ MeV}$

For Bi-214,

$\overline{T} = 1.44(19.7 \text{ min}) = 28.4 \text{ min}$
$N = 1.44(19.7 \text{ min})(100 \times 10^{-12} \text{ Ci})(3.7 \times 10^{10} \text{ dis/sec-Ci}) \times (60 \text{ sec/min})$
$\quad = 6298 \text{ dis}$
$E_{\text{alpha}} = 7.68 \text{ MeV/dis}$
$E_{\text{total}} = 7.68 \text{ MeV/dis} \times 6298 \text{ dis} = 4.84 \times 10^4 \text{ MeV}$

For Po-214,

$$\bar{T} = 1.44(1.0 \times 10^{-6} \text{ min}) = 1.4 \times 10^{-6} \text{ min}$$
$$N = 1.44(1.0 \times 10^{-6} \text{ min})(100 \times 10^{-12} \text{ Ci})(3.7 \times 10^{10} \text{ dis/sec-Ci}) \times$$
$$(60 \text{ sec/min})$$
$$= 3.2 \times 10^{-4} \text{ dis}$$
$$E_{\text{alpha}} = 7.68 \text{ MeV/dis}$$
$$E_{\text{total}} = 7.68 \text{ MeV/dis} \times 3.2 \times 10^{-4} \text{ dis} = 2.46 \times 10^{-3} \text{ MeV}$$

In summary, we have the following data:

Nuclide	Mean Lifetime	Alpha Energy (MeV)	Disintegrations per 100 pCi	Total Energy per 100 pCi
Po-218	4.4 min	6.00 + 7.68	976	1.34×10^4 MeV
Pb-214	38.6 min	7.68	8567	6.58×10^4 MeV
Bi-214	28.4 min	7.68	6298	4.84×10^4 MeV
Po-214	1.4×10^{-6} min	7.68	3×10^{-4}	0 MeV
			Total	1.3×10^5 MeV

The value 1.3×10^5 MeV of alpha energy completes the WL description.

Question 1.37

$C = (k_2 N)/(k_1 e f t_s)$
C = radon-222 concentration
k_1 = 150 dpm alpha/liter-WL
k_2 = (100 pCi/liter)/0.5 = 200 pCi/liter @ 50% equilibrium
N = total alpha counts per minute
  = 230 counts/1 minute = 230 cpm
e = counting efficiency = 0.3 cpm/dpm
f = sample pump flow rate = 10 liters/min
t_s = sample collection time = 5 min

$$C = \frac{(230 \text{ cpm})(200 \text{ pCi/WL})}{(0.3 \text{ cpm/dpm})(10 \text{ liters/min})(5 \text{ min})(150 \text{ dpm/WL})}$$
$$= 20.4 \text{ pCi/liter}$$

Question 1.38

Measurement for radon and its daughters is performed by using the following methods:

1. Track etch detectors
2. Charcoal absorption

3. Filter paper collection and analysis
4. Continuous working level monitors
5. Passive environmental radon monitors (PERMs)
6. Electrostatic or ELECTRET dosimeters

Scenario 12

Question 1.39

The "memory effect" is a potential source of error associated with measuring beta and gamma dose rates while moving in and out of a noble gas environment. While in the cloud, noble gas seeps inside the ion chamber. After leaving the noble gas environment, the instrument continues to respond to the gas within its chamber, which causes the meter to respond and yield a false-positive reading.

Question 1.40

Environmental conditions having an adverse effect on the accuracy of the instrument response include temperature, pressure, radiofrequency signals, and humidity. The temperature and pressure affect the density of the air within the chamber and therefore directly affect the number of ion pairs produced. Radiofrequency signals can interfere with the instrument's signal processing capability. Humidity can disrupt the flow of electricity, causing shorting of the anode/cathode which will lead to erratic readings.

Question 1.41

The geometry factor or the geometric relationship between the source and receptor is the most significant source of error associated with measuring true beta and gamma surface dose rates from contact measurements of small sources. The instrument response is related to the number of ions collected at the center of the detector, and not the dose rate at the surface of the source. This is also affected by the volume of the chamber exposed to the particle's radiation. The smaller the source, the smaller the volume irradiated and the larger the geometry effect.

Question 1.42

A source of error associated with measuring beta dose rates from large-area sources, with each source being a different beta emitter, is the energy of the emitted beta particle. Because beta particles are attenuated across the chamber depending upon their energy, higher-energy beta particles respond closer to the true beta dose rate. Lower-energy beta particles underrespond.

Question 1.43

For high-energy beta sources, beta penetration through the closed window is a source of error implicit in the application of the open and closed window readings. Normally, the open window (OW) measures both beta and gamma radiation, and the closed window (CW) measurement yields only gamma information:

$$OW = beta + gamma$$
$$CW = gamma$$

Therefore

$$OW - CW = beta$$

With a high-energy beta source, some of the beta particles penetrate the closed window. This results in an overestimate of the gamma contribution which causes the beta dose rate to be underestimated.

Scenario 13

Question 1.44: a

ANSI N13.11 does not apply to pocket dosimeters and extremity dosimeters.

Question 1.45:

Lens of the eye.

Question 1.46: a

The standard least adequately tests for low-energy beta particles and the spectrum of neutrons escaping from the reactor vessel in a power reactor.

Question 1.47: b

Passing the beta radiation performance standard does not guarantee accurate results for all beta sources.

Question 1.48: a

ANSI N13.11 forms the basis for the National Voluntary Laboratory Accreditation Program for dosimetry processors.

Scenario 14

Question 1.49

5—Proton recoil badge.

High-energy neutron detection is best accomplished with the proton recoil film badge.

Li-6-based dosimeters are preferred for lower-energy neutrons. Other dosimeters do not have good neutron detection capability.

Question 1.50

2—A TLD albedo containing both Li-6 and Li-7 elements.

The TLD albedo system has the best capacity for measuring both neutrons and gammas, but without the fogging which is often encountered with film systems.

A proton recoil film badge will also work, but not without interference.

Question 1.51

1—A common film badge.

The film badge has energy-compensating filters which provide satisfactory energy discrimination at the lower x-ray range.

Question 1.52

9–A calcium sulfate, dysprosium-activated TLD element in a tissue equivalent holder.

This system has a good response to high-energy gammas, low fading, and the tissue equivalent holder approximates conditions of electronic equilibrium.

Item 4, the manganese-activated TLD element, would also work, but its high-energy response is not as good.

Question 1.53

7—The four-element TLD with lithium borate phosphors, a thin mylar filter over one element, plastic filters over two elements, and an aluminum filter over the fourth element.

Lithium borate is a good photon and electron detector, and the multiple filters allow good energy determination.

Scenario 15

Question 1.54

Gamma calibration factor = 6000 TL units/500 mrem

$$GCF = 12 \text{ TL units/mrem}$$

$$\text{Beta calibration factor} = 750 \text{ TL units}/1000 \text{ mrem}$$
$$\text{BCF} = 0.75 \text{ TL units/mrem}$$
$$\begin{aligned}\text{Gamma dose} &= (\text{Chip 2} - \text{Control Chip 2})/\text{GCF}\\&= (11{,}520 - 120)\text{TL units}/12 \text{ TL units/mrem}\\&= 950 \text{ mrem}\end{aligned}$$
$$\begin{aligned}\text{Beta dose} &= (\text{Chip 1} - \text{Chip 2})/\text{BCF}\\&= (12{,}270 - 11{,}520)\text{TL units}/0.75 \text{ TL units/mrem}\\&= 1000 \text{ mrem}\end{aligned}$$
$$\text{Whole-body dose} = \text{gamma dose} = 950 \text{ mrem}$$
$$\text{Skin dose} = \text{beta dose} + \text{gamma dose} = 1950 \text{ mrem}$$

Question 1.55

The correct depth to use for the lens of the eye is 300 mg/cm². The beta dose equivalent at 300 mg/cm² is 25% of that at 7 mg/cm².

$$\begin{aligned}\text{Eye dose} &= \text{beta dose (300 mg/cm}^2) + \text{gamma dose}\\&= 0.25 \times 1000 \text{ mrem} + 950 \text{ mrem}\\&= 1200 \text{ mrem}\end{aligned}$$

Question 1.56

No regulatory limits (e.g., 10 CFR 20) were exceeded. The 1993 10 CFR 20 revision establishes a 15-rem/year limit for the eye dose equivalent.

Scenario 16

Question 1.57: b

The neutron quality factor is not constant between 0 and 20 MeV.

Question 1.58: a

The gamma dose measured on a phantom is greater than the gamma dose measured in air due to the H(n, γ)-D reaction in the phantom.

Question 1.59: e

TLD albedo dosimeters calibrated with a bare Cf-252 source will overrespond to soft neutron spectra. Corrections are needed for accurate results.

Question 1.60: c

Neutron bubble detectors are affected by temperature.

Question 1.61: a

For power reactor containment entries, a TLD albedo dosimetry system calibrated to D_2O-moderated Cf-252 will accurately measure the neutron dose equivalent.

Scenario 17

Question 1.62: Departments A, B, and C

The nuclear medicine, x-ray, and radiation therapy departments will require personnel monitoring for photons because they utilize photon or x-ray sources. The research department will not require photon monitoring because it only uses H-3 and C-14, which are low-energy beta emitters.

Question 1.63: None

Because no department utilizes neutron sources, neutron monitoring is not required.

Question 1.64: Department B

The x-ray department will benefit from the dual monitors because the apron will provide attenuation of the x-ray source.

Question 1.65: Department A

The nuclear medicine department should utilize ring badges because of the handling and preparation of Tc-99m.

Question 1.66: None

However, a misapplication could lead to significant skin doses that would require assessment. The dose factor (rem/hr per $\mu Ci/cm^2$) for a 7-mg/cm^2 skin depth should be considered. Factors for applicable nuclei are:

Nuclide	Dose factor
H-3	0.0
C-14	1.09
Co-60	4.13
Tc-99	3.49
Cs-137	6.46

230 SOLUTIONS FOR CHAPTER 1

Question 1.67: Department D

The use of H-3 and C-14 suggests that the research department should consider a routine bioassay program.

Question 1.68

Positive characteristics of film dosimeters:

1. Provides a permanent record
2. Rugged and durable
3. Responds to a wide exposure range
4. Provides a stable latent record
5. Reasonable energy discrimination
6. Relatively inexpensive

Question 1.69

Negative characteristics of film dosimeters:

1. No beta sensitivity for energies below about 200 keV
2. Sensitive to heat
3. Cannot be reused
4. Processing time causes delays in information retrieval
5. Not tissue-equivalent
6. Fades with time
7. Energy-dependent

Question 1.70

Positive characteristics of TLDs:

1. Rugged in construction
2. Tissue-equivalent
3. Energy-independent
4. Applicable over a wide exposure range
5. Sensitive to low doses
6. Can be reused
7. Can be immediately processed on site
8. Reasonable beta sensitivity

Question 1.71

Negative characteristics of TLDs:

1. No permanent record
2. Fades with heat
3. Sensitive to oil and water
4. Angular dependence
5. Supralinearity
6. Expensive
7. Maintenance costs are higher than those for film.

Scenario 18

Question 1.72

For cylindrical gas ionization chambers, the output current (I) is directly proportional to the volume of the detector:

$$I = kV$$

where k is a proportionality constant (A/cm^3). For a cylinder detector, the volume (V) is given by

$$V = \pi r^2 h$$

where r is the radius of the detector and h is the detector height. For the two detectors, we have

$$r_A = 0.25 \text{ cm}, \quad r_B = 0.50 \text{ cm}$$
$$h_A = 5.0 \text{ cm}, \quad h_B = 5.0 \text{ cm}$$

The current outputs of detectors A and B are related by

$$\frac{I_B}{I_A} = \frac{\pi r_B^2 h_B k}{\pi r_A^2 h_A k}$$

Because $h_A = h_B$ and $r_B = 2r_A$, we obtain

$$I_B = I_A(2)^2 = 4I_A = 4(1.0 \times 10^{-10} \text{ A}) = 4.0 \times 10^{-10} \text{ A}$$

The correct answer is b.

232 SOLUTIONS FOR CHAPTER 1

Question 1.73

The detector sensitivity (S) is directly proportional to the gas pressure (P) within the detector

$$S = kP$$

where k is a constant. The sensitivity obtained by changing the gas pressure may be obtained from the relationship

$$S_1/S_2 = P_1/P_2$$

or

$$S_2 = S_1(P_2/P_1) = (1.2 \times 10^{-10} \text{ A-hr/R})$$

$$\times \frac{11{,}400 \text{ torr}}{7600 \text{ torr}}$$

$$= 1.8 \times 10^{-10} \text{ A-hr/R}$$

Question 1.74

The detector current (I) may be written in terms of the conditions at STP, the detector volume (V), the detector gas density (p), ambient temperature (T), pressure (P), and the exposure rate ($\dot{X}$):

$$I = kpV \frac{T_{STP}}{T} \frac{P}{P_{STP}} \dot{X}.$$

where

$T_{STP} = 273°K \qquad T = 20°C$

$P_{STP} = 760 \text{ torr} \qquad P = 7600 \text{ torr}$

$p = 1.29 \text{ kg/m}^3$

$V = 100 \text{ cm}^3$

$I = 9.0 \times 10^{-14} \text{ A}$

$k = (2.58 \times 10^{-4} \text{ coulomb/kg-R})(1 \text{ hr}/3600 \text{ sec})(1 \text{ A sec/coulomb})$

$= 7.17 \times 10^{-8} \text{ A-hr/R-kg}$

SOLUTIONS FOR CHAPTER 1   233

$\dot{X}$ may be obtained from the current equation:

$$\dot{X} = I \bigg/ \left(kpV \frac{T_{STP}}{T} \frac{P}{P_{STP}} \right)$$

$$= \frac{9.0 \times 10^{-14} \text{ A}}{(7.17 \times 10^{-8} \text{ A-hr/R-kg})(1.29 \text{ kg/m}^3)(100 \text{ cm}^3)(1 \text{ m}/100 \text{ cm})^3}$$

$$\times \frac{1}{[273°K/(273°K + 20°K)][7600 \text{ torr}/760 \text{ torr}]}$$

$$= \frac{9.0 \times 10^{-14} \text{ A}}{8.62 \times 10^{-11} \text{ A-hr/R}} = 1.04 \times 10^{-3} \text{ R/hr} = 1.04 \text{ mR/hr}$$

Question 1.75

The detector was calibrated under the following conditions:

$$T_C = (273 + 20)°K = 293°K$$
$$P_C = 591.6 \text{ torr}$$

Measurements were made at sea level:

$$T = (273 + 36)°K = 309°K$$
$$P = 760 \text{ torr}$$
$$\dot{X} = 100 \text{ mR/hr}$$

From the first equation of this scenario, we have

$$I = K\dot{X}\frac{P}{T}$$

where K is a constant for a given detector. Because $I = I_C$, the exposure rate ($\dot{X}_C$) that will be read in a 100-mR/hr field is

$$\dot{X}_C = \dot{X}(P/P_C)(T_C/T)$$
$$= (100 \text{ mR/hr})(760 \text{ torr}/591.6 \text{ torr})(293°K/309°K)$$
$$= 121.8 \text{ mR/hr}$$

The correct answer is d.

Scenario 19

Question 1.76

The total gamma dose rate at 1 ft is

$$\dot{H} = (S/4\pi r^2)(u_{en}/p)E_{gamma}$$

where

S = gamma source strength = 2.2×10^6 gammas/sec
r = distance from the source = 1 ft = 30.48 cm
E_{gamma} = gamma-ray energy of source = 6.1 MeV
u_{en}/p = mass-energy absorption coefficient
   = 0.0178 cm^2/g for muscle

$\dot{H}$ = [(2.2×10^6 gammas/sec)/(4π)(30.48 cm)^2]
   $\times$ (0.0178 cm^2/g)
   $\times$ (6.1 MeV/gamma)(1.6×10^{-6} erg/MeV)(1 rad/100 erg/g)
   $\times$ (3600 sec/hr)(1 rem/rad)(1000 mrem/rem)
   = 1.18 mrem/hr

Question 1.77

The total neutron dose rate at 1 ft is given by

$$\dot{H}_n = (S_n/4\pi r^2)k$$

where

S_n = neutron source strength = 2.0×10^5 n/sec
k = flux-to-dose conversion factor
   = (2.5 mrem/hr)/(20 n sec^{-1} cm^{-2})

$$\dot{H}_n = \frac{(2.5 \text{ mrem/hr})(2.0 \times 10^5 \text{ n sec}^{-1})}{(20 \text{ n sec}^{-1} \text{ cm}^{-2})(4\pi)(30.48 \text{ cm})^2}$$

= 2.14 mrem/hr

Question 1.78: a

The best shielding arrangement is polyethylene followed by lead. The polyethylene thermalizes the neutrons, and lead attenuates both the capture gamma rays and the source's 6.1-MeV gammas.

Question 1.79

The required gamma shielding thickness is obtained as follows:

$$\dot{H}(x) = \dot{H}(0) B \exp(-ux)$$

$\dot{H}(0)$ = unshielded exposure rate

T = transmission factor = $\dot{H}(x)/\dot{H}(0) = 1/5 = 0.2$
$\quad = B \exp(-ux)$

From the table of buildup factors, a table of transmission factors can be assembled:

i	ux_i	B_i	T_i
1	1.0	1.18	0.434
2	2.0	1.40	0.189

When it is assumed that $\ln(T)$ versus ux is linear in this region of shield thickness, interpolation leads to the required ux value that corresponds to $T = 0.2$:

$$ux = ux_1 + (ux_2 - ux_1) \frac{\ln(T/T_1)}{\ln(T_2/T_1)}$$

$$= 1.0 + (2.0 - 1.0) \frac{\ln(0.2/0.434)}{\ln(0.189/0.434)}$$

$$= 1.0 + \frac{-0.7747}{-0.8313} = 1.93$$

$ux = 1.93$

$x = 1.93/u$

u_a/p = mass attenuation coefficient for lead
$\quad = 0.0435 \text{ cm}^2/\text{g}$

$p = 11.35 \text{ g/cm}^3$

$u = (u_a/p)p$

$$x = \frac{1.93}{(0.0435 \text{ cm}^2 \text{ g}^{-1})(11.35 \text{ g cm}^{-3})}$$

$\quad = 3.91 \text{ cm}$

Scenario 20

Question 1.80

Method	Advantages	Disadvantages
Time for decay	By waiting, there will be reduced exposures during decontamination and decomissioning (D&D).	Longer time will require controls throughout the storage period.
	Costs for D&D can be deferred to a future date.	There is a potential for continuing releases to the environment.
	Waiting will permit new technology to emerge to limit dose and cost.	Costs for future waste disposal are uncertain and will likely increase.
	More time is allowed for a D&D fund to accumulate.	Waiting will prohibit use of land, buildings, and facility resources.
Immediate removal	Costs are known.	During removal, the potential for releases to the environment are increased.
	Problem is solved, and resources may be used for other purposes.	Transportation risks will be increased during removal operations.
	Waste will be removed to a storage facility designed for this purpose.	Higher occupational exposures will be received compared to a delayed disposal.
Shielding	May be the lowest cost option.	A possession licence will be required.
	Some occupational exposure, but not as high as removal.	Future liability may include eventual removal.

Question 1.81

Based upon the data provided, the depth of excavation required to allow free release can be obtained from the relationship

$$I = I_0 \exp(-ux) \quad \text{(no buildup)}$$

Based upon the available information, excluding buildup is appropriate. The required excavation depth can be obtained by rearranging this equation and by inserting the given data into this equation:

$$x = -(1/u) \ln(I/I_0)$$

Relaxation length $= 1/u = 15$ cm

$$I_0 = 20 \ \mu R/hr$$
$$I = 5 \ \mu R/hr$$
$$x = -(15 \text{ cm}) \ln(5 \ \mu R/hr / 20 \ \mu R/hr) = 20.8 \text{ cm}$$

Question 1.82

$$A = A_0 \exp(-\lambda t)$$
$$t = (-1/\lambda) \ln(A/A_0)$$

For Co-60, we have

$$\lambda = 0.693/T_{1/2}$$
$$= 0.693/5.27 \text{ year} = 0.131 \text{ year}^{-1}$$
$$t = (-1/0.131 \text{ year}^{-1}) \ln(5/20) = 10.54 \text{ year}$$

Question 1.83

The mass attenuation coefficient is 0.06 cm^2/g and the density of concrete is 2.5 g/cm^3. Buildup is assumed to be a constant factor of 2.

To obtain the shield thickness, the following relationship is utilized:

$$I = I_0 B \exp(-uX) \text{ (buildup included)}$$
$$B = \text{buildup factor}$$
$$u = (u/p)(p)$$
$$u/p = 0.06 \text{ cm}^2/\text{g}$$
$$p = 2.5 \text{ g/cm}^3$$

Solving for X, we obtain

$$X = (-1/u) \ln(I/I_0 B)$$
$$X = (-1/0.06 \text{ cm}^2/\text{g} \times 2.5 \text{ g/cm}^3)$$
$$\times \ln(5 \ \mu R/hr / 2 \times 20 \ \mu R/hr)$$
$$X = 13.9 \text{ cm}$$

Scenario 21

Question 1.84

The reaction described in this problem is

$$\text{Incident particle} + \text{Al-27} \rightarrow \text{Na-24} + \text{ejectile}$$

with a production cross section of 20 mb.

The general equation for the buildup of Na-24 activity is

$$A = N\sigma\phi\,[1 - \exp(-\lambda t_{\text{irrad}})]$$
$$\times \exp(-\lambda t_{\text{decay}})$$

For this problem, the saturation activity is assumed and the activating flux is to be determined:

$A_{\text{sat}} = N\sigma\phi$

A_{sat} = saturation activity of Na-24

N = number of atoms of Al-27 in the target

σ = cross section for producing Na-24 from Al-27

   = 20 mb

ϕ = activating flux

Solving for the activating flux, we obtain

$\phi = A_{\text{sat}}/N\sigma$

$A_{\text{sat}} = 4.0 \times 10^7$ dis/sec

$N = (6.02 \times 10^{23} \text{ atoms/mole})(1 \text{ g})/(27 \text{ g/mole})$

   $= 2.23 \times 10^{22}$ atoms of Al-27

$\sigma = (20 \text{ mb})(1 \text{ b}/1000 \text{ mb})(1.0 \times 10^{-24} \text{ cm}^2/\text{barn})$

   $= 2 \times 10^{-26}$ cm^2

$$\phi = \frac{(4.0 \times 10^7 \text{ dis/sec})(1 \text{ particle/dis})}{(2.23 \times 10^{22} \text{ atoms})(2.0 \times 10^{-26} \text{ cm}^2/\text{atom})}$$

   $= 8.97 \times 10^{10}$ particles/cm^2-sec

Question 1.85

The general activation relationship can be written as

$$A = A_{sat}[1 - \exp(-\lambda t_{irrad})] \exp(-\lambda t_{decay})$$

Because we are interested in the time immediately after shutdown, $t_{decay} = 0$ and the activity relationship becomes

$$A = A_{sat}[1 - \exp(-\lambda t_{irrad})]$$

The activity after the 30-hr irradiation is

$$A = (4.0 \times 10^7 \text{ Bq}) \{1 - \exp[-(0.693/15 \text{ hr})(30 \text{ hr})]\}$$
$$= 3.0 \times 10^7 \text{ Bq}$$

Question 1.86

The dose-equivalent rate to a person standing a distance r from an unshielded point activation source is given by

$$\dot{D} = AEu/(4\pi r^2)$$
$$u = \text{mass attenuation coefficient } (\text{cm}^2/\text{g}) = u_{en}/p$$
$$= (2.3 \times 10^{-5} \text{ cm}^{-1})/(0.00129 \text{ g/cm}^3) = 0.018 \text{ cm}^2/\text{g}$$
$$E = \sum_i E_i Y_i = [1.4(1.0) + 2.8(1.0)] \text{ MeV} = 4.2 \text{ MeV}$$

A point source approximation is reasonable because the size of the 1-gram mass is small with respect to the 1-meter distance. With this approximation the dose rate is

$$\dot{D} = (4.0 \times 10^7 \text{ dis/sec})(4.2 \text{ MeV})(1.6 \times 10^{-6} \text{ erg/MeV})$$
$$\times (1 \text{ rad}/100 \text{ erg/g})(0.018 \text{ cm}^2/\text{g})/(4)(3.14)(100 \text{ cm})^2$$
$$= 3.85 \times 10^{-7} \text{ rad/sec} \times 1000 \text{ mrem/rem} \times 3600 \text{ sec/hr} \times 1 \text{ rem/rad}$$
$$= 1.39 \text{ mrem/hr}$$

Scenario 22

Question 1.87

$$\dot{X} = AG/r^2$$

240 SOLUTIONS FOR CHAPTER 1

where

$$A = \text{source activity} = 1.0 \times 10^7 \text{ Ci}$$
$$G = \text{gamma constant for Co-60}$$
$$r = \text{distance from the point source}$$
$$\dot{X} = (1.0 \times 10^7 \text{ Ci})(1.32 \text{ R-m}^2/\text{hr-Ci})/(3.0 \text{ m})^2$$
$$= 1.47 \times 10^6 \text{ R/hr}$$

Questions 1.88

The maximum photon fluence $I(r)$ at the exterior wall surface is given by

$$I(r) = I_0(r) \sum_i B_i f_i \exp(-u_i x)$$

B_i = buildup factor for energy E_i

u_i = attenuation factor for 1-m-thick concrete wall for energy E_i

$u_1 = 0.140/\text{cm}$ for 1.17 MeV

$u_2 = 0.130/\text{cm}$ for 1.33 MeV

x = thickness of the concrete wall = 1 m

E_1 = energy of Co-60 gamma ray #1 = 1.17 MeV

E_2 = energy of Co-60 gamma ray #2 = 1.33 MeV

f_i = fraction of I_0 contributed by E_i or the partial yield of the ith gamma ray

$f_1 = f_2 = 0.5$

Each term contributing to $I(r)$ will be calculated separately. The unattenuated flux is given by

$$I_0 = AkY/(4\pi r^2)$$

I_0 = unattenuated flux @ 3.5 m (gammas cm^{-2} sec^{-1})

S = source strength (gammas/sec)

  = AkY

A = source activity (Ci) = 1×10^7 Ci

k = conversion factor (3.7×10^{10} dis/sec-Ci)

Y = yield (gammas/dis) = 2 gammas/dis for Co-60

SOLUTIONS FOR CHAPTER 1   241

r = distance from the source (cm) = 350 cm

$I_0 = (1 \times 10^7 \text{ Ci})(3.7 \times 10^{10} \text{ dis/sec-Ci})(2 \text{ gammas/dis})/(4\pi)(350 \text{ cm})^2$

$= 4.81 \times 10^{11}$ gammas cm^{-2} sec^{-1}

Determine the attenuation provided by the concrete wall for each gamma ray:

$$1.17 \text{ MeV}: \quad i = 1$$
$$1.33 \text{ MeV}: \quad i = 2$$

$\exp[-(u_1 x)] = \exp(-0.140/\text{cm} \times 100 \text{ cm}) = 8.3 \times 10^{-7}$
$\exp[-(u_2 x)] = \exp(-0.130/\text{cm} \times 100 \text{ cm}) = 2.3 \times 10^{-6}$

Determine the buildup factors for each energy by interpolation:

$$u_1 x = (0.140/\text{cm})(100 \text{ cm}) = 14.0$$
$$u_2 x = (0.130/\text{cm})(100 \text{ cm}) = 13.0$$

	B	
i	$ux = 10$	$ux = 15$
1	17.5	30.6
2	16.3	28.2

$$B_1(14) = 17.5 + \frac{14 - 10}{15 - 10} \times (30.6 - 17.5) = 28.0$$

$$B_2(13) = 16.3 + \frac{13 - 10}{15 - 10} \times (28.2 - 16.3) = 23.4$$

Finally, apply these buildup factors and sum to obtain the total fluence rate:

$$I(r) = I_0(r) \sum_i B_i f_i \exp(-u_i x)$$

$I(r) = (4.8 \times 10^{11} \text{ gammas cm}^{-2} \text{ sec}^{-1})[(0.5)(8.3 \times 10^{-7})(28)$
$+ (0.5)(2.3 \times 10^{-6})(23.4)] = 1.9 \times 10^7$ gammas/cm^2-sec

Question 1.89

Buildup factors apply to broad-beam or poor geometry conditions. For these cases, scattering becomes important and the buildup factor is needed to correct for this scattering. The correct answer is d.

Scenario 23

Question 1.90

The linear attenuation coefficient (u) is defined by

$$I = I_0 \exp(-ux)$$

where

I = shielded radiation exposure rate (R/hr)
I_0 = unshielded radiation exposure rate (R/hr)
x = shield thickness (cm)

u may be obtained from its defining equation:

$$u = -\frac{1}{x} \ln(I/I_0)$$

Using the values in the data tables leads to the following values of $u(x)$ for the narrow beam:

$$u(1.0 \text{ cm}) = -\frac{1}{1.0 \text{ cm}} [-\ln(29.5 \text{ mR/hr}/127.0 \text{ mR/hr})]$$

$$= 1.46 \text{ cm}^{-1}$$

$$u(2.0 \text{ cm}) = -\frac{1}{2.0 \text{ cm}} [-\ln(7.7 \text{ mR/hr}/127.0 \text{ mR/hr})]$$

$$= 1.40 \text{ cm}^{-1}$$

$$u(3.0 \text{ cm}) = -\frac{1}{3.0 \text{ cm}} [-\ln(1.9 \text{ mR/hr}/127.0 \text{ mR/hr})]$$

$$= 1.40 \text{ cm}^{-1}$$

The linear attenuation coefficient for the given data is in the range of 1.40–1.46 1/cm. An average of these values yields 1.42/cm.

Question 1.91

The buildup factor (B) is defined for broad-beam conditions by the equation

$$I = I_0 B(x) \exp(-ux)$$

Solving for $B(x)$ leads to the result

$$B(x) = \frac{I}{I_0 B \exp(-ux)}$$

$B(2.5 \text{ cm})$ is requested. It can be obtained from the $B(2.0 \text{ cm})$ and $B(3.0 \text{ cm})$ values:

$$B(2.0 \text{ cm}) = \frac{I}{I_0 \exp(-ux)}$$

$$= \frac{13.0 \text{ mR/hr}}{(127 \text{ mR/hr}) \exp(-2.0/\text{cm} \times 2.0 \text{ cm})}$$

$$= 5.59$$

$$B(3.0 \text{ cm}) = \frac{I}{I_0 \exp(-ux)}$$

$$= \frac{4.0 \text{ mR/hr}}{(127 \text{ mR/hr}) \exp(-2.0/\text{cm} \times 3.0 \text{ cm})}$$

$$= 12.71$$

The value at 2.5-cm thickness is obtained by linear interpolation. Simple linear interpolation leads to the following value:

$$B(2.5 \text{ cm}) = (5.59 + 12.71)/2 = 9.15$$

The use of linear or logarithmic interpolation will be governed by the available data and the manner by which $B(x)$ scales with x.

Question 1.92

The mass-attenuation coefficient (u/p) is

$$u/p = \frac{2.0/\text{cm}}{18.9 \text{ g/cm}^3} = 0.11 \text{ cm}^2/\text{g}$$

Scenario 24

Question 1.93

The dose from the K-40 distributed throughout the body can be estimated by determining its activity in the body:

244 SOLUTIONS FOR CHAPTER 1

$A = \lambda N = (0.693/T_{1/2})N$
$T_{1/2}$ = half-life (years) = 1.2×10^9 years
N = number of atoms of K-40 in reference man
  = mA_0/M
m = mass of K-40 in the body = 140 g $\times$ 0.00012
M = gram atomic weight of K-40 = 40 g/GAW
A_0 = Avogadro's number = 6.02×10^{23} atoms/GAW
N = (140 g $\times$ 0.00012)(6.02×10^{23} atoms/GAW)/(40 g/GAW)
  = 2.53×10^{20} atoms
A = $(0.693/1.2 \times 10^9 \text{ years})(2.53 \times 10^{20})$ = 1.46×10^{11} dis/year
$\bar{e}$ = average energy deposited into the tissue by the K-40 decay
  = $E_{max} f/3$
E_{max} = maximum beta energy of the K-40 decay
  = 1.3 MeV
f = probability that a beta particle is emitted when K-40 decays
  = 0.9
$\bar{e}$ = (1.3 MeV)(0.9)/3 = 0.39 MeV

The dose rate delivered to the body from the K-40 ($\dot{D}$) is given by

$$\dot{D} = \frac{A\bar{e}}{m} k$$

m = mass of the whole body = 70,000 g

k = conversion factor = $(1.6 \times 10^{-6}$ erg/MeV$)/(100$ erg/g-rad$)$

$$\dot{D} = \frac{(1.46 \times 10^{11} \text{ dis/year})(0.39 \text{ MeV})(1.6 \times 10^{-6} \text{ erg/MeV})}{(70,000 \text{ g})(100 \text{ erg/g-rad})}$$

  = 0.013 rad/year $\times$ 1 year/52 weeks
  = 2.5×10^{-4} rad/week

Question 1.94: c

K-40 has no regulatory significance in whole-body counting, but it provides an important qualitative system check.

Question 1.95: e

ICRP-26 focuses upon minimizing the total effective dose equivalent, which is the sum of the external (deep dose equivalent) and internal (committed effective dose equivalent) doses. As such, the use of air samples and stay-time calculations instead of respirator usage is acceptable if the total dose is minimized.

SOLUTIONS FOR CHAPTER 1   245

Question 1.96: a

Electronic equilibrium is least likely at the surface of the skin.

Scenario 25

The reader should note that medical exposure is not included as part of the occupational exposure. This scenario reflects a radiation protection program that takes the diagnostic exposure into account in order to determine a worker's readiness for duty.

Question 1.97

Initial activity in the thyroid:

$$q = \text{administered activity } (\mu\text{Ci}) = 1.0 \ \mu\text{Ci}$$
$$A(0) = \text{initial activity in the thyroid}$$
$$= qf_2 = 0.3 \times 1.0 \ \mu\text{Ci} = 0.3 \ \mu\text{Ci}$$

Effective half-life:

$$T_{\text{phy}} = 8.08 \text{ days}$$
$$T_{\text{bio}} = 74 \text{ days}$$
$$T_{\text{eff}} = (8.08 \text{ days})(74 \text{ days})/(8.08 \text{ days} + 74 \text{ days})$$
$$= (7.28 \text{ days})(24 \text{ hr/day})$$
$$= 174.8 \text{ hr}$$

Cumulated activity:

$$\tilde{A} = A(0)(1.44)(T_{\text{eff}})$$
$$= (0.3 \ \mu\text{Ci})(1.44)(174.8 \text{ hr}) = 75.51 \ \mu\text{Ci-hr}$$

Thyroid dose:

$$D = \tilde{A}S(T \leftarrow S)$$
$$= 1.44A(0)T_{\text{eff}}S(T \leftarrow S)$$
$$D = (0.3 \ \mu\text{Ci})(1.44)(174.8 \text{ hr})(2.2 \times 10^{-2} \text{ rad}/\mu\text{Ci-hr})$$
$$= 1.66 \text{ rad}$$

Question 1.98

Because the ICRP-10 investigation level is 300 nCi, the HP program places a restriction on the worker at 10% of this value, or 30 nCi.

The worker reaches unrestricted status at 30 nCi. Therefore, you must find the time for the 300-nCi intake to decay to 30 nCi. This is given by a simple exponential decay using the effective half-life. Because the diagnostic procedure investigates a potential thyroid abnormality, this technique is only an estimate of the dose. The intake retention function would be required to provide a retrospective assessment of the actual dose. This can be accomplished by either *in vitro* or *in vivo* techniques.

$$A(t) = A(0) \exp[(-0.693/T_{eff})t]$$

$$A(0) = 300 \text{ nCi}$$

$$A(t) = 30 \text{ nCi}$$

Solving for t, we obtain

$$t = -(T_{eff}/0.693) \ln[A(t)/A(0)]$$

$$t = -(1.44)(7.28 \text{ days}) \ln(30/300)$$

$$= 24.2 \text{ days}$$

Question 1.99: e

Only statement e is correct per the ICRP-10 investigation level specification.

Scenario 26

Question 1.100

Assuming an instantaneous distribution of the 10 Ci and no dilution by the ventilation system, the dose to the individual in the room is given by

$$D = C(BR)t(DCF)$$

C = the tritium air concentration

$\quad = (10 \text{ Ci})/(10 \text{ ft} \times 20 \text{ ft} \times 30 \text{ ft} \times 0.02832 \text{ m}^3/\text{ft}^3)$

$\quad = 5.89 \times 10^{-2} \text{ Ci/m}^3$

BR = breathing rate = $3.5 \times 10^{-4} \text{ m}^3/\text{sec}$

t = exposure time = 30 min

DCF = dose conversion factor = 158 rem/Ci

SOLUTIONS FOR CHAPTER 1

$D = C(BR)t(DCF)$
$= (5.89 \times 10^{-2} \text{ Ci/m}^3)(3.5 \times 10^{-4} \text{ m}^3/\text{sec})(30 \text{ min} \times 60 \text{ sec/min})$
$\times (158 \text{ rem/Ci})$
$= 5.86 \text{ rem}$

Question 1.101

To calculate the dose to the individual at the site boundary, the following assumptions are made:

1. All H-3 is released to the environment in 30 min.
2. The person is stationed at the plume centerline during the entire accident.

$D = A(X/Q)(BR)(DCF)$
$A = \text{activity released} = 10 \text{ Ci}$
$X/Q = \text{atmospheric diffusion factor} = 1.0 \times 10^{-4} \text{ sec/m}^3$
$D = (10 \text{ Ci})(1.0 \times 10^{-4} \text{ sec/m}^3)(3.5 \times 10^{-4} \text{ m}^3/\text{sec})(158 \text{ rem/Ci})$
$= 5.53 \times 10^{-5} \text{ rem}$

Question 1.102

Assuming uniform distribution of the tritium and exponential removal by the ventilation system, the concentration of tritium activity $C(t)$ is given by

$$C(t) = C_0 \exp(-rt)$$

$r = \text{ventilation removal rate}$
$r = 3/\text{hr} \times 1 \text{ hr}/60 \text{ min} = 0.05/\text{min}$

The average concentration $(\overline{C})$ in the room during the time of the exposure ($t = 30$ min) is obtained by integrating $C(t)$ from $t = 0$ to t:

$$\overline{C} = \int_0^t C(t)\,dt \Big/ \int_0^t dt$$

$$\overline{C} = C_0[1 - \exp(-rt)]/(rt)$$

For $t = 30$ min,

$$\overline{C} = \frac{(5.89 \times 10^{-2} \text{ Ci/m}^3)[1.0 - \exp(-0.05/\text{min} \times 30 \text{ min})]}{(0.05/\text{min})(30 \text{ min})}$$

$$= 3.05 \times 10^{-2} \text{ Ci/m}^3$$

The dose is given by

$$D = \overline{C}(\text{BR})t(\text{DCF})$$
$$= (3.05 \times 10^{-2} \text{ Ci/m}^3)(3.5 \times 10^{-4} \text{ m}^3/\text{sec})(30 \text{ min} \times 60 \text{ sec/min})$$
$$\times (158 \text{ rem/Ci})$$
$$= 3.04 \text{ rem}$$

Question 1.103

The assumptions are the same as noted in question 1.101. The same quantity of tritium is released, and the individual is present during the entire release. Therefore, the dose is the same and independent of the room ventilation characteristics:

$$D = 5.53 \times 10^{-5} \text{ rem}$$

Question 1.104

The assumptions are the same as noted in question 1.102. To solve this problem, the average concentration to give a dose of 0.5 rem should be found.

$$D = \overline{C}(\text{BR})t(\text{DCF})$$
$$\overline{C} = D/[(\text{BR})t(\text{DCF})]$$
$$= \frac{0.5 \text{ rem}}{(3.5 \times 10^{-4} \text{ m}^3/\text{sec})(30 \text{ min} \times 60 \text{ sec/min})(158 \text{ rem/Ci})}$$
$$= 5.02 \times 10^{-3} \text{ Ci/m}^3$$

From question 1.102:

$$\overline{C} = C_0[1 - \exp(-rt)]/rt$$

The results of question 1.102 suggest that $r > 3/\text{hr}$, which corresponded to a dose of 3 rem. Because

$$1 - \exp(-rt) \to 1$$

as r increases, we can simplify the previous equations:

$$\overline{C} = C_0/rt$$

$$\begin{aligned} r &= C_0/\overline{C}t \\ &= (5.89 \times 10^{-2}\ \text{Ci/m}^3)/[(5.02 \times 10^{-3}\ \text{Ci/m}^3)(0.5\ \text{hr})] \\ &= 23.5/\text{hr} \end{aligned}$$

Scenario 27

Question 1.105

To calculate the CDE and weighted CDE or committed effective dose equivalent (CEDE), the data for Cs-137 and Ba-137m must be organized:

Cs-137

Targets (T)	SEE($T \leftarrow$ Lungs)	U_S	$U_S \times$ SEE
Gonads	0.0	1.9×10^4	0.0
Breast	0.0	1.9×10^4	0.0
Red marrow	0.0	1.9×10^4	0.0
Lungs	1.9×10^{-4}	1.9×10^4	3.6
Thyroid	0.0	1.9×10^4	0.0
Bone surfaces	0.0	1.9×10^4	0.0
SI wall	0.0	1.9×10^4	0.0
ULI wall	0.0	1.9×10^4	0.0
LLI wall	0.0	1.9×10^4	0.0
Uterus	0.0	1.9×10^4	0.0
Adrenals	0.0	1.9×10^4	0.0

Cs-137

Targets (T)	SEE($T \leftarrow$ Total Body)	U_S	$U_S \times$ SEE
Gonads	2.7×10^{-6}	7.7×10^6	2.1×10^1
Breast	2.7×10^{-6}	7.7×10^6	2.1×10^1
Red marrow	2.7×10^{-6}	7.7×10^6	2.1×10^1
Lungs	2.7×10^{-6}	7.7×10^6	2.1×10^1
Thyroid	2.7×10^{-6}	7.7×10^6	2.1×10^1
Bone surfaces	2.7×10^{-6}	7.7×10^6	2.1×10^1
SI wall	2.7×10^{-6}	7.7×10^6	2.1×10^1
ULI wall	2.7×10^{-6}	7.7×10^6	2.1×10^1
LLI wall	2.7×10^{-6}	7.7×10^6	2.1×10^1
Uterus	2.7×10^{-6}	7.7×10^6	2.1×10^1
Adrenals	2.7×10^{-6}	7.7×10^6	2.1×10^1

Ba-137m

Targets (T)	SEE($T \leftarrow$ Lungs)	U_S	$U_S \times$ SEE
Gonads	5.7×10^{-8}	1.8×10^4	1.0×10^{-3}
Breast	2.7×10^{-6}	1.8×10^4	4.9×10^{-2}
Red marrow	2.5×10^{-6}	1.8×10^4	4.5×10^{-2}
Lungs	9.5×10^{-5}	1.8×10^4	1.7
Thyroid	2.6×10^{-6}	1.8×10^4	4.7×10^{-2}
Bone surfaces	2.0×10^{-6}	1.8×10^4	3.6×10^{-2}
SI wall	5.9×10^{-7}	1.8×10^4	1.1×10^{-2}
ULI wall	8.0×10^{-7}	1.8×10^4	1.4×10^{-2}
LLI wall	1.7×10^{-7}	1.8×10^4	3.1×10^{-3}
Uterus	2.1×10^{-7}	1.8×10^4	3.8×10^{-3}
Adrenals	4.9×10^{-6}	1.8×10^4	8.8×10^{-2}

Ba-137m

Targets (T)	SEE($T \leftarrow$ Total Body)	U_S	$U_S \times$ SEE
Gonads	4.7×10^{-6}	7.3×10^6	3.4×10^1
Breast	3.9×10^{-6}	7.3×10^6	2.8×10^1
Red marrow	4.3×10^{-6}	7.3×10^6	3.1×10^1
Lungs	4.0×10^{-6}	7.3×10^6	2.9×10^1
Thyroid	3.9×10^{-6}	7.3×10^6	2.8×10^1
Bone surfaces	4.0×10^{-6}	7.3×10^6	2.9×10^1
SI wall	4.9×10^{-6}	7.3×10^6	3.6×10^1
ULI wall	4.8×10^{-6}	7.3×10^6	3.5×10^1
LLI wall	4.9×10^{-6}	7.3×10^6	3.6×10^1
Uterus	4.9×10^{-6}	7.3×10^6	3.6×10^1
Adrenals	5.2×10^{-6}	7.3×10^6	3.8×10^1

The following table summarizes the intermediate result:

$$\text{Sum} = H_{50,T}/1.6 \times 10^{-10} = \sum_S \sum_j \left[U_S \sum_i \text{SEE}(T \leftarrow S_i) \right]_j$$

$$= \left[U_S \sum_i \text{SEE}(T \leftarrow S_i) \right]_{\text{Cs-137}}$$

$$+ \left[U_S \sum_i \text{SEE}(T \leftarrow S_i) \right]_{\text{Ba-137m}}$$

$U_S \times$ SEE Summary and Combination of Parent (Cs-137) and Daughter (Ba-137m) Results

	Cs-137		Ba-137m		
Targets (T)	$(T \leftarrow L)^a$	$+ (T \leftarrow TB)^b +$	$(T \leftarrow L)$	$+ (T \leftarrow TB) =$	Sum
Gonads	0.0	2.1×10^1	1.0×10^{-3}	3.4×10^1	5.5×10^1
Breast	0.0	2.1×10^1	4.9×10^{-2}	2.8×10^1	4.9×10^1
Red marrow	0.0	2.1×10^1	4.5×10^{-2}	3.1×10^1	5.2×10^1
Lungs	3.6	2.1×10^1	1.7	2.9×10^1	5.5×10^1
Thyroid	0.0	2.1×10^1	4.7×10^{-2}	2.8×10^1	4.9×10^1
Bone surfaces	0.0	2.1×10^1	3.6×10^{-2}	2.9×10^1	5.0×10^1
SI wall	0.0	2.1×10^1	1.1×10^{-2}	3.6×10^1	5.7×10^1
ULI wall	0.0	2.1×10^1	1.4×10^{-2}	3.5×10^1	5.6×10^1
LLI wall	0.0	2.1×10^1	3.1×10^{-3}	3.6×10^1	5.7×10^1
Uterus	0.0	2.1×10^1	3.8×10^{-3}	3.6×10^1	5.7×10^1
Adrenals	0.0	2.1×10^1	8.8×10^{-2}	3.8×10^1	5.9×10^1

^{a}L = lungs.
^{b}TB = total body.

$H_{50,T}$ may be obtained from the previous table by multiplying the final column (Sum) by 1.6×10^{-10}.

$$\text{CDE} = H_{50,T} = 1.6 \times 10^{-10} \times \sum_S \sum_j \left[U_S \sum_i \text{SEE}(T \leftarrow S_i) \right]_j$$

Targets (T)	$H_{50,T}$ (Sv/Bq)	w_T	$W_T H_{50,T}$ (Sv/Bq)
Gonads	8.8×10^{-9}	0.25	2.2×10^{-9}
Breast	7.8×10^{-9}	0.15	1.2×10^{-9}
Red marrow	8.3×10^{-9}	0.12	1.0×10^{-9}
Lungs	8.8×10^{-9}	0.12	1.1×10^{-9}
Thyroid	7.8×10^{-9}	0.03	2.3×10^{-10}
Bone surfaces	8.0×10^{-9}	0.03	2.4×10^{-10}
SI wall	9.1×10^{-9}	0.06	5.5×10^{-10}
ULI wall	9.0×10^{-9}	0.06	5.4×10^{-10}
LLI wall	9.1×10^{-9}	0.06	5.5×10^{-10}
Uterus	9.1×10^{-9}	—^{a}	—
Adrenals	9.4×10^{-9}	—^{a}	—
Remainder	9.4×10^{-9}	0.12	1.1×10^{-9}

^{a}The ICRP chooses to not assign a weight of 0.06 to both the uterus and the adrenals. Instead, the largest $H_{50,T}$ value is selected (adrenals), and this value is assigned a weighting factor of 0.12. The organ with this weight is defined to be the remainder.

252   SOLUTIONS FOR CHAPTER 1

$$\text{CEDE} = \sum_T w_T H_{50,T} = 8.7 \times 10^{-9} \text{ Sv/Bq} \quad \text{(by summing the last column)}$$

The ALI may be determined from calculation of the stochastic (S) and nonstochastic (NS) results. The ALI is taken to be the larger of the two values:

$$\text{ALI}_S \leq \frac{0.05 \text{ Sv}}{\sum_T w_T H_{50,T}} = \frac{0.05 \text{ Sv}}{8.7 \times 10^{-9} \text{ Sv/Bq}} = 5.7 \times 10^6 \text{ Bq}$$

$$\text{ALI}_{NS} \leq \frac{0.5 \text{ Sv}}{H_{50,T}} = \frac{0.5 \text{ Sv}}{9.4 \times 10^{-9} \text{ Sv/Bq}} = 5.3 \times 10^7 \text{ Bq}$$

where the largest CDE value (remainder) is used to provide the smallest nonstochastic ALI value.

Comparing these two values yields an ALI value of 5.7×10^6 Bq. This is the value that satisfies both the stochastic and nonstochastic inequalities.

The value of the DAC follows from its definition:

$$\text{DAC} = \text{ALI}/2400 \text{ m}^3$$
$$= 5.7 \times 10^6 \text{ Bq}/2400 \text{ m}^3 = 2.4 \times 10^3 \text{ Bq/m}^3$$

By convention, the ICRP rounds these values to one significant figure. Therefore,

$$\text{ALI} = 6.0 \times 10^6 \text{ Bq (inhalation)}$$
$$\text{DAC} = 2.0 \times 10^3 \text{ Bq/m}^3$$

Scenario 28

Question 1.106

The total transfer rate constant $k^{(j)}$ for insoluble material in a given segment of the gastrointestinal (GI) tract is given by

$$k^{(j)} = \lambda + 1/T^{(j)}$$

where $T^{(j)}$ is the mean residence time for the segment, and $j = 1, 2, 3,$ and 4 for the stomach (S), small intestine (SI), upper large intestine (ULI), and lower larger intestine (LLI), respectively.

Of the total quantity of material entering a given segment of the GI tract, the fraction decaying in that segment will be given by $1/k^{(j)}$. The fraction that will be transported to the next segment is given by

$$\frac{1/T^{(j)}}{k^{(j)}} = \frac{1}{k^{(j)} T^{(j)}}$$

SOLUTIONS FOR CHAPTER 1

Using this information, the activity entering and decaying in each segment based on an intake q into the stomach is given by:

Segment	Activity Entering	Activity Decaying
S	q	$\dfrac{\lambda}{k^{(1)}} q$
SI	$\dfrac{q}{k^{(1)}T^{(1)}}$	$\dfrac{\lambda}{k^{(2)}} \dfrac{1}{k^{(1)}T^{(1)}} q$
ULI	$\dfrac{q}{k^{(1)}T^{(1)}k^{(2)}T^{(2)}}$	$\dfrac{\lambda}{k^{(3)}} \dfrac{q}{k^{(1)}T^{(1)}k^{(2)}T^{(2)}}$
LLI	$\dfrac{q}{k^{(1)}T^{(1)}k^{(2)}T^{(2)}} \times \dfrac{1}{k^{(3)}T^{(3)}}$	$\dfrac{\lambda}{k^{(4)}} \dfrac{q}{k^{(1)}T^{(1)}k^{(2)}T^{(2)}} \times \dfrac{1}{k^{(3)}T^{(3)}}$

The cumulated activity $\tilde{A}(j)$ for each segment is obtained by multiplying the activity decaying in each segment by the radiological mean lifetime $(1/\lambda)$ of Tc-99m.

$$\tilde{A}(1) = \frac{q}{k^{(1)}}$$

$$\tilde{A}(2) = \frac{q}{k^{(2)}k^{(1)}T^{(1)}}$$

$$\tilde{A}(3) = \frac{q}{k^{(3)}k^{(1)}T^{(1)}k^{(2)}T^{(2)}}$$

$$\tilde{A}(4) = \frac{q}{k^{(4)}k^{(1)}T^{(1)}k^{(2)}T^{(2)}k^{(3)}T^{(3)}}$$

where

$$k^{(j)} = \lambda + \frac{1}{T^{(j)}}$$

$$\lambda = 0.693/6.03 \text{ hr} = 0.1149/\text{hr}$$

$$q = 1 \ \mu\text{Ci}$$

$$T^{(1)} = 1 \text{ hr}$$

$$T^{(2)} = 4 \text{ hr}$$

$$T^{(3)} = 13 \text{ hr}$$

$$T^{(4)} = 24 \text{ hr}$$

Using these values and the model's mean residence times leads to the cumulated activity in each GI tract segment:

$$\tilde{A}(1) = \frac{1\ \mu\text{Ci}}{(0.1149 \times 1/1)/\text{hr}} = 0.897\ \mu\text{Ci-hr} \tag{S}$$

$$\tilde{A}(2) = \frac{1\ \mu\text{Ci}}{(1\ \text{hr})[(0.1149 + 1/1)/\text{hr}][(0.1149 + 1/4)/\text{hr}]}$$
$$= 2.458\ \mu\text{Ci-hr} \tag{SI}$$

$$\tilde{A}(3) = \frac{1\ \mu\text{Ci}}{(1\ \text{hr})(4\ \text{hr})[(0.1149 \times 1/1)/\text{hr}][(0.1149 + 1/4)/\text{hr}]}$$
$$\times \frac{1}{(0.1149 + 1/13)/\text{hr}} \tag{ULI}$$
$$= 3.204\ \mu\text{Ci-hr}$$

$$\tilde{A}(4) = \frac{1\ \mu\text{Ci}}{(1\ \text{hr})(4\ \text{hr})(13\ \text{hr})[(0.1149 + 1/1)/\text{hr}][(0.1149 + 1/4)/\text{hr}]}$$
$$\times \frac{1}{[(0.1149 + 1/13)/\text{hr}][(0.1149 + 1/24)/\text{hr}]} \tag{LLI}$$
$$= 1.574\ \mu\text{Ci-hr}$$

Question 1.107

The absorbed doses for each segment may be obtained by utilizing the tabulated S values (rad/μCi-hr) and the cumulated activities calculated in question 1.106. The dose to the target organ is obtained by summing over the four source organs comprising the GI tract model's organs:

$$D(T) = \sum_S S(T \leftarrow S)\tilde{A}(S)$$

Explicitly writing these sums permits the dose equivalent to the walls of each segment to be determined:

$$D(1) = S(1 \leftarrow 1)\tilde{A}(1) + S(1 \leftarrow 2)\tilde{A}(2)$$
$$+ S(1 \leftarrow 3)\tilde{A}(3) + S(1 \leftarrow 4)\tilde{A}(4)$$
$$= (1.3 \times 10^{-4}\ \text{rad}/\mu\text{Ci-hr})(0.897\ \mu\text{Ci-hr})$$
$$+ (3.7 \times 10^{-6}\ \text{rad}/\mu\text{Ci-hr})(2.458\ \mu\text{Ci-hr})$$
$$+ (3.8 \times 10^{-6}\ \text{rad}/\mu\text{Ci-hr})(3.204\ \mu\text{Ci-hr})$$
$$+ (1.8 \times 10^{-6}\ \text{rad}/\mu\text{Ci-hr})(1.574\ \mu\text{Ci-hr})$$
$$= 1.407 \times 10^{-4}\ \text{rad} \times 1\ \text{rem/rad} = 1.407 \times 10^{-4}\ \text{rem}$$

The last line is based upon the quality factor of the gamma radiation emitted from Tc-99m.

$$D(2) = S(2 \leftarrow 1)\tilde{A}(1) + S(2 \leftarrow 2)\tilde{A}(2)$$
$$+ S(2 \leftarrow 3)\tilde{A}(3) + S(2 \leftarrow 4)\tilde{A}(4)$$
$$= (2.7 \times 10^{-6} \text{ rad}/\mu\text{Ci-hr})(0.897 \ \mu\text{Ci-hr})$$
$$+ (7.8 \times 10^{-5} \text{ rad}/\mu\text{Ci-hr})(2.458 \ \mu\text{Ci-hr})$$
$$+ (1.7 \times 10^{-5} \text{ rad}/\mu\text{Ci-hr})(3.204 \ \mu\text{Ci-hr})$$
$$+ (9.4 \times 10^{-6} \text{ rad}/\mu\text{Ci-hr})(1.574 \ \mu\text{Ci-hr})$$
$$= 2.634 \times 10^{-4} \text{ rad} \times 1 \text{ rem/rad} = 2.634 \times 10^{-4} \text{ rem}$$

$$D(3) = S(3 \leftarrow 1)\tilde{A}(1) + S(3 \leftarrow 2)\tilde{A}(2)$$
$$+ S(3 \leftarrow 3)\tilde{A}(3) + S(3 \leftarrow 4)\tilde{A}(4)$$
$$= (3.5 \times 10^{-6} \text{ rad}/\mu\text{Ci-hr})(0.897 \ \mu\text{Ci-hr})$$
$$+ (2.4 \times 10^{-5} \text{ rad}/\mu\text{Ci-hr})(2.458 \ \mu\text{Ci-hr})$$
$$+ (1.3 \times 10^{-4} \text{ rad}/\mu\text{Ci-hr})(3.204 \ \mu\text{Ci-hr})$$
$$+ (4.2 \times 10^{-6} \text{ rad}/\mu\text{Ci-hr})(1.574 \ \mu\text{Ci-hr})$$
$$= 4.853 \times 10^{-4} \text{ rad} \times 1 \text{ rem/rad} = 4.853 \times 10^{-4} \text{ rem}$$

$$D(4) = S(4 \leftarrow 1)\tilde{A}(1) + S(4 \leftarrow 2)\tilde{A}(2)$$
$$+ S(4 \leftarrow 3)\tilde{A}(3) + S(4 \leftarrow 4)\tilde{A}(4)$$
$$= (1.2 \times 10^{-6} \text{ rad}/\mu\text{Ci-hr})(0.897 \ \mu\text{Ci-hr})$$
$$+ (7.3 \times 10^{-6} \text{ rad}/\mu\text{Ci-hr})(2.458 \ \mu\text{Ci-hr})$$
$$+ (3.2 \times 10^{-6} \text{ rad}/\mu\text{Ci-hr})(3.204 \ \mu\text{Ci-hr})$$
$$+ (1.9 \times 10^{-4} \text{ rad}/\mu\text{Ci-hr})(1.574 \ \mu\text{Ci-hr})$$
$$= 3.283 \times 10^{-4} \text{ rad} \times 1 \text{ rem/rad} = 3.283 \times 10^{-4} \text{ rem}$$

The ULI receives the largest exposure from the postulated 1-μCi stomach intake.

Question 1.108

Considering that the facility administrative organ limit is 15 rem, the ULI will be limited to this value. For insoluble Tc-99m, this dose equivalent would result from an intake of Q microcuries:

$$Q = \frac{15 \text{ rem/year}}{4.853 \times 10^{-4} \text{ rem}/\mu\text{Ci}} = 30{,}909 \ \mu\text{Ci/year}$$

256   SOLUTIONS FOR CHAPTER 1

Because the residence time in the GI tract is relatively short, the intake time period is somewhat arbitrary. Because we are interested in an annual limit, the annual average intake rate for occupational exposure is

$$P = \frac{30{,}900 \ \mu\text{Ci/year}}{(40 \ \text{hr/week})(50 \ \text{weeks/year})} = 15.45 \ \mu\text{Ci/hr}$$

This value can be converted into the maximum allowed occupational concentration limit (MAOC). Because the intake rate (1100 ml/day) is known, we obtain

$$\text{MAOC} = \frac{15.45 \ \mu\text{Ci/hr}}{(1100 \ \text{ml/day})(1 \ \text{day/8 hr})} = 0.11 \ \mu\text{Ci/ml}$$

Scenario 29

Question 1.109

The ICRP-26 risk coefficient is 2×10^{-4} excess cancer deaths/rem.

$$\begin{aligned}
\text{Risk} &= (\text{Risk Coefficient})(\text{Dose}) \\
&= (2.0 \times 10^{-4} \ \text{excess cancer deaths/rem})(100{,}000 \ \text{mrem}) \\
&\quad \times (1 \ \text{rem}/1000 \ \text{mrem}) \\
&= 0.02
\end{aligned}$$

There is about a 1 in 50 probability of developing a radiation-induced fatal cancer over the worker's lifetime.

Question 1.110

Normal: $0.2 \times 1000 = 200$
Radiation: 3 in 1000 = 3

203 in 1000 is the total probability of developing a fatal cancer.

Question 1.111: a

The ICRP-26 cancer risk model is based on an absolute risk model.

Question 1.112: b

The PC tables are based on a relative risk model.

SOLUTIONS FOR CHAPTER 1   257

Question 1.113: b

Smoking history is considered when using the PC tables to estimate risk. Items a, c, d, and e are true.

Scenario 30

Question 1.114: c

Tritiated thymidine will cause a greater biological effect because it is incorporated into the cell's nucleus.

Question 1.115:

The correct matches are:

Strontium-90 (soluble)	Bone
Cesium-137	Total body
Plutonium-239 (soluble)	Bone
Uranium-238 (insoluble)	Lung/kidney
Radon-222	Lung

Question 1.116

The listed order (a, b, c, d, e) is from most to least radiosensitive.

Question 1.117: d

Occupational exposure to x-rays does not account for 1% of the cataracts in x-ray technicians.

Question 1.118: c

Statements a, b, d, and e are correct.

Scenario 31

Question 1.119: e

Most soluble radioactive material is removed from the body through excretion via urine.

Question 1.120: d

Prior to 1 January 1993, the requirements of 10CFR20 are based upon ICRP-2. Thereafter, ICRP-26/30 will form the basis of the revised 10CFR20.

258   SOLUTIONS FOR CHAPTER 1

Question 1.121: d

The ICRP-26 annual dose equivalent limit to prevent nonstochastic effects in tissue is 0.5 Sv (50 rem).

Question 1.122: a

The stochastic limit is based on uniform whole-body radiation. On the principle that the risk of a stochastic effect should be equal whether the whole body is uniformly irradiated or the radiation dose is distributed in a nonuniform manner, the ICRP recommended that stochastic effects for occupational exposures be limited to 50 mSv (5 rem) effective dose equivalent in 1 year.

Question 1.123: d

ICRP-26 replaced the critical organ with the concept of tissue region or target tissue.

Question 1.124: c

Large sets of occupational radiation exposure data are fit by a log-normal distribution.

Question 1.125: a

The annual dose limit for the eye was changed from 0.3 Sv (30 rem) to 0.15 Sv (15 rem).

Question 1.126: a

NCRP-91 recommended that cumulative occupational exposure be limited to N rem, where N is the worker's age in years.

Question 1.127: c

Note that when this question was asked on the 1989 ABHP Comprehensive Examination, it was based on the ICRP-26 risk coefficient of 2×10^{-4} excess cancer deaths/rem:

$$r = (2 \times 10^{-4} \text{ ecd/rem})(0.5 \text{ rem})$$
$$= 1 \times 10^{-4} \text{ ecd}$$

BEIR V and its 8×10^{-4} coefficient yields a risk of 4×10^{-4} ecd.

SOLUTIONS FOR CHAPTER 1   259

Question 1.128: b

Safe industries in the United States have an annual average fatal accident rate of 1×10^{-4}.

Scenario 32

Question 1.129: e

The skin response to acute radiation exposure in correct chronological order is erythema, dry desquamation, moist desquamation, and recovery.

Question 1.130: c

Factor	Affects Severity of the Reaction
Skin pigmentation	Yes
Fractionation of dose	Yes
Charged particle equilibrium at the basal cell layer	No
Dose rate	Yes
LET	Yes

Question 1.131: b

The basal cell layer is the most radiosensitive.

Question 1.132

$$T_{physical} = 8 \text{ days}, \quad T_{biological} = 5 \text{ days}$$

$$T_{eff} = (8 \text{ days})(5 \text{ days})/(8 \text{ days} + 5 \text{ days}) = 3.08 \text{ days}$$

$$A(t) = A(0) \exp(-0.693t/T_{eff})$$

$$A(t)/A(0) = 0.1$$

$$t = \ln[A(t)/A(0)](-T_{eff}/0.693)$$

$$t = \ln(0.1)(-3.08 \text{ days}/0.693) = 10.2 \text{ days}$$

Question 1.133: a

The weighting factor is used to relate the effective dose equivalent to skin dose. Effective dose equivalent = $0.01 \times$ skin dose. The radiogenic skin cancer risk is low.

Scenario 33

Question 1.134

The types of information from animal studies that can be useful in human risk estimation include:

1. Predicting the effects of high-LET radiations including neutrons, heavy ions, and high-energy particles
2. Estimating the effects at low dose rates
3. Gauging effects for dose fractionation sequences
4. Predicting the effects of radionuclide intakes
5. Predicting radiation-induced mutations
6. Estimating the relative effects of doses as a function of the radiation type
7. Gauging the effects of the presence of mitigating, protective, or synergistic agents

Question 1.135

The radiobiological factors which affect either the onset or the development of malignant tumors in experimental animals include:

Radiation type	Radiation dose
Radiation dose rate	Dose fractionation
Linear energy transfer	Sex
Age at exposure	General health
Diet	Genetic predisposition toward
Presence of other agents	tumor formation

Question 1.136: b

$$\begin{aligned} \text{ecd} &= \text{excess cancer deaths} \\ &= (\text{BEIR V risk coefficient})(\text{acute dose equivalent}) \\ &= (8 \times 10^{-4} \text{ ecd/rem})(0.1 \text{ Sv})(100 \text{ rem/Sv}) \\ &= 8 \times 10^{-3} \text{ ecd} \end{aligned}$$

Question 1.137: d

The additive risk model has been dropped in the BEIR V Committee Report in favor of the multiplicative risk model.

Question 1.138: a

The constant (1) does not ensure positive values of the excess risk estimate. All other items—b c, d, and e—are correct.

SOLUTIONS FOR CHAPTER 2

Scenario 34

Question 2.1

NCRP-37 recommends that patients in a hospital receive no more than 100 mR per stay from other patients. Because a patient could occupy an adjacent room for 168 hr each week, the average exposure to adjacent patients must not exceed 100 mR/168 hr = 0.6 mR/hr.

Question 2.2

Exposure Rate to Adjacent Patient:

$$D_0 = m_{eqRa} G_{Ra} (1 - P)/r^2$$

where

$$G_{Ra} = 8.25 \text{ mR-ft}^2/\text{mg-hr}$$

For this application, the dose to the adjacent patient, assuming no shielding in wall A, is

$$D_0 = (70 \text{ mg})(8.25 \text{ mR-ft}^2/\text{hr-mg})(1 - 0.3)(1/4 \text{ ft})^2$$
$$= 25.3 \text{ mR/hr}$$

SOLUTIONS FOR CHAPTER 2

Shielding Required to Meet NCRP-37 Recommendations: The number of half-value layers (N) required to reduce D_0 to the NCRP-37 value is

$$(1/2)^N = D_{NCRP\,37}/D_0$$

where

$$D_{NCRP\,37} = 0.6 \text{ mR/week} \quad \text{(NCRP-37 recommendation)}$$
$$(1/2)^N = (0.6 \text{ mR/hr}/25.3 \text{ mR/hr}) = 0.024$$
$$N \ln(1/2) = \ln(0.024) \text{ HVL}$$
$$N = 5.4 \text{ HVL}$$

The required thickness is just N times the half-value thickness:

$$t = N t_{HVL}$$

Because one HVL = 0.65 cm (for Cs-137 in lead), the thickness of lead required to reduce the dose rate to the NCRP-37 value is

$$t = 5.4 \text{ HVL} \times 0.65 \text{ cm/HVL} = 3.5 \text{ cm Pb}$$

This type of problem could also be solved from barrier transmission versus shield thickness tables or curves.

Question 2.3

The linear attenuation coefficient (u) for concrete is

$$u = (0.06 \text{ cm}^2/\text{g})(2.4 \text{ g/cm}^3) = 0.144 \text{ cm}^{-1}$$

The linear attenuation coefficient can be related to the HVL as follows:

$$I = I_0 \exp(-ux)$$

If $x = 1$ HVL, then $I/I_0 = 1/2$. Solving for the HVL leads to the relationship

$$HVL = 0.693/u$$

For the values given in this problem,

$$HVL = 0.693/0.144 \text{ cm}^{-1} = 4.81 \text{ cm concrete}$$

The dose rate for transmission through the concrete floor is solved using the exposure rate equation:

$$D_0 = m_{eq\,Ra} G_{Ra}(1 - P)/r^2$$

$$D_0 = (70 \text{ mg})(8.25 \text{ mR-ft}^2/\text{hr-mg})(1 - 0.3)/(7 \text{ ft})^2$$

$$= 8.25 \text{ mR/hr}$$

Using the same NCRP dose guideline for the nursery as for the adjacent patient and the half-value layer equation permits the ceiling concrete thickness to be determined:

$$(1/2)^N = (0.6 \text{ mR/hr})/(8.25 \text{ mR/hr}) = 0.073$$
$$N \ln(1/2) = \ln(0.073) \text{ HVL}$$
$$N = 3.78 \text{ HVL} \times 4.81 \text{ cm concrete/HVL} = 18.2 \text{ cm}$$
$$= 18.2 \text{ cm concrete } (1 \text{ in.}/2.54 \text{ cm})$$
$$= 7.2 \text{ in. concrete}$$

Question 2.4

Radiation protection differences between Cs-137 and I-125 brachytherapies include:

Iodine-125

1. Permanent implant.
2. Seeds inserted directly into the tissue.
3. Lower external exposure rate at the surface of the patient.
4. The window or thickness of the radiation instrument is critical to the measured radiation level.

Cesium-137

1. Temporary implant.
2. Afterloading devices are utilized to load the seeds.
3. Higher external exposure rate at the surface of the patient.
4. The dose rate can be detected by most gamma instrumentation.

Question 2.5

The dose equivalent (D) is given in terms of the relationship

$$D = BtSw_T = D_T w_T$$

where

w_T = thyroid weighting factor

264   SOLUTIONS FOR CHAPTER 2

The effective half-life is needed in order to calculate the dose equivalent to the thyroid from the uptake:

$$T_{\text{eff}} = \frac{(60 \text{ days})(138 \text{ days})}{(60 \text{ days} + 138 \text{ days})} = 41.8 \text{ days} \times 24 \text{ hr/day} = 1004 \text{ hr}$$

The dose to the thyroid is given by

$$D_T = BtS$$

where

B = organ burden = 300 nCi
t = mean lifetime = $T_{\text{eff}}/\ln(2)$
S = I-125 S-factor = 3×10^{-3} rad/(μCi-hr)
D_T = (1004 hr/0.693)(300 nCi × 1 μCi/1000 nCi)
       × (3×10^{-3} rad/μCi-hr) = 1.30 rad

The dose equivalent is obtained from the thyroid weighting factor:

$$D = 0.03 \times 1.30 \text{ rad} = 39.1 \text{ mrad}$$

Question 2.6

The sealed source is slowly leaking iodine, and the thyroid burden will continue to increase. This is not a statistical variation in the thyroid counting.

Scenario 35

Question 2.7

Because C-14 is long-lived, the contribution of the glucose and exhaled CO_2 can be handled individually. The requisite relationship for the dose to tissue T due to an activity deposition (A) is

$$D_T = (A/\lambda)S$$

where

$\lambda = 0.693/T_{1/2}$

Glucose: The half-life for glucose being metabolized into CO_2 is

$$T_{1/2} = 67 \text{ hr}$$

The dose to the liver from the glucose injection is written as

$$D_{liver} = [200\ \mu Ci/(0.693/67\ hr)](5.8 \times 10^{-5}\ rad/\mu Ci\text{-}hr)$$
$$= 1.12\ rad$$

The contribution of the glucose to the whole-body dose equivalent is

$$D_{whole\,body}(glucose) = (0.06)(1.12\ rad) = 0.067\ rad$$

CO_2: CO_2 is released from the whole body via the lungs with a half-life of

$$T_{1/2} = 1.2\ hr$$

The whole-body dose from the exhaled CO_2 is obtained by using the S value:

$$D_{whole\,body} = [200\ \mu Ci/(0.693/1.2\ hr)](1.5 \times 10^{-5}\ rad/\mu Ci\text{-}hr)$$
$$= 0.005\ rad$$

Total Committed Effective Dose:

$$D = \text{total committed effective dose (rad)}$$
$$= D_{whole\,body}(CO_2) + w_{liver}D_{liver}(glucose)$$
$$D = 0.005\ rad + 0.06 \times 1.12\ rad = 0.072\ rad$$

Question 2.8

$$\dot{V} = Vrk$$
$$k = \text{conversion factor}$$
$$= (24\ hr/day)(365\ day/year)(2.832 \times 10^4\ ml/ft^3)$$
$$V = \text{room volume}$$
$$\dot{V} = (18\ ft \times 20\ ft \times 8\ ft)(3/hr)(24\ hr/day)(365\ day/year)$$
$$\times (2.832 \times 10^4\ ml/ft^3)$$
$$= 2.14 \times 10^{12}\ ml/year$$
$$\dot{A} = ANf$$
$$f = \text{loss factor} = 0.2$$
$$\dot{A} = (200\ \mu Ci/subject)(20\ subjects/year)(0.2)$$
$$= 800\ \mu Ci/year$$

The average room concentration may be determined from the information derived above:

$$\overline{C} = \dot{A}/\dot{V}$$

$$\overline{C} = (800\ \mu\text{Ci/year})/(2.14 \times 10^{12}\ \text{ml/year}) = 3.73 \times 10^{-10}\ \mu\text{Ci/ml}$$

Because this is much less than the hospital's administrative limit, no action is warranted.

Question 2.9

Based upon the volume, one could reasonably conclude that the administrative limit (AL) will not present a problem.

$$V_a = VNk$$

V_a = air volume (ml) in the hospital

N = number of room volumes in the total volume = 50

$k = 2.832 \times 10^4\ \text{ml/ft}^3$

$$V_a = (18\ \text{ft} \times 20\ \text{ft} \times 8\ \text{ft})(50)(2.832 \times 10^4\ \text{ml/ft}^3)$$

$$= 4.08 \times 10^9\ \text{ml}$$

The average room concentration may be determined from the information derived above:

$$\overline{C} = \dot{A}/\dot{V}$$
$$= (800\ \mu\text{Ci})/(4.08 \times 10^9\ \text{ml})$$
$$= 1.96 \times 10^{-7}\ \mu\text{Ci/ml}$$

Any amount of ventilation would reduce the activity; and because it is 1/10 the AL, there is no problem. You could also conclude from the previous calculation that if the air in the room is much less than the AL, then any recirculated air would further reduce the concentration.

Question 2.10

A = quantity of $^{14}CO_2$ released into the room per patient

$$A = \int_0^t \frac{dN_{CO_2}}{dt} = \int_0^t \lambda_{CO_2} N_{CO_2}(t)\ dt$$

Let C replace CO_2 for simplicity and let G denote glucose:

$$\lambda_G = 0.693/67 \text{ hr} = 0.0103/\text{hr}$$

$$\lambda_C = 0.693/1.2 \text{ hr} = 0.5775/\text{hr}$$

The serial decay relationship is used to write

$$A = \int_0^t \frac{\lambda_C \lambda_G}{\lambda_C - \lambda_G} N_G(0) [\exp(-\lambda_G t) - \exp(-\lambda_C t)]$$

Performing the integration from $t = 0$ to $t = 6$ hr (the release period) yields

$$A = \frac{\lambda_C \lambda_G}{\lambda_C - \lambda_G} N_G(0)$$

$$\times \left[\frac{1 - \exp(-\lambda_G t)}{\lambda_G} - \frac{1 - \exp(-\lambda_C t)}{\lambda_C} \right]$$

$$A = \frac{(0.0103/\text{hr})(0.5775/\text{hr})}{0.5775/\text{hr} - 0.0103/\text{hr}} (200 \ \mu\text{Ci})$$

$$\times \left\{ \frac{[1 - \exp(-0.0103/\text{hr} \times 6 \text{ hr})]}{0.0103/\text{hr}} \right.$$

$$\left. - \frac{[1 - \exp(-0.5775/\text{hr} \times 6 \text{ hr})]}{0.5775/\text{hr}} \right\}$$

$$= 8.68 \ \mu\text{Ci}$$

Scenario 36

Question 2.11

$$W = E N_v N_p k$$

$E = 120$ mA-sec/view

$N_v = 4$ views/patient

$N_p = 40$ patients/week

k = conversion factor (1 min/60 sec)

$W = (120 \text{ mA-sec/view})(4 \text{ views/patient})(40 \text{ patients/week})$

$\times (1 \text{ min}/60 \text{ sec})$

$= 320$ mA-min/week

Question 2.12

The correct answer is d. NCRP-49 recommends 10 mR/week to be conservative.

Question 2.13

Orientation	Number of Views per Patient	Number of Images Directed Toward Control Booth
Cephalocaudal	2	0
Mediolateral	2	1
Total	4	1

Only the mediolateral images on one side would be directed at this wall.

$$U_{CB} = V_{CB}/N_v$$
$$U_{CB} = 1/4 = 0.25$$

The correct answer is b. Only 1 of 4 images is directed at the control booth.

Question 2.14

The correct answer is e. Based on NCRP 49, the occupancy of the control booth would be 1.00.

Question 2.15

$$\dot{X}_{CB}^{unshielded}(r) = O_{PB}(T_{CB})(U_{CB})(W)k(r_0/r)^2$$

O_{PB} = 1.0 mR/mA-sec @ 100 cm (primary beam output)

T_{CB} = 0.5

U_{CB} = 0.5

W = 160 mA-min/week

k = conversion factor (60 sec/min)

r = distance from x-ray source to point of interest (mediolateral position) = 200 cm

r_0 = location of measured primary beam output (cm)

$$\dot{X}_{CB}^{unshielded}(r) = (1.0 \text{ mR/mA-sec})(100 \text{ cm}/200 \text{ cm})^2(0.5)(0.5)$$
$$\times (160 \text{ mA-min/week})(60 \text{ sec/min})$$
$$= 600 \text{ mR/week}$$

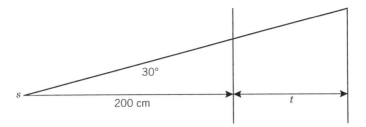

Fig. S2.1 Relative location of x-ray source (s) and the gypsum wall of thickness t.

The 600-mR/week value must be reduced to 5 mR/week. These values permit the calculation of the transmission factor (T):

$$T = \frac{\dot{X}(\text{NCRP-49})}{\dot{X}_{CB}^{\text{unshielded}}} = \frac{5 \text{ mR/week}}{600 \text{ mR/week}} = 8.33 \times 10^{-3}$$

The required shielding is obtained from the gypsum transmission curve which yields a thickness of 1.1 cm. The 30° angle reduces the required shielding thickness as illustrated in Fig. S2.1.

$$\cos 30° = t/1.1 \text{ cm}$$
$$t = 0.866 \times 1.1 \text{ cm} = 0.95 \text{ cm}$$

which is the required thickness.

Question 2.16

$$\dot{X}_{DO,\text{unshielded}}^{\text{scatt}}(r) = O_{PB}(W) k f_{\text{scatter}} (\text{OF}_{DO}) (r_0/r)^2$$

$\dot{X}_{DO,\text{unshielded}}^{\text{scatt}}(r) = $ exposure rate in the doctor's office without gypsum board addition (mR/week) due to the scattered radiation

$O_{PB} = 1.0$ mR/mA-sec @ 100 cm

$W = 160$ mA-min/week

$k = (60$ sec/min$)$

$r = $ distance from x-ray source to point of interest (doctor's office) $= 150$ cm

$r_0 = 100$ cm $= $ location of the primary beam measurement

$f_{\text{scatter}} = 0.0003$

$\text{OF}_{DO} = 0.6$

$$\dot{X}^{\text{scatt}}_{\text{DO, unshielded}} = (1.0 \text{ mR/mA-sec})(100 \text{ cm}/150 \text{ cm})^2$$
$$\times (0.0003)(160 \text{ mA-min/week})(60 \text{ sec/min})$$
$$\times (0.6)$$
$$= 0.77 \text{ mR/week}$$

Because this is less than the 5-mR/week criterion, no shielding is needed.

Question 2.17

$$\dot{X}^{\text{unshielded}}_{\text{DO, leakage}}(r) = (\dot{X}_L(r_0)/i)(W)k(T_{\text{DO}})(r_0/r)^2$$

$\dot{X}_L(r_0) =$ leakage dose rate $= 100$ mR/hr at 100 cm (r_0)

$r_0 =$ location of leakage measurement

$i = 7$ mA

$W =$ workload $= 640$ mA-min/week

$k =$ conversion factor (1 hr/60 min)

$r =$ distance from x-ray source to point of interest (doctor's office) is 200 cm

$T_{\text{DO}} =$ occupancy factor for the doctor's office $= 0.5$

$$\dot{X}^{\text{unshielded}}_{\text{DO, leakage}} = (100 \text{ mR}/7 \text{ mA-hr})(100 \text{ cm}/200 \text{ cm})^2$$
$$\times (640 \text{ mA-min/week})(1 \text{ hr}/60 \text{ min})$$
$$\times (0.5)$$
$$= 19.0 \text{ mR/week}$$

Because this is more than the 5-mR/week design criterion, gypsum board shielding is needed. The required thickness can be calculated from the following relationship:

$$\dot{X}^{\text{shielded}}_{\text{DO, leakage}} = \dot{X}^{\text{unshielded}}_{\text{DO, leakage}} \exp(-0.693t/\text{HVL})$$

$\dot{X}^{\text{shielded}}_{\text{DO, leakage}} =$ design exposure rate (mR/week) $= 5$ mR/week

HVL $=$ half-value layer $= 0.5$ cm for 30 kVp

$t =$ shield thickness required to reduce the unshielded exposure rate to the criterion (5 mR/week)

$$5 \text{ mR/week} = (19.0 \text{ mR/week}) \exp(-0.693t/0.5 \text{ cm})$$

Solving for t yields

$$\ln(5/19.0) = -0.693t/0.5 \text{ cm}$$
$$t = -(0.5 \text{ cm}/0.693)\ln(5/19.0)$$
$$t = 0.96 \text{ cm of gypsum board}$$

Scenario 37

Question 2.18

General Formulation:

$$a = \text{Mo-99}, \quad T_{1/2} = 66 \text{ hr}$$
$$b = \text{Tc-99m}, \quad T_{1/2} = 6 \text{ hr}$$

$$a \xrightarrow{\lambda_a} b \xrightarrow{\lambda_b} c$$

$$N_b = \frac{\lambda_a N_{a0}}{\lambda_b - \lambda_a}[\exp(-\lambda_a t) - \exp(-\lambda_b t)]$$

$$A_b = \frac{\lambda_b A_{a0}}{\lambda_b - \lambda_a}[\exp(-\lambda_a t) - \exp(-\lambda_b t)]$$

where

N_i = number of atoms of isotope i
A_i = activity of isotope i

Using the relationship, $\lambda_i = \ln(2)/T^i_{1/2}$, the activity equation can be written in terms of the half-lives:

$$A(\text{Tc-99m}) = A(\text{Mo-99})[T(\text{Mo-99})/(T(\text{Mo-99}) - T(\text{Tc-99m}))]$$
$$\times [\exp(-0.693t/T(\text{Mo-99})) - \exp(-0.693t/T(\text{Tc-99m}))]$$
$$= 1000 \text{ mCi } [66/(66 - 6)][\exp(-0.693 \times 24/66)$$
$$- \exp(-0.693 \times 24/6)]$$
$$= 1000(1.1)(0.7772 - 0.0625) \text{ mCi}$$
$$= 786 \text{ mCi}$$

Question 2.19

The specific activity $C(t = 8 \text{ hr})$ at 1600 hours (4:00 p.m.) is

$$C(t_2) = C(t_1)[\exp -0.693t/T(\text{Tc-99m})]$$
$$C(1600 \text{ hours}) = 80 \text{ mCi/cm}^3 [\exp(-0.693 \times 8/6)]$$
$$= 31.75 \text{ mCi/cm}^3$$

V_a = volume of Tc-99m to be added (cm³)

V_l = volume of sulfur colloid liquid with the kit = 5 cm³
(*Note*: This is not radioactive material.)

C = concentration at 1600 (mCi/cm³)

C_d = desired concentration = 10 mCi/cm³

The volume of added material is obtained from an activity balance prior to and following the addition of the sulfur colloid liquid:

$$A = V_a C = (V_a + V_l) C_d$$

which may be solved for the added volume:

$$V_a C = V_a C_d + V_l C_d$$
$$V_a (C - C_d) = V_l C_d$$
$$V_a = \frac{V_l C_d}{C - C_d}$$
$$= \frac{(5 \text{ cm}^3)(10 \text{ mCi/cm}^3)}{31.75 \text{ mCi/cm}^3 - 10 \text{ mCi/cm}^3}$$
$$= 2.3 \text{ cm}^3$$

Question 2.20

$$\dot{X} = AG \exp(-ux)/r^2$$

$\dot{X}$ = exposure rate (mR/hr)

G = gamma constant = 0.56 R-cm²/mCi-hr

u = attenuation coefficient = 3.25/cm

A = 50 mCi

r = 50 cm

x = 0.5 cm

Buildup is ignored; that is, assume $B = 1$:

$$\dot{X} = \frac{(50 \text{ mCi})(0.56 \text{ R-cm}^2/\text{mCi-hr})}{(50 \text{ cm})^2}$$
$$\times \exp(-3.25/\text{cm} \times 0.5 \text{ cm})(1000 \text{ mR/R})$$
$$= 2.2 \text{ mR/hr}$$

Question 2.21

The remaining activity of Mo-99 is determined from

$$A(t) = A_0 \exp[-0.693 \times t/T_{1/2}(\text{Mo-99})]$$
$$= (1000 \text{ mCi}) \exp\{-[(0.693/66 \text{ hr})(90 \text{ days} \times 24 \text{ hr/day})]\}$$
$$= 1.41 \times 10^{-7} \text{ mCi} \times 1 \text{ Ci}/10^3 \text{ mCi} \times 10^{12} \text{ pCi/Ci} = 141 \text{ pCi}$$

This activity may also be expressed in dpm:

$$A(t) = 1.41 \times 10^{-7} \text{ mCi} \times 1 \text{ Ci}/10^3 \text{ mCi} \times 3.7 \times 10^{10} \text{ dis/sec-Ci}$$
$$\times 60 \text{ sec/min}$$
$$= 314 \text{ dpm}$$

Question 2.22

Because 3700 dpm is much larger than the remaining activity from Mo-99 (314 dpm), there must be another source. This could be from long-lived radiochemical impurities from the original Mo-99 production. Tc-99 is probably not the source of the additional counts unless the generator column is fractured.

Assuming that Ru-106 ($T_{1/2}$ = 372 days) has been seen in old generator columns at a fraction of microcurie levels, it is a likely candidate for explaining the 3700-dpm value. One should also evaluate that Tc-99 is the source of the contamination.

Tc-99 ($T_{1/2}$ = 213,000 years) is also present in the column. Its weak beta (293 keV) should not penetrate the glass wall of the column. It is also unlikely that there would be enough Tc-99 contamination on the outside of the column to account for the 3700 dpm, or enough bremsstrahlung from the inside to explain the 3700-dpm value.

Scenario 38

Question 2.23

The beamline exposure rate ($\dot{X}$) is given by

$$\dot{X} = k \frac{SE}{AW} (u_{en}/p)_{air}$$

where

S = photon source strength = 4.0×10^{11} photons/sec
E = photon energy = 30 keV/photon
A = beam area = 0.05 cm × 12.3 cm
W = 34 eV/ion pair

u_{en}/p = energy absorption coefficient = 0.15 cm²/g-air
k = conversion factor

$$\dot{X} = \frac{(4.0 \times 10^{11} \text{ photons/sec})(30 \text{ keV/photon})}{(0.05 \text{ cm} \times 12.3 \text{ cm})(34.0 \text{ eV/ip})}$$
$\times (0.15 \text{ cm}^2/\text{g-air})(1000 \text{ eV/keV})(1.6 \times 10^{-19} \text{ coulomb/ip})$
$\times (1000 \text{ g/kg})(1 \text{ R-kg-air}/2.58 \times 10^{-4} \text{ coulomb})$
= 53.4 R/sec

Question 2.24

For simplicity assume no scattered radiation enters the beamline. This assumption permits the attenuation relationship to be utilized:

$$\dot{X}(t) = \dot{X}_0 \exp(-ut)$$

where

$\dot{X}(t) = 2.0$ mR/hr (by design)
$\dot{X}_0 = 53.4$ R/sec (exposure rate in beamline)
$u = (u/p)(p)$
  $= (1.12 \text{ cm}^2/\text{g})(2.7 \text{ g/cm}^3) = 3.02/\text{cm}$

Solving for the desired thickness (t) yields

$$t = -(1/u) \ln [\dot{X}(t)/\dot{X}_0]$$
$= -(1/3.02/\text{cm}) \ln [(2.0 \text{ mR/hr} \times 1 \text{ hr}/3600 \text{ sec})/$
  $53.4 \text{ R/sec} \times 1000 \text{ mR/R}]$
= 6.08 cm

Question 2.25

For simplicity assume that the exposure rate is constant across the area of the beam.

The time required for the beam to move past any point is

$$t = \frac{\text{beam width}}{\text{traveling rate}} = \frac{0.5 \text{ mm}}{60 \text{ mm/sec}} = 8.33 \times 10^{-3} \text{ sec}$$

The exposure is just the exposure rate times the time:

$$X = \dot{X}t$$
$= (53.4 \text{ R/sec})(8.33 \times 10^{-3} \text{ sec}) = 0.445 \text{ R}$

Question 2.26

The transmission ionization chamber will not provide a good estimate of the surface dose. This ion chamber would measure the exposure rate over the area of the beam. Because the patient is moving relative to the beam, the exposure is delivered over a larger area than the beam area. Therefore, the average exposure rate delivered over the entire area irradiated is less than the beamline exposure rate. The two rates are the same only if the patient is stationary.

Question 2.27

The beamline exposure rate and the vertical velocity of the patient must be monitored to ensure that the patient exposure is maintained below the 1.0-R criteria.

Scenario 39

Question 2.28

$$X = \text{exposure per view at 40 in.}$$
$$S = 4 \text{ mR/mA-sec for 70 kVp at 40 in.}$$
$$I = \text{time/current} = 25 \text{ mA-sec/view}$$
$$X = SI$$
$$= (4 \text{ mR/mA-sec})(25 \text{ mA-sec/view})$$
$$= 100 \text{ mR/view}$$

Assuming that the thickness of the child is about 4 in., the exposure at the surface of the child (X_0) is

$$X_0 = X(d/d_0)$$

where

$d = 40$ in. $=$ source image distance
$d_0 =$ distance to surface of child $= 40$ in. $- 4$ in. $= 36$ in.

$$X_0 = 100 \text{ mR/view} \frac{(40 \text{ in.})^2}{(40 \text{ in.} - 4 \text{ in.})^2} = 123 \text{ mR/view}$$

In order to estimate the abdomen dose, the distance between the primary beam and the woman's abdomen (d_1) must be determined. From the problem statement, this distance is 18 in. The abdomen skin dose (X_1) may be written in terms of these parameters:

$$X_1 = X_0 f_{scatter} N (d_0/d_1)^2$$

where

$f_{scatter}$ = scattered radiation intensity from the primary beam at 1 m normalized to the primary beam intensity
    = 0.001
N = number of views = 5 views
X_1 = (123 mR/view)(0.001)(5 views)(36 in./18 in.)^2
    = 2.46 mR

Question 2.29

In order to estimate the fetal exposure, recall that the woman was wearing a lead apron. Assuming the lead apron is 0.5 mm equivalent lead, it would lead to a transmission factor (TF) of 10%. A depth dose factor (DDF) of about 30% would also be a reasonable assignment.

The fetal exposure (X_F) can be written in terms of these factors

$$X_F = X_1 (DDF)(TF)$$
$$= (2.46 \text{ mR})(0.3)(0.1)$$
$$= 0.074 \text{ mR}$$

Question 2.30

NRC and NCRP guidance is to limit fetal exposure to 500 mR or less during its gestation.

Question 2.31

The exposure calculations, assumptions, and relevant regulatory guidance should be presented to the attending physician. Dose effects from fetal exposures should also be discussed with the physician. The final responsibility to provide consultation with the patient resides with the physician.

Question 2.32

The practice of excluding the mother appears to be overly conservative when the risk to the fetus is compared to the benefit to the child being x-rayed. The use of mechanical restraints is an option that should also be considered.

Scenario 40

Question 2.33

I-131 poses both an internal and external radiation hazard due to its associated beta and gamma radiation. From an internal viewpoint, each microcurie taken up by the thyroid leads to a dose commitment of about 6.6 rem. The gamma constant for I-131 is 2.2 R hr^{-1} mCi^{-1} cm^2, which indicates that gamma exposure must be evaluated. The short radiological half-life of 8 days suggests that the hazards will be relatively short-lived, but still long enough that good radiological controls practices are warranted.

Question 2.34

The exposure rate a distance r from a point source of activity A is

$$\dot{X}(r) = AG/r^2$$

$$\dot{X}(\text{contact}) = (200 \text{ mCi})\left(2.2 \; \frac{\text{R-cm}^2}{\text{hr-mCi}}\right) \Big/ (1.0 \text{ cm})^2$$

$$= 440 \text{ R/hr}$$

The contact exposure rate suggests that source shielding and good exposure control practices are warranted.

$$\dot{X}(1.0 \text{ m}) = (200 \text{ mCi})\left(2.2 \; \frac{\text{R-cm}^2}{\text{hr-mCi}}\right) \Big/ (100 \text{ cm})^2$$

$$= 0.044 \text{ R/hr} \times 1000 \text{ mR/R}$$

$$= 44 \text{ mR/hr}$$

Exposure control for hospital personnel and other patients should be considered following this administration because fields around the patient will be on the order of 10–100 mR/hr.

Question 2.35

The following health physics practices are recommended:

1. *Minimize Airborne Radioactivity.* Administration of the I-131 should be in a basic pH solution to minimize the evolution of airborne I-131. Good room ventilation is recommended to minimize the airborne concentration of any I-131 that volatilizes.

2. *Minimize Radiation Exposures.* It is recommended that the 200-mCi solution be contained in a shielded vial to minimize radiation exposures to the

technician administering the I-131. The use of bedside shields should also be considered.

Family members and visitors should be instructed to maintain their distance from the patient (10 feet) during their visit to minimize the radiation exposure. Similar guidance will be provided to the hospital staff.

Following the oral administration of I-131 to the patient, fluids are to be administered to rinse the mouth and esophagus. This action minimizes exposure to these organs. Intravenous injection would eliminate the exposures to the mouth and esophagus.

3. *Implement Contamination Control Practices.* The technician wears protective clothing and gloves during the administration process and during the subsequent cleanup. The room setup prior to the administration should minimize the contamination of the patient's clothing, bedding, and room, and any patient excretion including vomiting should be easily controlled. The collection of bedding should also include monitoring for contamination resulting from patient sweating.

4. *Posting of Radiological Areas.* The room must be properly posted and controlled to ensure that exposures are minimized and all applicable standards are met.

5. *Periodic Surveys.* Radiation and contamination surveys should be performed to properly assess the radiological conditions in the room. This information will also impact access to the patient.

6. *Post-administration Cleanup and Waste Disposal.* Post-administration follow-up should properly control any waste material. Good contamination control practices and dose control methods are to be utilized.

All urine will be collected and stored in waste containers in a waste storage area to permit decay prior to disposal in the sanitary waste. This procedure also minimizes the exposure to downstream populations after the waste water is recycled for public use. *Note*: There is no requirement to store urine, and it may be directly discharged.

# SOLUTIONS FOR CHAPTER 3

Scenario 41

Question 3.1

$$e(E) = N(E)/S \times t$$

As an example of the efficiency calculation, consider the 60- and 1408-keV data:

Gamma Energy (keV)	Calibration Source (gammas/sec) Decay Corrected	Ge(Li) Detector Net Counts (1000-second count)
60	2000	83,762
1408	1170	8215

$$e(60 \text{ keV}) = \frac{83{,}762 \text{ counts}/1000 \text{ sec}}{2000 \text{ gammas/sec}}$$

$$= 0.042 \text{ counts/gamma}$$

$$e(1408 \text{ keV}) = \frac{8215 \text{ counts}/1000 \text{ sec}}{1170 \text{ gammas/sec}}$$

$$= 0.007 \text{ counts/gamma}$$

Following a similar procedure leads to the efficiency results for the other gamma energies considered in this problem:

E(keV):	60	88	121	344	768	963	1408
$e(E)$	0.042	0.052	0.050	0.021	0.012	0.009	0.007

Question 3.2

The values of $e(E)$ versus E provided in question 3.1 will lead to the desired curve.

Question 3.3

From the curve or by interpolating the tabular results, we obtain the value $e(2000 \text{ keV}) = 0.004$.

The 2000-keV efficiency can be obtained by assuming that the tail of the $e(E)$ versus E curve is linear:

$$e(E) = mE + b$$

$$963 \text{ keV}: \quad 0.009 = 963m + b \quad [1]$$

$$1408 \text{ keV}: \quad 0.007 = 1408m + b \quad [2]$$

$$[1] - [2]: \quad 0.002 = (963 - 1408)m$$

$$m = -4.494 \times 10^{-6}$$

Inserting this value for m leads to b:

$$[1]: \quad 0.009 = 963(-4.494 \times 10^{-6}) + b$$

$$b = 0.0133$$

Therefore,

$$e(E) = -4.494 \times 10^{-6} E + 0.0133$$

Using this equation permits the efficiency at 2000 keV to be determined:

$$E = 2000 \text{ keV}: e(2000 \text{ keV}) = (-4.494 \times 10^{-6})(2000) + 0.0133$$

$$= 0.004$$

Question 3.4: a

Between 100 and 300 keV, the counting efficiency decreases because photoelectric absorption decreases.

SOLUTIONS FOR CHAPTER 3   281

Question 3.5: a

At low gamma energies (<80 keV), counting efficiency decreases because detector and housing attenuation become significant.

Question 3.6

The 400.7-keV peak only contains a contribution from Se-75, and the activity of Se-75 can be obtained from its analysis. Once the Se-75 activity is obtained, the Co-57 activity can be derived from the 121.5-keV peak. Finally, the Hg-203 activity is obtained from the 279.2-keV peak once the Se-75 activity is known.

Peak 1: 400.7 keV

$$A_i = C(E)k/(t)\,[e(E)]\,[Y(E)]$$

$$e(400 \text{ keV}) = 0.018 \text{ counts/gamma}$$

A(Se-75, 400.7-keV peak)

$$= \frac{(629 \text{ counts})(2.22 \times 10^6 \text{ dpm}/\mu\text{Ci})^{-1}}{(10 \text{ min})(0.018 \text{ counts/gamma})(0.116 \text{ gammas/dis})}$$

$$= 1.36 \times 10^{-2} \, \mu\text{Ci}$$

Peak 2: 122.1 keV

Given the activity of Se-75 determined from the 400.7-keV peak, the Se-75 contribution to the 121.1-keV peak can be determined. The Co-57 activity can be determined from the 122.1-keV sample count results once the Se-75 contribution is subtracted.

The Se-75 count rate contributing to the 121.1-keV peak is obtained from the previous relationship:

$$A_i = C(E)k/(t)\,[e(E)]\,[Y(E)]$$

Solving for $C(E)$ provides the count rate contribution:

$$C(E) = A_i t e(E) Y(E)/k$$

C(Se-75, 121.1 keV) $= (1.36 \times 10^{-2} \, \mu\text{Ci})(10 \text{ min})(0.05 \text{ counts/gamma})$

$\qquad \times (0.173 \text{ gammas/dis})(2.22 \times 10^6 \text{ dpm}/\mu\text{Ci})$

$\qquad = 2611.6 \text{ counts}$

A(Co-57, 122.1 keV)

$$= \frac{(7266 \text{ counts} - 2612 \text{ counts})(2.22 \times 10^6 \text{ dpm}/\mu\text{Ci})^{-1}}{(10 \text{ min})(0.05 \text{ counts/gamma})(0.859 \text{ gammas/dis})}$$

$$= 4.88 \times 10^{-3} \, \mu\text{Ci}$$

Peak 3: 279.2 keV

A similar approach is utilized to obtain the Hg-203 activity:

$$e(279.5 \text{ keV}) = 0.025 \text{ counts/gamma}$$

$$\begin{aligned}C(\text{Se-75, 279.5 keV}) = &\ (1.36 \times 10^{-2}\ \mu\text{Ci})(10 \text{ min})(0.025 \text{ counts/gamma}) \\ &\times (0.252 \text{ gammas/dis})(2.22 \times 10^6 \text{ dpm}/\mu\text{Ci}) \\ = &\ 1902.1 \text{ counts}\end{aligned}$$

$$A(\text{Hg-203, 279.2 keV})$$

$$= \frac{(2279 \text{ counts} - 1902 \text{ counts})(2.22 \times 10^6 \text{ dpm}/\mu\text{Ci})^{-1}}{(10 \text{ min})(0.025 \text{ counts/gamma})(0.815 \text{ gammas/dis})}$$

$$= 8.33 \times 10^{-4}\ \mu\text{Ci}$$

Scenario 42

Question 3.7

- P-32   is not volatile, even when heated, and can be ignored as an airborne contaminant.
- HTO   is an airborne hazard.
- I-125 is an airborne hazard.

Question 3.8

Air sampling would be performed as follows:

- HTO   Bubble air through distilled water
         Pass air through desiccant (silica gel)
         Cold trap/finger
         Ion chamber
- I-125 Charcoal filters with a vacuum pump impinger

Question 3.9

Because neither HTO nor sodium iodide resides in dust, anisokinetic sampling can be used.

Question 3.10

An intake (I_i) will be possible for I-125 and H-3. Based upon its properties, P-32 is not volatilized. Because both students are present for the hour, they will each receive uptakes of I-125 and H-3.

$$I_i = (BR)tA_i/V$$
$$V = (20 \text{ ft} \times 20 \text{ ft} \times 10 \text{ ft})(28.3 \text{ liters/ft}^3)$$
$$= 113{,}200 \text{ liters}$$

I-125. For I-125, the intake relationship is sufficiently accurate to estimate the activity in the students:

$$I(\text{I-125}) = (20 \text{ liters/min})(60 \text{ min})(5000 \ \mu\text{Ci})/(113{,}200 \text{ liters})$$
$$= 53.0 \ \mu\text{Ci I-125/student}$$

HTO. Tritium oxide can enter the body via both inhalation and skin absorption in approximately equal amounts. Therefore, skin absorption must be included in the HTO intake estimate. A factor of 2 will be used to account for the skin absorption.

$$I(\text{HTO}) = 2(20 \text{ liters/min})(60 \text{ min})(10{,}000 \ \mu\text{Ci})/(113{,}200 \text{ liters})$$
$$= 212.0 \ \mu\text{Ci HTO/student}$$

Question 3.11

The critical organs for HTO and I-125 are

HTO   Whole body
I-125 Thyroid

The amount of HTO in the whole body is given in question 3.10.

The amount of I-125 in the thyroid will be 20–30% of the quantity calculated previously in question 3.10.

Question 3.12

H-3 and I-125 can be counted simultaneously by using liquid scintillation counting (LSC). Samples could be counted in a sodium iodide well counter or other gamma counter to assess only the I-125. To obtain the tritium activity subtract the absolute activity, obtained from the gamma counter, from the gross absolute activity obtained from the LSC.

P-32 could be ignored because it is not volatile. Its presence or absence could be confirmed by counting in an LSC with a wide window.

Question 3.13

Borotritide chemical reactions lead to the evolution of tritium gas. Therefore, tritiated water collection techniques, which are not efficient collection mechanisms for tritium gas, would not be appropriate. The tritium gas could be oxidized via dry combustion using a palladium catalyst. Alternatively, ion chamber techniques could be employed to detect the tritium gas.

Question 3.14

The intakes noted above were caused by the laboratory exhaust fan motor being turned off for maintenance without the knowledge of the laboratory workers. The following measures would prevent recurrence of this accident:

1. Develop a procedure for maintenance personnel or outside contractors who repair exhaust/ventilation equipment. The procedure should include posting the hood to warn the researcher that the equipment is inoperable.
2. Install a warning light or horn at the hood to notify users when the hood fan is off or not functioning correctly.
3. Improve worker training and include operational checks of hood ventilation or a requirement for workers to contact maintenance prior to operating the hoods.
4. Institute a red tag system for activities that interfere with hood operability.

Question 3.15

Factors that would affect the intake of radioactive material by the workers utilizing the nonoperable hood include:

1. Room air changes or ventilation system characteristics
2. Uniformity of radionuclide distribution in the air
3. Location of the workers
4. Plateout of radioactive material on room or hood structures
5. Metabolism of the workers
6. Actual quantity of material volatilized versus the 100% assumption

Scenario 43

Question 3.16: e

Use a 5-in. × 0.06-in. NaI crystal probe to survey all areas suspected of being contaminated. The NaI detector of this configuration will have a good response to the low-energy x-rays. It is also more sensitive than the 1.5-in. × 1-in. NaI probe.

Poorer Answers

a. Take wipes in all suspected areas and count them on a shielded gas-flow proportional counter. You would have to smear every square inch and

document the smear location. This would take considerable time, and the wipes would not necessarily remove contamination from all surfaces such as carpeting.
b. Use a pancake GM probe to survey all suspected areas of contamination. This probe will respond, but its efficiency is not as good as the NaI probe.
c. Use a ZnS-coated/photomultiplier-tube-based portable alpha probe to survey all suspected areas. Because the ZnS probe detects only alphas, it will miss significant levels of contamination due to self-shielding by carpet fibers.
d. Use a 1.5-in. × 1.0-in. NaI crystal probe to survey all suspected contaminated areas. The 1.5-in. × 1-in. NaI crystal geometry is not as sensitive as the 5-in. × 0.06-in. NaI probe.

Question 3.17: d

Use a 1.5-in. × 1.0-in. NaI crystal probe to survey all areas suspected of being contaminated. This size and thickness of NaI detector will have the best sensitivity to the 662-keV Cs-137 gammas.

Question 3.18: e

Analyze a 24-hr fecal sample collected on day 2 with a germanium detector/multichannel analyzer. Clearance of insoluble Am-241 particles from the lung will route through the gastrointestinal tract to feces pathway.

Poorer Answers
a. Count nasal swipes on a shielded gas-flow proportional counter to estimate the inhaled activity. Although swipe counting will detect the Am-241, there is no established method to allow a quantitative determination of lung dose based on nasal wipe data.
b. Whole-body counting in a shielded facility via a 3 × 3 NaI detector/MCA. Whole-body counting for low-energy x-rays would have a very low sensitivity due to their attenuation by the body.
c. Whole-body counting in a shielded facility via a coaxial germanium detector/MCA. Whole-body counting by GeLi for low-energy x-rays has a greater sensitivity than a NaI detector, but it will be less sensitive than fecal counting.
d. Analysis of a 24-hr urine sample collected on day 2 by liquid scintillation counting. A negligible amount of insoluble Am-241 is expected to appear in the urine.

Question 3.19: b

Whole-body counting in a shielded facility via a NaI detector/MCA. This system would easily detect the 662-keV Cs-137 gammas and quantify the Cs-137 uptake.

Poorer Answers

a. Count nasal swipes on a shielded gas-flow proportional counter to estimate the inhaled activity. Although swipe counting will detect the Cs-137, there are too many uncertainties in using this technique to quantify the uptake.

c. Analysis of activity exhaled in the breath. Insignificant activity and calibration uncertainties complicate the use of this approach.

d. Analysis of a 24-hr urine sample collected on day 2 by liquid scintillation counting. The activity going into the urine would vary greatly, and the retention function would be uncertain within the first 2 days.

e. Analyze a 24-hr fecal sample collected on day 2 with a germanium detector/multichannel analyzer. This technique would allow an estimate, but there is more uncertainty than with whole-body counting.

Question 3.20

$\dot{D}_0 = 2.13 PE(QF)/m$

P = activity deposited into the lung (μCi)

E = 5.57 MeV (energy of alphas and recoil atoms deposited into the lung)

m = lung mass (2) = 1000 g

QF = quality factor for alphas = 20 rem/rad

$\dot{D}_0 = (2.13)[(\text{rad/hr})/(\mu\text{Ci-MeV/g})](1000\ \mu\text{Ci})(5.57\ \text{MeV})$

$\quad \times (20\ \text{rem/rad})/(1000\ \text{g})$

$\quad = 237\ \text{rem/hr}$

Question 3.21: c

Over a 50-year period, the doses for both cases would be essentially the same. For the 50-year period, details of the intake over the first 100 days are insignificant when compared to the lung retention time.

Question 3.22: b

ICRP-26 specifies a lung weighting factor of 0.12.

Question 3.23: d

Am-241's most common production mode is from the beta decay of Pu-241.

Scenario 44

Question 3.24

Dose rate information, a knowledge of the fetal–source distance, and the geometry of the x-ray system can be used to calculate patient exposures including the dose to a fetus.

In order to use the tabular data, the beam quality HVL must be determined. From the available survey data, the measured dose can be plotted, on semilog graph paper, as a function of the filtration (mm Al) to determine a HVL value of 3.0 mm of Al. Interpolation can also be used to obtain the HVL thickness:

4.5-mm Al filtration corresponds to 180 mR
2.5-mm Al filtration corresponds to 280 mR
0.0-mm Al filtration corresponds to 500 mR
HVL of Al filtration corresponds to 250 mR

$$\frac{\log(4.5) - \log(2.5)}{180 \text{ mR} - 280 \text{ mR}} = \frac{\log(4.5) - \log(\text{HVL})}{180 \text{ mR} - 250 \text{ mR}}$$

$$\log(\text{HVL}) = -\frac{180 \text{ mR} - 250 \text{ mR}}{180 \text{ mR} - 280 \text{ mR}} [\log(4.5) - \log(2.5)] + \log(4.5)$$

$$= 0.6532 - 0.7(0.2553)$$

$$= 0.4745$$

$$\text{HVL} = 2.98 \text{ mm Al}$$

Distances of interest are illustrated in Fig. S3.1.

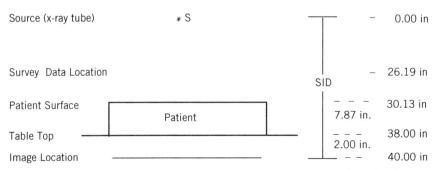

Fig. S3.1 Location of the patient, x-ray source (*S*) and image plane for the unanticipated exposure event.

SOLUTIONS FOR CHAPTER 3

The following dimensions are relevant to the fetal exposure assessment:

X-ray source location = 0.00 in.

The patient thickness is 20.0 cm = 7.87 in.

The location of the survey data is 30 cm above the table top, which is the following distance from the source: (40.00 in. − 2.00 in. − 30.00 cm/ 2.54 cm/in. = 26.19 in.

The location of the top of the patient relative to the source: (40.00 in. − 2.00 in. − 7.87 in.) = 30.13 in.

The information given in the problem statement indicates that the patient's skin is 30.13 in. from the tube. This is illustrated above.

In order to obtain the patient's entrance skin exposure, assume:

1. Linearity with respect to mA-sec.
2. The inverse square law applies.
3. The entrance skin dose can be obtained from the given tabular data.

The embryo/fetus dose can be determined from the tabular data for the entrance skin exposures. To use the tables, we must first determine the absorbed dose for AP and PA films. In the x-ray series, 2 AP and PA films were taken. The total entrance skin exposure (X) for both posterior–anterior (PA) and anterior–posterior (AP) films is provided by the following equation, where all distances are measured from the x-ray source:

$$X = \frac{(26.19 \text{ in.})^2}{(30.13 \text{ in.})^2} \times \frac{35 \text{ mA-sec}}{30 \text{ mA-sec}} \times 500 \text{ mR} \times 2$$

$$= 882 \text{ mR} = 0.882 \text{ R}$$

In the tables under the 3.0-mm Al column for AP film, an absorbed dose of 330 mrad/R at the skin entrance is obtained. Therefore, for both AP films the embryo absorbed dose is

$$D_{AP} = X_{AP} \times \text{(tabular mrad/R conversion factor)}$$

$$= 0.882 \text{ R} \times 330 \text{ mrad/R}$$

$$= 291 \text{ mrad}$$

For the PA exposure, the 3.0-mm Al column yields 174 mrad/R to the embryo at the skin entrance. Therefore, for both PA films we obtain

$$D_{PA} = X_{PA} \times \text{(tabular mrad/R conversion factor)}$$

$$= 0.882 \text{ R} \times 174 \text{ mrad/R}$$

$$= 153 \text{ mrad}$$

Therefore, the total embryo absorbed dose (D) is

$$D = D_{AP} + D_{PA}$$
$$D = 291 \text{ mrad} + 153 \text{ mrad}$$
$$= 444 \text{ mrad}$$

Question 3.25: b

ICRP 49 indicates that prior to the eighth week following fertilization there is apparently little risk of severe mental retardation for radiation dose *in utero*.

Question 3.26: e

The NCRP-91 recommendations for pregnant radiation workers are to limit the dose equivalent to 500 mrem to the embryo/fetus over the entire pregnancy.

Scenario 45

Question 3.27

$$D = 73.8(C)(f)(\text{AF}) \sum_i E_i Y_i (T_{\text{eff}})$$

where

D = dose in rads
C = activity concentration in the thyroid (μCi/gram of tissue)
f = mean number of photons emitted/disintegration
AF = absorbed fraction
i = number of photons emitted
E_i = mean photon energy (MeV) of the ith photon
Y_i = yield of the ith photon of energy E_i
T_{eff} = effective half-life (days)
$T_{\text{eff}} = (T_{\text{phy}} \times T_{\text{bio}})/(T_{\text{phy}} + T_{\text{bio}})$
$T_{\text{phy}} = 60$ days, $T_{\text{bio}} = 130$ days
$T_{\text{eff}} = (60 \text{ days} \times 130 \text{ days})/(60 \text{ days} + 130 \text{ days})$
= 41.05 days

The activity (2.4 μCi) at 14 days is known. This information and the effective half-life can be used to determine the initial uptake of I-125 in the thyroid:

$$A(t) = A(0) \exp(-0.693t/T_{\text{eff}})$$
$$A(0) = 2.4 \ \mu\text{Ci} \exp(+0.693 \times 14 \text{ days}/41.05 \text{ days})$$
$$= 3.04 \ \mu\text{Ci}$$

$$D = 73.8\,(3.04/30)\,(1.398 \times 0.028 + 0.067 \times 0.035)$$
$$\times\ (0.7)\,(41.05)\ \text{rad}$$
$$= 8.92\ \text{rad}$$

Question 3.28

$$H = D + w_T H_T$$
$$= 1.25\ \text{rem} + 0.03 \times 25\ \text{rem} = 2.0\ \text{rem}$$

Question 3.29

No overexposure occurred. The ICRP organ dose limit is 50 rem (nonstochastic), and the total effective dose equivalent limit is 5 rem (stochastic).

Scenario 46

Question 3.30

The total activity (A) of Co-60 in the column may be obtained by assuming that there is no decay during the sampling period. This assumption is valid in view of the Co-60 (5.3 years) half-life.

$$A = VCe$$
$$= (5000\ \text{liters})\,(1000\ \text{ml/liter})\,(1.92 \times 10^{-2}\ \mu\text{Ci/ml})\,(0.95)$$
$$= 9.12 \times 10^4\ \mu\text{Ci} = 91.2\ \text{mCi}$$

The exposure rate from the column may be obtained from a line source approximation:

$$\dot{D} = GC_L\,\theta/h$$

Determination of the angle θ is illustrated in Fig. S3.2.

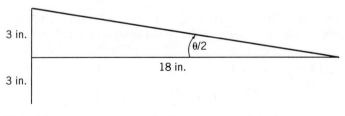

Fig. S3.2 Line source geometry for determination of the included angle (θ).

$$\tan \theta/2 = 3/18$$

$$= 0.1667$$

$$\theta = \tan^{-1}(0.3333333) = 18.92° = 0.33 \text{ rad}$$

$$C_L = 91.2 \text{ mCi}/(6.0 \text{ in.} \times 2.54 \text{ cm/in.})$$

$$= 5.98 \text{ mCi/cm}$$

$$\dot{D} = 13.2 \frac{\text{R-cm}^2}{\text{mCi-hr}} (5.98 \text{ mCi/cm})(0.33)$$

$$\times (1000 \text{ mR/R})/(18 \text{ in.} \times 2.54 \text{ cm/in.})$$

$$= 570 \text{ mR/hr}$$

A reasonable approximation to this dose rate may be obtained from a point source approximation:

$$\dot{D} = GA/r^2$$

$$\dot{D} = 13.2 \frac{\text{R-cm}^2}{\text{mCi-hr}} (91.2 \text{ mCi}) \times 1000 \text{ mR/R}/(18 \text{ in.} \times 2.54 \text{ cm/in})^2$$

$$= 576 \text{ mR/hr}$$

Question 3.31

Safety concerns regarding the storage of the material are as follows:

1. Radiolytic decomposition of water results in the generation and accumulation of hydrogen and oxygen gases, which present an explosive hazard. To assess this concern, the gas generation must be quantified.

Gas Generation Assumptions

a. Storage time is 1 year.
b. Co-60 activity will be assumed to be constant during the interval (91.2 mCi).
c. Anion resin is used for I-129, and cation resin is used for Co-60.
d. All beta energy emitted by radioactive decay is absorbed for gas generation. One-half is absorbed by the anion resin and one-half by the cation resin. No gamma absorption is assumed.

The energy absorbed per year (E) is given by

$$E = (91.2 \times 10^{-3} \text{ Ci})(3.7 \times 10^{10} \text{ dis/sec-Ci})(0.096 \text{ MeV/dis})$$

$$\times (365 \text{ days/year})(24 \text{ hr/day})(3600 \text{ sec/hr})$$

$$= 1.02 \times 10^{16} \text{ MeV/year}$$

The quantity of hydrogen generation (V) is obtained from the absorbed energy and radiolytic rates:

$$V = (1.02 \times 10^{16} \text{ MeV/year})(1.0 \times 10^6 \text{ eV/MeV})(0.5)$$
$$\times\ [(0.6\ H_2\ \text{molecules}/100\ \text{eV})$$
$$+\ (0.13\ H_2\ \text{molecules}/100\ \text{eV})](22.4\ \text{liters/g-mole})$$
$$\times\ (1000\ \text{cm}^3/\text{liter})/6.023 \times 10^{23}\ \frac{H_2\ \text{molecules}}{\text{g-mole}}$$
$$=\ 1.38\ \text{cm}^3/\text{year}$$

This gas generation rate is insignificant. Moreover, the careful researcher will have vented his storage column, which further reduces the potential of the buildup of an explosive mixture.

2. Radiation damage to the plexiglass resin column resulting in its fracture or eventual failure with a subsequent release of its contents.

Material Damage Assumptions

a. Co-60 activity will be assumed to be constant during the interval (91.2 mCi).
b. All the beta energy is absorbed in the resin column. This is reasonable based upon the beta energies encountered in this scenario.
c. A minimum amount of gamma energy is absorbed in the resin column. Attenuation coefficients suggest that this is a valid assumption.

In order to assess the dose to the resin column walls, we first calculate the dose to the resin contained within the column. The average dose is the total energy absorbed (E) divided by the volume of the resin column (v) contents:

$$D_{\text{resin}} = E/v$$
$$D_{\text{resin}} = (1.02 \times 10^{16}\ \text{MeV/year})(1\ \text{year})(1.6 \times 10^{-8}\ \text{g-rad/MeV})$$
$$\times\ (1\ \text{g/cm}^3)^{-1}$$
$$\times\ \frac{1}{(3.14)(0.5\ \text{in.})^2(6\ \text{in.})(2.54\ \text{cm/in.})^3}$$
$$=\ 2.11 \times 10^6\ \text{rad}$$

The dose to the wall of the resin column will be a fraction of the dose delivered to the contents of the column. This fraction will vary between 0 and 1, depending on the nature of the contents and the container. A guide to the correct

SOLUTIONS FOR CHAPTER 3   293

factor is derived from the ICRP-30 ingestion model; this suggests that for water contents in tissue, the dose to the walls is about 0.5 times the dose delivered to the contents.

$$D_{\text{column}} = (0.5) D_{\text{resin}}$$
$$= (0.5)(2.11 \times 10^6 \text{ rad}) = 1.06 \times 10^6 \text{ rad}$$

Radiation damage to the column walls is possible in this dose range. The damage will involve discoloration of the plexiglass. Some fine cracking is also possible. Rupture of the container is not likely during the first year.

Question 3.32

Type of Control	Control Measure
Administrative	1. Perform a radiological safety evaluation to assess potential hazards resulting from the storage. 2. The operation and usage of the column should be specified in a written procedure. 3. An ALARA (as low as reasonably achievable) evaluation should be performed to limit radiation exposures.
Operational	1. Access controls including high radiation control should be imposed. 2. Temporary shielding should be added to limit radiation exposures. 3. Appropriate radiological postings and warning signs should be installed.

Question 3.33

The definition of the lower limit of detectability (LLD) is

$$\text{LLD} = \frac{4.66 \, S_b}{evY \exp(-\lambda t)}$$

All factors defining the LLD remain constant except for the sample volume (v). Therefore, the reduction in the LLD is related to the ratio of sample volumes:

$$\text{LLD} = 3.0 \times 10^{-10} \, \mu\text{Ci/ml} \, (1.0 \text{ liter}/5000 \text{ liters})$$
$$= 6.0 \times 10^{-14} \, \mu\text{Ci/ml}$$

Scenario 47

Question 3.34

The transport index of a package is defined as the maximum dose rate at a distance of 1.0 from the surface of the package.

The transport index is specified to be 3. Therefore, the maximum dose at 1 m from the surface of the package is 3 mrem/hr.

Question 3.35

The package dimensions are 60 × 60 × 60 cm. From the transport index definition, it is known that the dose equivalent rate 1.0 m from the surface of the package or 130 cm (100 cm + 30 cm) from the source is 3.0 mrem/hr. Assuming that the source resides at the center of the package and that the additional air and carton offer minimal attenuation and buildup, the point source approximation can be used to calculate the dose rate at the package surface:

$$\dot{D}(r) = S/r^2$$

where S is the source strength. Using this equation, we can write

$$\dot{D}(a)r_a^2 = \dot{D}(b)r_b^2$$

Letting a = 130 cm and b = 30 cm (the distance to the package surface), the dose rate at 30 cm is obtained:

$$\dot{D}(30\ cm) = 3.0\ mrem/hr\ (130\ cm/30\ cm)^2$$
$$= 56.3\ mrem/hr$$

Question 3.36

A medical package is legal for shipment on a passenger aircraft if the following conditions are met:

1. The transport index (TI) does not exceed 3.0.
2. The highest surface dose equivalent rate does not exceed 200 mrem/hr.

For the packages in question, the TI = 3.0 and the maximum surface dose rate is 56.3 mrem/hr. Therefore, this is a legal package.

Question 3.37

The shipment contains four packages as described in the scenario. The TI of four packages is just four times the TI for a single package, or 4 × 3 (mrem/hr) = 12 (mrem/hr).

Question 3.38

The TLD package is 50 cm from the package surface or 80 cm (30 cm + 50 cm) from the source assumed to reside at the center of the package. The dose delivered to the TLD package from the four generator packages at the end of a 6-hr exposure is

$$D = (3 \text{ mrem/hr-package})(130 \text{ cm}/80 \text{ cm})^2$$
$$\times (4 \text{ packages})(6 \text{ hr})$$
$$= 190.1 \text{ mrem}$$

Question 3.39

Provisions made by commercial vendors to compensate for transient exposures include:

1. The package will contain control dosimeters. These dosimeters were exposed to a known exposure and are used to determine the transit dose that is to be subtracted from the TLD shipment:

   Transit dose = Total TLD dose − Known exposure

2. The TLD package will contain a warning label stating that the package is sensitive to x-rays to preclude unnecessary x-ray exposure.

Question 3.40

The package is returned to its manufacturer 2 weeks after its initial labeling. Assuming that the Tc-99m and Mo-99 are in secular equilibrium, the activity decays with an effective half-life of the parent. The TI at the end of the 2-week period is

$$TI(t) = TI_0 \exp(-\lambda t)$$
$$= (3)[(-0.693/67 \text{ hr})(14 \text{ days} \times 24 \text{ hr/day})]$$
$$= 0.093$$

Question 3.41

Another package has the following properties:

$$TI = 2.0$$
Surface dose equivalent rate = 15.0 mrem/hr

Because the TI exceeds 1, a radioactive type III label is required.

Scenario 48

Question 3.42

Actions taken to evaluate and correct the spread of radioactive material are as follows:

1. Secure the area and control access.
2. Contain the leaking source and store it in a radiologically controlled area.
3. Determine the contamination levels on the source.
4. Survey the area to determine the extent and degree of contamination. Appropriately post the area based on the survey results.
5. Determine which and how many workers were contaminated.
6. Survey and decontaminate the affected personnel.
7. Obtain nasal swabs and bioassay on the workers involved in the incident.
8. Contact the source manufacturer to determine if the source was manufactured with biologically inert microspheres or other material to limit their biological impact.
9. Notify university, state, and Federal officials as required by your health physics procedures.

Question 3.43

Pertinent polonium-210 decay and hazard parameters are as follows:

Primary mode of decay: Alpha 5.3 MeV @ 100%
                        Gamma 0.8 MeV @ 0.001%

Half-life: 138 days
Hazard type: Internal

Question 3.44

Measurement	Method	Locations	Concerns/Precautions
Air	Particulate High-volume grab sample	Lab involved Adjacent labs	Interfering nuclides Representativeness of the sample Measurement sensitivity
Smears	Filters	Lab benches, walls, and floors	All locations not easily accessible
	Sanitary napkins	Remainder of building	Radon/thoron interference
	Mops	Worker residences	

Measurement	Method	Locations	Concerns/Precautions
Surveys	Alpha probe Large-area detectors	Campus buildings Parking Lots	Interfering nuclides Measurement sensitivity
		Personal vehicles Worker residences	Surveys on rough surfaces Po-210 may be masked by asphalt, soil, vegetation, and carpet
		Personal property Roadways	

Question 3.45

Areas that should receive priority attention for decontamination are as follows:

Area	Cleanup Motivation
University buildings (excluding the affected lab)	Permit the university to function normally.
Parking lots and roads	Ensure that uncontrolled areas are returned to a normal status.
Worker residences and vehicles	Ensure that uncontrolled areas are returned to a normal status.

Question 3.46

Items for inclusion in the incident report are as follows:

1. Root cause and contributing factors for the incident.
2. Quantification of the degree and extent of the contamination.
3. Listing of personnel contaminated, their contamination levels, and the associated internal exposures.
4. Recommendations to preclude recurrence of this type of event.
5. Program modifications suggested by this event (changes to survey frequency, personnel frisking requirements, changes to training program, and source inspection requirements).
6. Corrective actions taken and planned with responsible person and due date identified.
7. Incident recovery cost.
8. Sequence of events of the incident and recovery, including dates, times, and personnel involved.
9. Identification of regulatory consequences and required reporting requirements.
10. Personnel statements provided during the investigation.

Scenario 49

Question 3.47

Relative Hazards	P-32	I-125	HTO
Skin dose potential	High	Moderate	Low
Bioassay requirement	Low	High	High
Eye hazard	High	Low	None
Personnel dosimetry requirement	High	High	None
Air sampling requirement	Low	High to moderate	High to moderate

Question 3.48

Low-atomic-number material should be utilized to shield the 1.71-MeV beta particle from P-32. One choice would be lucite or plexiglass with a density of about 1.18 g/cm^3.

The range of the P-32 beta particle may be approximated by the beta-range relationship:

$$R = 412 \, E^{1.265 - 0.0954 \ln(E)}$$

where R is the range of the beta particle in mg/cm^2 and E is the maximum beta energy in MeV.

$$R = 412 (1.71)^{1.265 - 0.0954 \ln(1.71)}$$
$$= 790.2 \text{ mg/cm}^2$$

The thickness of material is given by

$$t(\text{cm}) = R(\text{mg/cm}^2)/p(\text{g/cm}^3)$$
$$= (790.2 \text{ mg/cm}^2)(1 \text{ g}/1000 \text{ mg})/(1.18 \text{ g/cm}^3)$$
$$= 0.670 \text{ cm} \times 1 \text{ in.}/2.54 \text{ cm}$$
$$= 0.264 \text{ in. or about 3/8 in. to be conservative}$$

Question 3.49

The shielding thickness does not change. All P-32 betas are stopped by the 3/8-in. thickness that will stop the maximum beta energy particle.

Question 3.50

P-32 waste material should be stored for 10 half-lives to permit radioactive decay to background levels. After documented surveys, the P-32 waste can be discarded as common trash.

Question 3.51

KI should not be taken before the procedure. The researcher could have an allergic reaction to the KI, and there is no need for this practice if the fume hood is working properly. IF there were an incident, KI would still be effective if it were taken within hours of the uptake. It would be advantageous if experimenters were cleared for KI use in advance by a university physician. A listing of those qualified for KI use should be maintained by the RSO.

Question 3.52

A liquid scintillation counter should be utilized. Both beta particles and gamma rays will produce light in a liquid scintillation cocktail. Appropriate calibration standards and laboratory procedures would be required to ensure that sufficient accuracy is maintained.

SOLUTIONS FOR CHAPTER 4

Scenario 50

Question 4.1

The intake based on the CAM concentration estimate is

$$I = kt(\text{BR})C$$
$$I = (80 \text{ sec})(20 \text{ liters/min})(1 \text{ min}/60 \text{ sec})(1000 \text{ ml/liter})$$
$$\times (2.0 \times 10^{-5} \text{ }\mu\text{Ci/ml})$$
$$= 0.53 \text{ }\mu\text{Ci}$$

Calculation of organ doses:

$$D = I \sum_i f_i (\text{DCF}_i)$$

$$\begin{aligned}
D(\text{lung}) = 0.53 \text{ }\mu\text{Ci}[&(1.2 \times 10^{-2})(7.0 \times 10^{-5} \text{ Sv/Bq}) \\
&+ (1.4 \times 10^{-1})(6.7 \times 10^{-5} \text{ Sv/Bq}) \\
&+ (3.1 \times 10^{-2})(6.7 \times 10^{-5} \text{ Sv/Bq}) \\
&+ (8.1 \times 10^{-1})(7.0 \times 10^{-9} \text{ Sv/Bq}) \\
&+ (1.5 \times 10^{-3})(1.8 \times 10^{-5} \text{ Sv/Bq})] \\
&\times (3.7 \times 10^4 \text{ Bq/}\mu\text{Ci})(100 \text{ rem/Sv}) \\
= 24.2 \text{ rem}&
\end{aligned}$$

$$D(\text{liver}) = 0.53 \; \mu\text{Ci}[(1.2 \times 10^{-2})(9.7 \times 10^{-7} \; \text{Sv/Bq})$$
$$+ (1.4 \times 10^{-1})(9.2 \times 10^{-7} \; \text{Sv/Bq})$$
$$+ (3.1 \times 10^{-2})(9.2 \times 10^{-7} \; \text{Sv/Bq})$$
$$+ (8.1 \times 10^{-1})(1.1 \times 10^{-9} \; \text{Sv/Bq})$$
$$+ (1.5 \times 10^{-3})(1.5 \times 10^{-6} \; \text{Sv/Bq})]$$
$$\times (3.7 \times 10^4 \; \text{Bq}/\mu\text{Ci})(100 \; \text{rem/Sv})$$
$$= 0.34 \; \text{rem}$$

$$D(\text{bone surfaces}) = 0.53 \; \mu\text{Ci}[(1.2 \times 10^{-2})(3.5 \times 10^{-6} \; \text{Sv/Bq})$$
$$+ (1.4 \times 10^{-1})(3.5 \times 10^{-6} \; \text{Sv/Bq})$$
$$+ (3.1 \times 10^{-2})(3.5 \times 10^{-6} \; \text{Sv/Bq})$$
$$+ (8.1 \times 10^{-1})(3.8 \times 10^{-9} \; \text{Sv/Bq})$$
$$+ (1.5 \times 10^{-3})(5.4 \times 10^{-5} \; \text{Sv/Bq})]$$
$$\times (3.7 \times 10^4 \; \text{Bq}/\mu\text{Ci})(100 \; \text{rem/Sv})$$
$$= 1.42 \; \text{rem}$$

$$D(\text{red bone marrow}) = 0.53 \; \mu\text{Ci}[(1.2 \times 10^{-2})(3.0 \times 10^{-7} \; \text{Sv/Bq})$$
$$+ (1.4 \times 10^{-1})(2.7 \times 10^{-7} \; \text{Sv/Bq})$$
$$+ (3.1 \times 10^{-2})(2.7 \times 10^{-7} \; \text{Sv/Bq})$$
$$+ (8.1 \times 10^{-1})(2.7 \times 10^{-10} \; \text{Sv/Bq})$$
$$+ (1.5 \times 10^{-3})(4.3 \times 10^{-6} \; \text{Sv/Bq})]$$
$$\times (3.7 \times 10^4 \; \text{Bq}/\mu\text{Ci})(100 \; \text{rem/Sv})$$
$$= 0.11 \; \text{rem}$$

Question 4.2

$$H = \sum_T W_T H_T$$
$$= (24.2 \; \text{rem})(0.12)_{\text{lung}} + (0.34 \; \text{rem})(0.06)_{\text{liver}}$$
$$= (1.42 \; \text{rem})(0.03)_{\text{bone surfaces}} + (0.11 \; \text{rem})(0.12)_{\text{RBM}}$$
$$= 2.98 \; \text{rem}$$

Question 4.3

Stochastic Limit: 5 rem/year total effective dose equivalent
Nonstochastic Limit: 50 rem/year committed dose equivalent for the organs of interest for this intake.

Neither limit was exceeded.

Question 4.4

Extensive bioassay follow-up will be required. This will include urinalysis, fecal analysis, and *in vivo* lung counting. All voids should be collected the first week for Pu and Am analysis. A routine frequency should be established for several months thereafter until elimination functions are established.

No clinical symptoms are expected for these dose levels.

Scenario 51

Question 4.5

The intake activity (A) can be defined in terms of the count rate on the filter paper:

$$A = CR/ek$$

Knowing the activity permits the determination of the average air concentration ($\overline{C}$) inhaled by the worker:

$$\overline{C} = A/ft$$

The intake I can be estimated using the relationship

$$I = (CR)(BR)/ekf(PF)$$

which may be written in terms of the available information given in the problem:

$$I = (\overline{C})(BR)t/PF$$

Using these relationships, the intakes for workers "A" and "B" can be determined:

Worker A:

$$A = \frac{30{,}000 \text{ cpm}}{(0.3 \text{ cpm/dpm})(2.2 \times 10^6 \text{ dpm}/\mu\text{Ci})} = 4.55 \times 10^{-2} \ \mu\text{Ci on the filter}$$

$$\bar{C} = \frac{4.55 \times 10^{-2} \, \mu\text{Ci}}{(2 \text{ liters/min})(10 \text{ min})}$$

$= 2.3 \times 10^{-3} \, \mu\text{Ci/liter}$ average inhaled concentration

$I = (2.3 \times 10^{-3} \, \mu\text{Ci/liter})(20 \text{ liters/min})(10 \text{ min})/(10)$

$= 4.5 \times 10^{-2} \, \mu\text{Ci}$ intake

Worker B:

$$A = \frac{20,000 \text{ cpm}}{(0.3 \text{ cpm/dpm})(2.2 \times 10^6 \text{ dpm}/\mu\text{Ci})} = 3.03 \times 10^{-2} \, \mu\text{Ci on the filter}$$

$$\bar{C} = \frac{3.0 \times 10^{-2} \, \mu\text{Ci}}{(2 \text{ liters/min})(10 \text{ min})}$$

$= 1.5 \times 10^{-3} \, \mu\text{Ci/liter}$ average inhaled concentration

$I = (1.5 \times 10^{-3} \, \mu\text{Ci/liter})(20 \text{ liters/min})(10 \text{ min})/(1.0)$

$= 3.0 \times 10^{-1} \, \mu\text{Ci}$ inhaled

Question 4.6

The material is insoluble, and therefore class Y values for retention and elimination functions and dose conversion factors are appropriate.

For class Y material, the deposition in the pulmonary lung will result in a long-term dose contribution. The long-term activity in the pulmonary lung will be

$A_L = I D_P d_Y$

$A_L =$ long-term pulmonary deposition (μCi)

$I =$ intake (μCi)

$D_P =$ deposition factor which depends on the particle size and the location of the deposition (For pulmonary deposition of 1-μm particles, $D_p = 0.25$.)

$d_Y =$ class Y long-term retention function

$=$ sum of long-term compartment F values (i.e., those with $T = 500$ days)

$= F_e + F_g + F_h = 0.05 + 0.4 + 0.15 = 0.60$

From ICRP 30, for the Pulmonary Region

		Class Y	
Region	Compartment	T(day)^a	F^b
Pulmonary	e	500	0.05
	f	1.0	0.4
	g	500	0.4
	h	500	0.15

$^a T$ = compartment removal half-times.
$^b F$ = compartmental fractions.

The long-term pulmonary dose is given by

$$D = A_L \sum_i f_i (\text{DCF}_i)$$

D = long-term pulmonary dose (rem)

i = Number of isotopes in the air mixture inhaled by the worker

f_i = isotopic fraction in the mixture

DCF_i = inhalation lung dose equivalent conversion factors (rem/μCi)

For this problem, the following are to be used

Isotope	f_i	DCF$_i$ (rem/μCi) (first year)	DCF$_i$ (rem/μCi) (50 years)
U-234	0.10	230	1100
U-235	0.02	210	1000
U-238	0.88	200	980

Using these relationships, the desired information may be obtained.

Worker A:

$A_L = (4.55 \times 10^{-2} \, \mu\text{Ci})(0.25)(0.6) = 6.8 \times 10^{-3} \, \mu\text{Ci}$

$D = [(0.1)(1100) + (0.02)(1000) + (0.88)(980)] \, \text{rem}/\mu\text{Ci}$

$\times (6.8 \times 10^{-3} \, \mu\text{Ci}) = 6.75$ rem committed dose equivalent

Worker B:

$A_L = (3.03 \times 10^{-1} \, \mu\text{Ci})(0.25)(0.6) = 4.5 \times 10^{-2} \, \mu\text{Ci}$

$$D = [(0.1)(1100) + (0.02)(1000) + (0.88)(980)] \text{ rem}/\mu\text{Ci}$$
$$\times (4.5 \times 10^{-2} \ \mu\text{Ci}) = 44.6 \text{ rem committed dose equivalent}$$

Question 4.7

The percentage is represented by a ratio of DCF values:

$$\% \text{ CDE} = \frac{\sum_i f_i \text{ DCF}_i(\text{first year})}{\sum_i f_i \text{ DCF}_i(50 \text{ years})}$$

$$= \frac{(0.1)(230) + (0.02)(210) + (0.88)(200)}{(0.1)(1100) + (0.02)(1000) + (0.88)(980)} = 20.5\%$$

Question 4.8

$$H_E = \sum_T W_T H_T \leq 5 \text{ rem effective dose (stochastic dose limit)}$$

Using this information, worker doses can be determined:

Worker A: $(6.75 \text{ rem})(0.12) = 0.81 \text{ rem} < 5 \text{ rem}$
Worker B: $(44.6 \text{ rem})(0.12) = 5.35 \text{ rem} > 5 \text{ rem}$

Only Worker B exceeds the stochastic limit for the lung (5 rem/0.12 = 41.6 rem). The nonstochastic limit is 50 rem, which is not exceeded by either worker.

Question 4.9

Extensive bioassay follow-up is required. This will include urinalysis, fecal analysis, and *in vivo* lung counting. All voids should be collected for the first week (isotopic uranium, thorium, and radium analyses), and a routine frequency is established for several months thereafter. *In vivo* lung counting should be performed as soon as possible with repeated analysis every few months until elimination functions are determined.

Scenario 52

Question 4.10

Ionizing Radiation Hazards

1. X-rays are produced from the high-voltage (30 and 50 kV) power supplies. Extremity dosimeters and low-energy x-ray sensitive dosimeters would be part of a radiation monitoring program.

306 SOLUTIONS FOR CHAPTER 4

2. Airborne alpha and beta hazards exist during maintenance operations.
3. The dust/debris resulting from cleaning the ionization chamber may present disposal problems due to the quantity of TRU present.
4. Criticality concerns from the generation of highly enriched product cause potential gamma and neutron hazards.
5. Alpha/beta hazards exist from the raw feed material and enriched product.

Nonionizing Radiation Hazards

1. High-intensity laser beams can damage the skin and eyes.
2. Unshielded high-strength electromagnetic fields may cause biological injury.
3. Reflected laser light from the photoexcitation process may injure the eyes.

Other Hazards

1. Heat buildup from the interaction of the electron beam with the uranium feed material.
2. Noise hazards due to the high-energy systems transporting significant quantities of matter.

Question 4.11

Engineering design features to minimize ionizing radiation hazards include the following:

1. X-ray hazards can be minimized with proper shielding.
2. Facility design should position high-voltage equipment away from location where personnel will be routinely located.
3. Particulate alpha and beta airborne concentrations can be minimized with proper airflow and ventilation system design.
4. Air should be HEPA filtered and not recirculated from higher airborne concentration areas to lower concentration areas.
5. Proper sizing and arrangement of transfer lines and storage containers will reduce the probability of a criticality accident.
6. Alarming air monitors should be installed in areas where uranium dust is or may be present.
7. Criticality alarms should be installed with detectors properly located in enriched materials storage areas or near enriched material transport areas.
8. Glove boxes and fume hoods should be designed to facilitate chamber operations, maintenance, and cleaning.

SOLUTIONS FOR CHAPTER 4   307

Question 4.12

Engineering design features to minimize nonionizing radiation hazards include the following:

1. Beam tubes and optical transport systems should be designed to keep laser radiation out of occupied areas.
2. Interlocks should be used in all areas of high-intensity laser radiation.
3. Areas with high-strength electromagnetic fields should be contained in separate rooms with strict access controls.

Question 4.13

In order to determine the dose to the technician, the following assumptions are made:

1. The buildup from the polyethylene shield is negligible.
2. The average energy of the photons and neutrons is representative of the spectrum, and therefore monoenergetic energies can be used in the calculation.
3. The use of dose conversion factors is appropriate for the total dose and dose rate for the neutron and gamma energies of this problem.
4. A point source approximation is applicable due to the small size of the critical mass (source) and the distance of the technician from the source.

Neutron Dose: The unattenuated neutron dose is

$$D_n^0 = \frac{(1.0 \times 10^{16} \text{ fissions})(3 \text{ neutrons/fission})}{(4)(3.14)(20 \text{ ft} \times 30.48 \text{ cm/ft})^2}$$
$$\times [(2.5 \text{ mrad/hr})/(20 \text{ n/cm}^2\text{-sec})](1 \text{ hr/3600 sec})$$
$$= 2.23 \times 10^5 \text{ mrad}$$

Because the neutron attenuation is 0.3, for 2.5-MeV neutrons passing through 4 in. of polyethylene, the neutron dose to the technician is

$$D_n = (0.3)(2.23 \times 10^5 \text{ mrad})(1 \text{ rem}/1000 \text{ mrad}) = 66.9 \text{ rad}$$

Gamma Dose: The unattenuated gamma dose is

$$D_g^0 = \frac{(1.0 \times 10^{16} \text{ fissions})(8 \text{ gammas/fission})}{(4)(3.14)(20 \text{ ft} \times 30.48 \text{ cm/ft})^2}$$
$$\times [(1.0 \text{ R/hr})/(5.5 \times 10^5 \text{ gammas/cm}^2\text{-sec})](1 \text{ hr}/3600 \text{ sec})$$
$$= 8.65 \text{ R}$$

The attenuated gamma dose is

$$D_g = D_g^0 \exp(-ut)$$
$$= 8.65 \text{ R} \exp[-(0.0727 \text{ cm}^2/\text{g})(4 \text{ in.})(2.54 \text{ cm/in.})(1.4 \text{ g/cm}^3)]$$
$$= 3.08 \text{ R} \times 0.869 \text{ rad/R} = 2.67 \text{ rad}$$
$$D_{\text{total}} = 66.9 \text{ rad} + 2.7 \text{ rad} = 69.6 \text{ rad}$$

Scenario 53

Question 4.14

Assumptions for questions 4.14 and 4.15:

1. The criticality is adequately represented by a point source.
2. The buildup from the polyethylene shield is small; that is, $B = 1.0$.
3. The neutron and gamma spectra are adequately represented by their mean energies.
4. The neutron and gamma dose conversion factors are representative of their associated spectra.

The general equation for the neutron dose is

$$D_n = N_{\text{fissions}} \, vk \exp(-ut) B/(4\pi) r^2$$

k = conversion factors

$$= (2.5 \text{ mrem/hr}/20 \text{ neutrons/cm}^2\text{-sec}) \times (1 \text{ hr}/3600 \text{ sec})$$

Unattenuated Neutron Dose (D_n^0):

$$D_n^0 = \frac{(1.0 \times 10^{16} \text{ fissions})(3 \text{ neutrons/fission})}{(4)(3.14)(10 \text{ ft} \times 12 \text{ in./ft} \times 2.54 \text{ cm/in.})^2}$$

$$\times \frac{2.5 \text{ mrem/hr}}{20 \text{ neutrons/cm}^2\text{-sec}}$$

$$\times \frac{1 \text{ hr}}{3600 \text{ sec}} \frac{1 \text{ rem}}{1000 \text{ mrem}} = 892.7 \text{ rem}$$

Attenuated Dose (D_n): Because $B \exp(-ut)$ may be replaced by the neutron attenuation factor, the attenuated dose is written as

$$D_n = D_n^0 \times \text{(neutron attenuation for 2.5 MeV and 12 in. of polyethylene)}$$
$$D_n = (892.7 \text{ rem})(0.005) = 4.46 \text{ rem}$$

Question 4.15

The general equation for the gamma dose is

$$D_g = N_{\text{fissions}}\ vk \exp(-ut) B/4\pi r^2$$

Unattenuated Gamma Exposure (X_g^0)

$$X_g^0 = \frac{(1.0 \times 10^{16}\ \text{fissions})(8\ \text{gammas/fission})}{(4)(3.14)(10\ \text{ft} \times 12\ \text{in./ft} \times 2.54\ \text{cm/in.})^2}$$

$$\times \frac{1\ \text{hr}}{3600\ \text{sec}} \frac{1\ \text{R/hr}}{5.5 \times 10^5\ \text{gammas/cm}^2\text{-sec}} = 34.6\ \text{R}$$

Attenuated Exposure (X_g):

$$X_g = X_g^0 e^{-ut}$$

$$X_g = (34.6\ \text{R})\exp(-0.0727\ \text{cm}^2/\text{g} \times 12\ \text{in.} \times 2.54\ \text{cm/in.} \times 1.4\ \text{g/cm}^3)$$

$$= 1.56\ \text{R}$$

Gamma dose (D_g):

$$D_g = (1.56\ \text{R})(0.869\ \text{rad/R})(1\ \text{rem/rad}) = 1.36\ \text{rem}$$

Question 4.16

Because the alarm set point is 500 mR/hr and the detector responds to 1/2500 of the actual gamma exposure rate, the actual exposure rate which must exist at the detector to trip an alarm is X_t:

$$X_t/2500 = 500\ \text{mR/hr}$$

$$X_t = 1250\ \text{R/hr}$$

Given that 10^{15} fissions yields a gamma exposure of 2.5 R at 6.0 ft, 1.0×10^{16} fissions will yield an exposure of 25 R at 6.0 ft. Assuming that the 1.0×10^{16} fissions occur uniformly in time during the 1.0-msec transient, the effective exposure rate at 6.0 ft is

$$\dot{X} = 25\ \text{R}/0.001\ \text{sec} = 2.5 \times 10^4\ \text{R/sec} \times 3600\ \text{sec/hr}$$

$$= 9.0 \times 10^7\ \text{R/hr @ 6.0 ft}$$

Assuming that the criticality is adequately represented by a point source at the detector, the inverse square law is applicable:

$$(\dot{X})r^2 = \text{constant}$$
$$(9.0 \times 10^7 \text{ R/hr})(6.0 \text{ ft})^2 = (1250 \text{ R/hr})(d^2)$$

where d is the maximum distance at which the detector will alarm at the alarm set point. Solving the equation yields $d = 1610$ ft.

Scenario 54

Question 4.17

Characteristics of raw oils that make collecting a representative sample a difficult task include:

1. Water contamination
2. Oils having different density or stratification
3. Sludge
4. Various solubilities for different chemical compositions and radionuclides
5. Contaminants or other liquids discarded in oil collection drums

Question 4.18

Acceptable methods for obtaining representative samples from the tank are as follows:

1. Recirculation of the tank such that its volume is recirculated at least three times
2. Collecting a composite of repetitive dip samples taken at various depths

Question 4.19

Analysis methods for characterizing the waste oil include:

1. Radiochemical analysis to determine the radionuclide content.
2. Chemical toxicity testing to determine the presence of hazardous materials

Question 4.20

General survey techniques for trash release from the RCA include:

1. Beta and gamma surface contamination monitoring with state-of-the-art thin window detectors
2. Gamma monitoring of the disposal package with state-of-the-art low-background instrumentation

SOLUTIONS FOR CHAPTER 4   311

Question 4.21

The regulatory accepted limit for release is that no detectable activity above background is present.

Question 4.22

The radiation exposure pathways include:

1. Direct exposure from surface deposition
2. Consumption of fish living in the lagoon
3. Well-water consumption
4. Inhalation of resuspended dust
5. Consumption of garden vegetables
6. Consumption of dairy cow's milk from the grass–cow pathway
7. Consumption of beef from the cattle grazing pathway

Question 4.23

The exposure to the maximum exposed individual should be less than 5 mrem/year.

Question 4.24

Six general environmental categories that impact approval of alternate disposal methods are as follows:

1. Topographical
2. Geological
3. Meteorological
4. Hydrological
5. Ground- and surface-water usage
6. Nature and location of other industrial facilities

Question 4.25

Other types of information which must be included in the application are as follows:

1. Radiation control procedures used to minimize exposure
2. Description of licensed radioactive material
3. Other types of nonlicensed radioactive material
4. Quantity, type, and chemical form of radioactive material

5. Levels of radioactivity
6. Proposed manner and conditions of disposal
7. Alternative methods of disposal including their economic and radiological impacts

Scenario 55

Question 4.26

The committed dose equivalent to the thyroid from the I-131 intake is determined from the definition of the ALI:

$$H_T = I_{\text{I-131}} \frac{50 \text{ rem}}{\text{ALI(I-131)}}$$

where

$$I_{\text{I-131}} = \text{intake of I-131} = 19.5 \ \mu\text{Ci}$$

$$\text{ALI(I-131)} = \text{annual limit of intake for I-131}$$
$$= 50 \ \mu\text{Ci}$$

$$H_T = 19.5 \ \mu\text{Ci} \ \frac{50 \text{ rem}}{50 \ \mu\text{Ci}} = 19.5 \text{ rem}$$

Question 4.27

The committed effective dose equivalent (CEDE) from the I-131 and Cs-137 intakes is determined from the sum of the dose equivalents from each isotope:

$$H_E = 5 \text{ rem} \left(\frac{I_{\text{I-131}}}{\text{SALI(I-131)}} + \frac{I_{\text{Cs-137}}}{\text{SALI(Cs-137)}} \right)$$

$$= 5 \text{ rem} \left(\frac{19.5 \ \mu\text{Ci}}{200 \ \mu\text{Ci}} + \frac{80.0 \ \mu\text{Ci}}{200 \ \mu\text{Ci}} \right)$$

$$= 2.49 \text{ rem}$$

where the following assumptions are made:

1. The CEDE is evaluated over a 50-year period.
2. The 27-year-old worker has the characteristics of reference man.
3. The 1-μm AMAD particle size is applicable to this uptake.
4. Class D aerosols are representative of the inhaled particles.

Question 4.28

To determine if the NCRP-91 occupational dose limits have been exceeded, the limit and the corresponding quantity for this event must be determined.

The NCRP-91 annual dose recommendations include:

1. A total effective dose equivalent (TEDE) less than or equal to 5 rem
2. A 50-year committed dose equivalent (CDE) of 50 rem or less.

TEDE:

$$\text{TEDE} = H_{DDE} + H_E(50 \text{ year})$$

where

H_{DDE} = the deep dose equivalent = $(0.7 + 1.2 + 0.8 + 0.3)$ rem = 3.0 rem.
$H_E(50 \text{ year})$ was obtained in question 4.27.
TEDE = 3.0 rem + 2.49 rem = 5.49 rem.
The TEDE exceeds the NCRP-91 recommendation.

CDE: The dose equivalent (H_T) assigned to the thyroid is as follows:

$H_T = H_{DDE(thyroid)} + H_T(50 \text{ year}) = 3.0 \text{ rem} + 19.5 \text{ rem} = 22.5 \text{ rem}$.
H_T does not exceed the NCRP-91 recommendations.

The external dose contribution to the thyroid has not accounted for tissue attenuation prior to radiation reaching the thyroid.

Scenario 56

Question 4.29

The worker exposure is most easily determined using the thin disk source approximation:

$$\dot{X}(h) = \pi G C_a \ln \frac{r^2 + h^2}{h^2}$$

where

G = gamma constant = $0.33 \dfrac{\text{R-m}^2}{\text{hr-Ci}}$ (only Cs-137 contributes significantly to the external exposure)

C_a = gamma emitter isotopic concentration per unit area

$\quad = A/a$

A = total activity (Ci)

$\quad = C(\text{Cs-137}) \times V$

$C(\text{Cs-137})$ = Cs-137 waste concentration (Ci/liter)

$\quad = 0.2$ Ci/liter

V = spill volume = 100 liters

$A = 0.2$ Ci/liter $\times$ 100 liters = 20 Ci

a = spill area (m²)

$\quad = \pi r^2 = (3.14)(2.5 \text{ m})^2 = 19.625 \text{ m}^2$

$C_a = 20 \text{ Ci}/19.625 \text{ m}^2 = 1.02 \text{ Ci/m}^2$

r = radius of the circular spill = 2.5 m

h = distance above spill central axis = 1.0 m

Using these data permits the determination of the gamma dose:

$$\dot{X} = (3.14)\left(0.33 \frac{\text{R-m}^2}{\text{hr-Ci}}\right)(1.02 \text{ Ci/m}^2)$$

$$\times \ln \frac{(2.5 \text{ m})^2 + (1.0 \text{ m})^2}{(1.0 \text{ m})^2} = 2.09 \text{ R/hr}$$

Assuming 1 R/hr = 1 rem/hr:

$$\dot{D} = 2.09 \text{ R/hr} \frac{\text{rem/hr}}{\text{R/hr}} = 2.09 \text{ rem/hr}$$

The external dose equivalent is easily calculated because the exposure time is known;

$$D = \dot{D}t = (2.09 \text{ rem/hr})(1.5 \text{ hr}) = 3.14 \text{ rem}$$

Question 4.30

The effective dose equivalents may be obtained from a knowledge of the isotopic air concentration, breathing rate, exposure time, and dose conversion factors.

Isotopic Air Concentrations:

$$C_i = \frac{k(\text{CR})_i}{(\text{FR})T}$$

where

C_i = isotopic concentration of the ith radionuclide

$i = 1$, gamma count rate (dpm) for Cs-137

$ = 2$, beta count rate (dpm) for Sr-90

$ = 3$, alpha count rate (dpm) for Pu-238

FR = flow rate of air sampler = 2 liters/min

T = exposure time = 90 min

k = conversion factor = $1/2.2 \times 10^6$ dpm/μCi

Cs-137: $C_1 = \dfrac{1.0 \times 10^5 \text{ dpm}}{(2.2 \times 10^6 \text{ dpm}/\mu\text{Ci})(2 \text{ liters/min})(90 \text{ min})}$

$ = 2.53 \times 10^{-4}\ \mu\text{Ci/liter}$

Sr-90: $C_2 = \dfrac{(5.0 \times 10^3 \text{ dpm})}{(2.2 \times 10^6 \text{ dpm}/\mu\text{Ci})(2 \text{ liter/min})(90 \text{ min})}$

$ = 1.26 \times 10^{-5}\ \mu\text{Ci/liter}$

Pu-238: $C_3 = \dfrac{5.0 \times 10^3 \text{ dpm}}{(2.2 \times 10^6 \text{ dpm}/\mu\text{Ci})(2 \text{ liter/min})(90 \text{ min})}$

$ = 1.26 \times 10^{-5}\ \mu\text{Ci/liter}$

The first-year dose equivalent D_1 may be determined from the following relationship:

$$D_1 = \sum_i C_i(\text{BR})T\,\text{DCF}_i$$

where

D_1 = first-year dose equivalent

$ = (2.53 \times 10^{-4}\ \mu\text{Ci/liter})(20 \text{ liters/min})(90 \text{ min})(2.9 \times 10^{-2} \text{ rem}/\mu\text{Ci})$

$ + (1.26 \times 10^{-5}\ \mu\text{Ci/liter})(20 \text{ liters/min})(90 \text{ min})(0.32 \text{ rem}/\mu\text{Ci})$

$ + (1.26 \times 10^{-5}\ \mu\text{Ci/liter})(20 \text{ liters/min})(90 \text{ min})(30 \text{ rem}/\mu\text{Ci})$

$ = 0.013 \text{ rem} + 0.007 \text{ rem} + 0.680 \text{ rem} = 0.700 \text{ rem}$

Question 4.31

The 50-year committed effective dose equivalent is obtained in an analogous manner:

H_{50} = 50-year committed effective dose equivalent

$= (2.53 \times 10^{-4} \, \mu\text{Ci/liter})(20 \text{ liter/min})(90 \text{ min})(3.2 \times 10^{-2} \text{ rem}/\mu\text{Ci})$
$+ (1.26 \times 10^{-5} \, \mu\text{Ci/liter})(20 \text{ liter/min})(90 \text{ min})(1.30 \text{ rem}/\mu\text{Ci})$
$+ (1.26 \times 10^{-5} \, \mu\text{Ci/liter})(20 \text{ liter/min})(90 \text{ min})(310 \text{ rem}/\mu\text{Ci})$
$= 0.015 \text{ rem} + 0.029 \text{ rem} + 7.031 \text{ rem} = 7.075 \text{ rem}$

Question 4.32

Standards

10CFR20 (1/1/93). Per the January 1, 1993 revision, the committed effective dose equivalent (CEDE) plus annual deep dose equivalent (external dose) (DDE) or total effective dose equivalent (TEDE) is limited to 5.0 rem.

$$\text{TEDE} = \text{CEDE} + \text{DDE}$$
$$\text{TEDE} = 3.14 \text{ rem} + 7.08 \text{ rem} = 10.22 \text{ rem}$$

The 10 CFR 20 TEDE limit is exceeded.

10CFR20 (old). The old version 10 CFR 20 had two separate banks—that is, one for external and one for internal exposure:

External: 3 rem/quarter if $5(N - 18)$ was not exceeded (N = worker age in years). $D = 3.14$ rem, so this limit was exceeded.

Internal: 520 MPC-hr/quarter

$N_{\text{MPC-hr}} = (1.5 \text{ hr}) \left(\dfrac{2.53 \times 10^{-7} \, \mu\text{Ci/ml}}{1.0 \times 10^{-8} \, \mu\text{Ci/ml-MPC}} + \dfrac{1.26 \times 10^{-8} \, \mu\text{Ci/ml}}{1.0 \times 10^{-9} \, \mu\text{Ci/ml-MPC}} \right.$

$\left. + \dfrac{1.26 \times 10^{-8} \, \mu\text{Ci/ml}}{2.0 \times 10^{-12} \, \mu\text{Ci/ml-MPC}} \right)$

$= 38.0 \text{ MPC-hr} + 18.9 \text{ MPC-hr} + 9450 \text{ MPC-hr}$

$= 9507 \text{ MPC-hr}$

The 520 MPC-hr limit was exceeded.

SOLUTIONS FOR CHAPTER 4

DOE Order 5480.11. The 5-rem DOE limit is the sum of the external dose and the internal dose equivalent for the year of interest.

$$D_{DOE} = 3.14 \text{ rem} + 0.7 \text{ rem} = 3.84 \text{ rem}$$

The DOE 5480.11 criteria is not exceeded.

DOE Radiological Controls Manual. This manual specifies the use of the TEDE limited to 5 rem. The 50-year CEDE is added to the external deep dose equivalent and is compared to the 5-rem limit. The limit was exceeded. See the 10 CFR 20 (1/1/93) discussion.

10CFR835. Limits are the same as stated in the *DOE Radiological Controls Manual*.

Question 4.33

Instructions that should be given to the shift supervisor by the health physicist are as follows:

1. Stop the spill/release in the most dose effective manner.
2. Once the release is terminated, rope the area and keep unauthorized personnel out of the area.
3. Warn others via plant page to stay out of the area.
4. Minimize exposure in the recovery.
5. Check affected personnel for contamination and decontaminate as necessary.
6. Assess radiological conditions.
7. Notify appropriate personnel (NRC/DOE, plant management, and response teams as warranted).
8. Survey adjacent areas and decontaminate as necessary.
9. Clean up the spill area and minimize radioactive waste generation.
10. Stabilize plant conditions.
11. Establish plant shift schedules for long-term recovery.

Question 4.34

Cs-137 rapidly clears from the body. Most of it is eliminated during the first year post intake.

Pu-238 clears more slowly than Cs-137. A significant portion of the initial intake remains after the first year. Some will still remain after 50 years.

These comments are qualitatively understood by examining the ratio of the DCFs for 1 and 50 years:

Cs-137:

$$\frac{DCF_1}{DCF_{50}} = \frac{2.9 \times 10^{-2}}{3.2 \times 10^{-2}} = 0.91 \qquad (91\% \text{ of the dose in the first year})$$

Pu-238:

$$\frac{DCF_1}{DCF_{50}} = \frac{30}{310} = 0.10 \qquad (10\% \text{ of the dose in the first year})$$

SOLUTIONS FOR CHAPTER 5

Scenario 57

Question 5.1

$$A = (C_i)(F)(e)[1 - \exp(-\lambda t)]/(\lambda)$$

For Co-60 and the demineralizer utilized in this problem, we obtain

$\lambda = 0.693/(5.26 \text{ year} \times 365 \text{ days/year}) = 3.61 \times 10^{-4}/\text{day}$

$F = 350 \text{ liters/min} \times 10^3 \text{ ml/liter} = 3.5 \times 10^5 \text{ ml/min}$

$C_i = 6.0 \times 10^{-4} \ \mu\text{Ci/ml}$

$t = 200 \text{ days}$

$e = 1.0$

$$A = \frac{(6.0 \times 10^{-4} \ \mu\text{Ci/ml})(3.5 \times 10^5 \ \text{ml/min})(1.0)}{(3.61 \times 10^{-4}/\text{day})(1 \ \text{day}/1440 \ \text{min})} \ \frac{1.0 \ \text{Ci}}{10^6 \ \mu\text{Ci}}$$
$$\times \{1 - \exp[-(3.61 \times 10^{-4}/\text{day} \times 200 \ \text{days})]\} = 58.3 \ \text{Ci}$$

To compute the gamma dose, the gamma constant (G) is utilized.

$R = 2.0 \text{ ft}$ (the radius of the resin bed)

$h = 1.0 \text{ ft}$ (the distance from disk source to dose location)

SOLUTIONS FOR CHAPTER 5

The dose rate is to be calculated using the equation provided in the data section:

$$D(h) = 3.14 GC \ln \frac{h^2 + R^2}{h^2}$$

$$\ln(R^2 + h^2)/h^2 = \ln[(4+1)/1] = 1.609$$

Area of demineralizer bed = $3.14 \times (2 \text{ ft} \times 12 \text{ in./ft} \times 1 \text{ m}/39.37 \text{ in.})^2$
$$= 1.17 \text{ m}^2$$

$$D(1 \text{ ft}) = 3.14 \times 1.609 \times \left[(1.3 \text{ R-m}^2/\text{hr-Ci}) \times \frac{58.3 \text{ Ci}}{1.17 \text{ m}^2} \right]$$

$$= 327.3 \text{ R/hr} \times 0.877 \text{ rad/R} \times 1 \text{ rem/rad} = 287.0 \text{ rem/hr}$$

Note: The use of the point source approximation is not applicable.

$$D_{\text{Point}} = GA/r^2$$

$$D_{\text{Point}} = 1.3 \text{ R-m}^2/\text{hr-Ci} \times 58.3 \text{ Ci}$$

$$\times \frac{1}{(1 \text{ ft} \times 12 \text{ in./ft} \times 1 \text{ m}/39.37 \text{ in.})^2} = 815.8 \text{ R/hr}$$

Question 5.2

The shielded dose calculation requires the decay of demineralizer activity to a time of 6 months post shutdown:

$$A(\text{Co-60}) = 80 \text{ Ci} \times \exp\{-[(0.693)(0.5 \text{ year})/(5.26 \text{ years})]\} = 74.9 \text{ Ci}$$
$$R = 2.0 \text{ ft}$$
$$h = 25.0 \text{ ft}$$

$$\ln \frac{R^2 + h^2}{h^2} = \ln \frac{2^2 + 25^2}{25^2} = 6.38 \times 10^{-3}$$

The unshielded dose may be calculated based upon this information:

$$D(\text{unshielded}) = (3.14)(6.38 \times 10^{-3})$$
$$\times [(74.9 \text{ Ci} \times 1.3 \text{ R-m}^2 \text{ hr}^{-1} \text{ Ci}^{-1}/1.17 \text{ m}^2)]$$
$$= 1.67 \text{ R/hr}$$

Note: The point source approximation may be used because the field point distance is at least three times larger than the largest source dimension.

It should also be recognized that the beta dose component is effectively negated by distance and the demineralizer shell.

$$D_{\text{Point}} = (1.3 \text{ R-m}^2/\text{hr-Ci}) \times (74.9 \text{ Ci})$$

$$\times \frac{1}{(25 \text{ ft} \times 12 \text{ in./ft} \times 1 \text{ m}/39.37 \text{ in.})^2}$$

$$= 1.68 \text{ R/hr} \quad \text{(essentially the same result)}$$

Co-60 has two gamma rays (1.173 and 1.332 MeV) each with a yield of 100%. Because the average energy of these gammas is about 1.25 MeV, it is sufficient to use the 1.25-MeV attenuation coefficient for lead.

Interpolation of buildup factors could be performed, but is not necessary; that is, the steel shell has been neglected. The use of slightly conservative buildup factors is also justified based upon dose limitations.

$$t = 2.0 \text{ in.} \times 2.54 \text{ cm/in.} = 5.08 \text{ cm} \quad \text{(lead shield thickness)}$$
$$u(\text{Co-60}) = 0.65 \text{ cm}^{-1} \quad \text{(from the 1.25-MeV data)}$$

Calculate the optical thickness (ut) values:

$$\text{Co-60:} \quad ut = .65 \text{ cm}^{-1} \times 5.08 \text{ cm} = 3.30$$

The buildup factor is obtained as follows:

$$\text{Co-60:} \quad B(3.30) = B(ux = 4.0, @ 1.0 \text{ MeV}) = 2.26$$

The shielded dose rate may be obtained from the preceding factors and the relationship

$$D = D_0 B \exp(-ut)$$

$$D(2.0 \text{ in. Pb}) = (3.14)(6.38 \times 10^{-3})$$
$$\times (1.3 \text{ R-m}^2 \text{ Ci}^{-1} \text{ hr}^{-1})(74.9 \text{ Ci})(2.26)/1.17 \text{ m}^2$$
$$\times \exp(-3.3) \times (1000 \text{ mR/R}) = 139.0 \text{ mR/hr}$$

The total dose during the repair operation is

$$D_{\text{total}} = 3 \text{ hr} \times 139 \text{ mR/hr}$$
$$= 417 \text{ mR} \times 0.877 \text{ mrad/mR} \times \text{mrem/mrad} = 365.7 \text{ mrem}$$

This exceeds the 300-mrem limit. Therefore the 2 in. of lead is not acceptable. If no additional shielding were utilized, the worker would exceed his annual limit by (366 − 300) mrem = 66 mrem.

Scenario 58

Question 5.3: e

For a hot particle, the beta radiation emitted from the particle is the principal contributor to the skin dose.

Question 5.4: a

The increased use of more sensitive instrumentation is the major reason for the increased detection frequency of hot particles at nuclear power plants.

Question 5.5: d

Prior to 1988, neither the NCRP nor the ICRP provided explicit recommendations for hot-particle skin exposure. NCRP 106 (1989) provided this guidance.

Question 5.6: a

Distributed contamination on the skin is more likely to cause skin cancer when compared to the same isotopic activity residing within a hot particle. Hot particles are more likely to kill cells due to their high localized doses.

Question 5.7: c

For a 2.0-cm Sr-90 disk source, a detector located 1 cm from the source should have a detection efficiency of about 35%.

Question 5.8: c

The beta dose to the worker's finger may be determined from the relationship

$$D = A_i \mathrm{DF}_i t / S$$

where

D = beta dose (rad)
A_i = particle activity (μCi) = 0.39 μCi of Co-60
DF_i = 4.13 rad-cm^2/μCi-hr
t = time particle resides on the skin (hr) = 15.75 hr
S = area over which the dose is averaged (cm^2) = 1 cm^2
D = (3.9 × 10^{-1} μCi)(4.13 rad-cm^2/μCi-hr)
    × (15.75 hr)/(1 cm^2) = 25.37 rad

SOLUTIONS FOR CHAPTER 5   323

Question 5.9: d

Prior to January 1, 1993, this dose would exceed the 18.75-rem quarterly limit for extremities as outlined in 10 CFR 20.

After January 1, 1993, this dose would not exceed the 50-rem annual limit for skin dose as outlined in 10 CFR 20.

Question 5.10: e

Thyroid monitoring and urinalysis are not appropriate for hot-particle contamination residing on the skin because no internal hazard is likely. Moreover, hot particles are normally insoluble, and an intake is unlikely.

Question 5.11: a

An energy-compensated GM probe would provide the best sensitivity for detecting hot particles at the exit station.

Question 5.12: b

NCRP 106 established a 75 μCi-hr limit for hot-particle exposures.

Scenario 59

Question 5.13: a

Choice "a" results in higher doses. The other choices have been identified in various NRC and EPRI publications as reasons for the downward trend. See NUREG/CR-5158 for example.

Question 5.14: c

Co-60, a neutron activation product, has repeatedly been identified as the radionuclide responsible for at least 80% of these exposures.

Question 5.15: d

Co-60 is the major dose contributor to shutdown radiation fields.

Question 5.16: a

Hydrogen water chemistry increases doses by increasing N-16 in BWR turbines. The other choices are ways of reducing the BWR radiation field.

324   SOLUTIONS FOR CHAPTER 5

Question 5.17: g

Lack of knowledge of NRC Regulations by the radiological protection technician has not been a common error in unplanned exposures at nuclear power plants.

Question 5.18: a

The number of certified health physicists at a power reactor is not a useful indicator of radiological performance.

Question 5.19: a

The average annual exposure at nuclear power plants is in the 400 to 800-mrem range.

Question 5.20: b

Choice "b" is the only choice that is exactly the same as the previous requirements. Therefore, it cannot result in program changes. The other choices are changed requirements and should result in some program changes.

Question 5.21: c

The NRC dose limit for protection of the embryo–fetus applies once the female worker has voluntarily made her pregnancy known to her employer.

Question 5.22: b

ICRP-60 (1990) recommends that a 2-rem/year average or 10 rem over 5 years be adopted. Other recommendations suggest that 1 rem/year may be appropriate.

Scenario 60

Question 5.23

The activity derived from the irradiation is given by

$$A = N\sigma\phi[1 - \exp(-\lambda t_{irrad})] \exp(-\lambda t_{decay})$$

$$N = (10.0 \text{ g})(6.02 \times 10^{23} \text{ atoms/mole})/(59 \text{ g/mole})$$

$$= 1.02 \times 10^{23} \text{ atoms}$$

$$\sigma = 37b \times 1.0 \times 10^{-24} \text{ cm}^2/\text{b-atom} = 3.7 \times 10^{-23} \text{ cm}^2/\text{atom}$$

$$\phi = 1.0 \times 10^{10} \text{ n/cm}^2\text{-sec} \quad \text{(activation is by the thermal flux only)}$$

t_{irrad} = 10 years

t_{decay} = 0.5 years

λ = 0.693/5.27 years = 0.131 year^{-1}

A = (1.02 × 10^{23} atoms)(3.7 × 10^{-23} cm^2/atom)
× (1.0 × 10^{10} n/cm^2-sec)(1 dis/n)
× [1.0 − exp (−0.131 year^{-1} × 10 year)]
× exp (−0.131 year^{-1} × 0.5 year)
= 2.58 × 10^{10} dis/sec × 1 Ci/(3.7 × 10^{10} dis/sec) = 0.70 Ci

The exposure rate at a point 2.0 m from the small-particle source can be obtained from the point source approximation. Assuming no attenuation due to air, the exposure rate is

$$\dot{X}(N) = AG/r^2$$

$$= \frac{(0.70 \text{ Ci}) \times 1.3 \frac{\text{R-m}^2}{\text{hr-Ci}}}{(2.0 \text{ m})^2}$$

$$= 0.23 \text{ R/hr}$$

Question 5.24

The exposure rate at point P a distance of 2.0 m from the small-particle source can be obtained from the point source approximation:

$$\dot{X}(P) = AG/r^2$$

$$= \frac{(3.0 \text{ Ci}) \times 1.3 \frac{\text{R-m}^2}{\text{hr-Ci}}}{(2.0 \text{ m})^2}$$

$$= 0.98 \text{ R/hr}$$

Question 5.25

The general line source equation may be utilized for this question:

$$\dot{X}(Q) = GC_L\theta/h$$

The exposure rate at point Q, a distance of 2.0 m from the end of the sample line source, can be obtained by using a line source approximation as illustrated in Fig. S5.1.

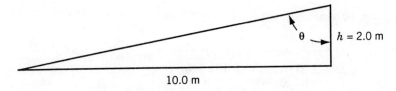

Fig. S5.1 Determination of the included angle (θ) for line source geometry.

$$\dot{X} = \frac{GC_1\theta}{h}$$

$$G = 1.3 \ \frac{\text{R-m}^2}{\text{hr-Ci}}$$

$$\tan \theta = 10.0 \text{ m}/2.0 \text{ m} = 5.0$$

$$\theta = \tan^{-1}(5.0) = 78.69°$$

$$= 78.69° \times 3.14 \text{ rad}/180.0°$$

$$= 1.37 \text{ rad}$$

$$C_1 = 3 \text{ Ci}/10 \text{ m}$$

$$= 0.3 \text{ Ci/m}$$

$$h = 2.0 \text{ m}$$

$$\dot{X} = \frac{1.3 \ \frac{\text{R-m}^2}{\text{hr-Ci}} \times \frac{0.3 \text{ Ci}}{\text{m}} \times 1.37}{2.0 \text{ m}}$$

$$= 0.27 \text{ R/hr}$$

Question 5.26

The exposure rate at point X, a distance of 2.0 m from the thin disk source, may be determined from the relationship:

$$\dot{X}(X) = \pi G C_a \ln \frac{r^2 + h^2}{h^2}$$

The geometric relationships utilized in the application of the disk source relationship are illustrated in Fig. S5.2.

$$G = 1.3 \ \frac{\text{R-m}^2}{\text{hr-Ci}}$$

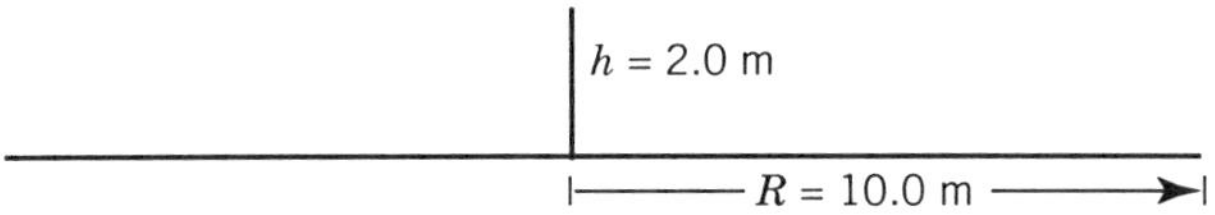

Fig. S5.2 Disk source geometry illustrating the 10.0-m-radius spill and the point of interest lying 2.0 m above the centerline of the spill.

$$C_a = 3 \text{ Ci}/(3.14)(10 \text{ m})^2 = 9.55 \times 10^{-3} \text{ Ci/m}^2$$

$$R = 10.0 \text{ m}$$

$$h = 2.0 \text{ m}$$

$$\dot{X} = 1.3 \frac{\text{R-m}^2}{\text{hr-Ci}} \times \frac{9.55 \times 10^{-3} \text{ Ci}}{\text{m}^2} \times 3.14$$

$$\times \ln \frac{(10.0 \text{ m})^2 + (2.0 \text{ m})^2}{(2.0 \text{ m})^2} = 0.13 \text{ R/hr}$$

Scenario 61

Question 5.27

The hollow pole can produce a radiation streaming path and could result in high dose rates in a narrow beam. The pole needs to have free-flooding holes or shield plugs inserted into the pole.

The installation of holes permits the possibility of trapping debris inside the pole and is the source of additional radioactive waste. Another option would be the use of a solid pole that could be easily decontaminated. This would eliminate both the radiation streaming and radioactive waste generating concerns.

The surfaces of the pole should also be polished to a smooth finish. Surfaces that could trap particles or debris must be avoided.

Question 5.28

The cutting head should be designed as a single disposable unit with quick disconnects to hydraulic or power lines. If this has not been part of the vendor supplied tool, then the replacement of cutter jaws will likely require that the tool be removed from the pool. The problem of trapped fragments, debris, or pieces of irradiated hardware in the tool must be considered as part of the radiological evaluation. The tool should be designed to minimize the possibility of trapping irradiated hardware fragments or debris. Large clearances or gaps are desirable. Ease of underwater inspection and cleanout should be available. Appropriate flush connections are to be provided to facilitate cleanout.

Question 5.29

The possibility of lifting an irradiated component out of the water or near the surface of the water must be addressed. The design of the crane must incorporate physical means to limit the travel of the crane to preclude the loss of the shielding provided by the water in the pool. Limit switches, special rigging, or lanyards should be used to limit the upward vertical travel of the crane.

Alarming radiation monitors or limit switch alarms should be provided to warn the operator of abnormal radiological conditions.

Question 5.30

The ARM detector is at a height of 6 ft above the pool surface and 20 ft from the edge of the pool railing.

Using this height and distance permits the associated angle to be determined as illustrated in Fig. S5.3:

$$\tan a = \frac{6 \text{ ft}}{20 \text{ ft}} = 0.3$$

Because the surface of the pool is parallel to the surface of the refueling floor, angle a may be used to determine the distance from the edge of the pool (d). This is illustrated in Fig. S5.4.

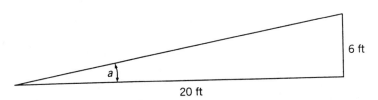

Fig. S5.3 Determination of the angle (a) describing the location of the radiation monitor relative to the edge of the spent-fuel pool.

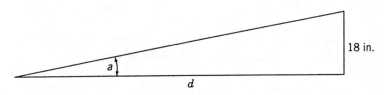

Fig. S5.4 Geometry for determining the distance (d) from the edge of the spent-fuel pool. The distance between the surface of the spent-fuel pool and the refueling deck is 18 in.

The distance between the pool and the refueling floor is 18 in.

$$\tan a = 18 \text{ in.}/d = 0.3$$
$$d = 18 \text{ in.}/0.3 = 60 \text{ in.} = 5 \text{ ft}$$

Question 5.31

The distance (d_1) from the center of the pool to the ARM is

$$d_1 = [(15 \text{ ft} + 20 \text{ ft})^2 + (6 \text{ ft} + 1.5 \text{ ft})^2]^{1/2}$$
$$= 35.8 \text{ ft}$$

The distance (d_2) from the center of the pool to the edge of the railing is

$$d_2 = [(15 \text{ ft})^2 + (1.5 \text{ ft} + 3.5 \text{ ft})^2]^{1/2}$$
$$= 15.8 \text{ ft}$$

The alarm setpoint of the installed ARM is adjusted so that it will alarm if a point source at the center of the pool yields an exposure rate of 1 R/hr at the pool railing ($\dot{X}_2$). Using these results, the correct ARM setpoint ($\dot{X}_1$) is given by the inverse-square law:

$$\dot{X}_1 d_1^2 = \dot{X}_2 d_2^2$$
$$\dot{X}_1 = \dot{X}_2 (d_2/d_1)^2$$
$$\dot{X}_1 = 1000 \text{ mR/hr } (15.8 \text{ ft}/35.8 \text{ ft})^2$$
$$= 194.8 \text{ mR/hr}$$

Question 5.32

Characterization Technique	Advantages	Disadvantages
Physical sampling	Allows for the most precise determination of radionuclide composition. Calculational conservatism is minimized.	Sample handling requires strict contamination control practices. Sampling techniques may produce hot particles that increase the risk of contamination events or unanticipated exposures.

Characterization Technique	Advantages	Disadvantages
Physical sampling (continued)		Sample must be shown to be representative of the material. Irradiation may not have been uniform, which requires multiple samples to be taken.
Underwater surveys	Direct handling of activated material is not required	Material composition uncertainties may require calculations to be overly conservative. Uncertainties in the measurement technique, such as the probe position, will cause significant errors.

Question 5.33

Characterization Technique	Sources of Error
Physical sampling	Assumptions regarding the uniform irradiation of the component. A sufficient number of samples are required to ensure that they are representative. Only surface conditions are included in the sample. Interior samples are not feasible with the proposed method.
Underwater surveys	The chemical composition of the material may be incomplete. This includes the fractional quantity of cobalt impurities in the material. The irradiation history of the material will be incomplete. Core power levels and the activating neutron fluence levels may be incomplete. This is complicated by not knowing the exact core location during the irradiation period. The position of the probe will be uncertain. This may be complicated by water clarity in the pool and difficulties in accurately positioning the probe relative to the equipment being measured. These uncertainties complicate and reduce the accuracy of the subsequent model calculations.

SOLUTIONS FOR CHAPTER 5   331

Question 5.34

Because of their composition and age, the filters are expected to be in a degraded condition and not have their initial structural integrity. Disintegration of the filters during movement is possible. The disintegration of these filters could lead to the following radiological controls concerns:

1. Release of numerous hot particles
2. Generation of debris that will float on the surface of the spent fuel pool. Without the water shielding, this debris presents a significant external radiation hazard.

Question 5.35

Appropriate radiation control measures to deal with the hazards outlined in question 5.34 include:

1. Minimize the distance that the filters are transported. An underwater transfer device, supplying structural integrity, could be employed.
2. Utilize tooling and transfer equipment that minimize the possibility of damaging the filters during transport.

Scenario 62

Question 5.36

Primary responsibilities of the RCM are as follows:

1. Perform timely calculations of the projected doses from facility releases.
2. Ensure the radiation safety of the personnel at the EOF.
3. Formulate/update Protection Action Recommendations (PARs).
4. Coordinate the movement of off-site sampling teams in order to assess the release.
5. Direct the overall radiological response to the event.
6. Determine if the use of thyroid blocking agents are warranted.
7. Discuss plant radiological conditions, dose projections, and PARs with senior State and Federal Officials.
8. Authorize exposure extensions, emergency exposure requests, and thyroid blocking agents.

Question 5.37

The three fission product barriers are:

1. Fuel element cladding
2. Reactor coolant system
3. Containment building

Question 5.38

Fission Product Barrier Status

1. *Intact*. The fission product barrier is capable of preventing the release of fission products to the environment. This is the normal condition.
2. *In Jeopardy*. The barrier is currently intact, but plant conditions are such that the barrier will likely be breached.
3. *Breached*. The fission product barrier is not able to prevent the release of fission products to the environment.

Question 5.39

Based on the data, we present the status of the three fission product barriers:

Fission Product Barrier	Status
Fuel cladding	Breached as noted by the high activity in the letdown (primary) system.
Reactor coolant system	Breached as evidenced by the high activity in the steam generator blowdown (secondary) sample.
Containment	In jeopardy as related to the increasing containment pressure, but more importantly due to the primary to secondary leakage and the possibility of a release if a relief valve lifts due to increases in secondary system pressure.

Question 5.40

Factors affecting the calculation of off-site doses include:

1. Release magnitude and isotopic composition
2. Release rate
3. Meteorological conditions (atmospheric stability class, wind speed, and precipitation)
4. Release height
5. Release duration
6. Changing plant conditions due to repairs or equipment failures

Question 5.41

The projected thyroid dose rate ($\dot{H}$) at 2 miles will be due to the I-131 released, and this value must consider any partitioning from the secondary side of the plant to the environment. The release concentration (C) of I-131 is determined from the I-131 concentration in the steam generator blowdown sample and the iodine partitioning factor:

$$C = (0.015)(6.3 \times 10^{-1} \ \mu Ci/cm^3)$$
$$= 9.45 \times 10^{-3} \ \mu Ci/cm^3$$

The release rate (Q) can be obtained from the release concentration by assuming that the release is through the atmospheric relief valve having a flow rate of 1.4×10^7 cm^3/sec:

$$Q = (1.47 \times 10^7 \ cm^3/sec)(9.45 \times 10^{-3} \ \mu Ci/cm^3)$$
$$= 1.32 \times 10^5 \ \mu Ci/sec = 0.132 \ Ci/sec$$

$$\dot{H} = \frac{X u}{Q} Q \frac{1}{u} (DCF) \quad @ \ 2 \ miles$$

$$= \frac{2.69 \times 10^{-3}}{m^2} (0.132 \ Ci/sec)(3600 \ sec/hr)$$

$$\times \frac{1}{(15 \ mi/hr)(1609.36 \ m/mi)(1 \ hr/3600 \ sec)}$$

$$\times 77.2 \ \frac{rem/sec}{\mu Ci/cm^3} \times \frac{1 \ m^3}{1 \times 10^6 \ cm^3} \times \frac{1 \times 10^6 \ \mu Ci}{Ci}$$

$$= 14.7 \ rem/hr$$

Question 5.42

Sheltering should be recommended at this time. Because a long-term release is not expected, evacuation is not warranted.

Question 5.43

A conservative estimate of the dose in the town can be performed by using the dose information calculated at the 2.0-mile distance:

$$D = 14.7 \ rem/hr \times 0.25 \ hr = 3.7 \ rem$$

Again, evacuation is not warranted. The population could not reasonably be expected to be evacuated in a 15-min period. Sheltering is the recommended protective action.

Scenario 63

Question 5.44

In order to determine the Rb-88 activity at the beginning of the counting interval, one must know the net filter counts, disintegration constant, counting time interval, and counting efficiency.

334   SOLUTIONS FOR CHAPTER 5

The activity at the beginning of the counting interval is

$$A = \lambda N$$

The number of atoms N is given by

$$N = \frac{C_S}{[1 - \exp(-\lambda T_{count})]e}$$

where C_S is the number of filter counts in the counting interval (30 min), e is the efficiency, and $1 - \exp(-\lambda T_{count})$ is the fraction of the atoms expected to decay during the counting interval (T_{count}). Therefore, the activity of Rb-88 is written as

$$A_{Rb\text{-}88} = \frac{\lambda C_S}{(1 - \exp[-\lambda T_{count}])e}$$

where
$\lambda = \ln(2)/T_{1/2} = 0.693/17.7 \text{ min} = 0.0392/\text{min}$
C_S = net counts = 1.5×10^5 counts
T_{count} = counting time = 30 min
e = counting efficiency = 0.1 count/disintegration

$$A_{Rb\text{-}88} = \frac{(0.0392/\text{min})(1.5 \times 10^5 \text{ counts})}{[1 - \exp(-0.0392/\text{min} \times 30 \text{ min})](0.1 \text{ count/disintegration})}$$

$$= \frac{8.503 \times 10^4 \text{ disintegration/min}}{(3.7 \times 10^{10} \text{ disintegration/sec-Ci})(60 \text{ sec/min})(1 \text{ Ci}/10^6 \, \mu\text{Ci})}$$

$$= 0.0383 \, \mu\text{Ci}$$

Question 5.45

To calculate the activity concentration $U_{Rb\text{-}88}$ a knowledge of the counting and sampling times is needed:

$$T_S \text{ (sampling time)} = 30 \text{ min}$$
$$T_T \text{ (transit time)} = 10 \text{ min}$$
$$T_C \text{ (counting time)} = 30 \text{ min}$$

The net counts on the filter paper C_S is a function of the total number of Rb-88 atoms (N) in the sampled air volume:

$$C_S = N(\text{FR}) \frac{1 - \exp(-\lambda T_S)}{\lambda T_S} \exp(-\lambda T_T)$$
$$\times [1 - \exp(-\lambda T_C)]e$$

where

N = number Rb-88 atoms in the sampled air volume (V)

$$= \frac{U_{\text{Rb-88}}}{\lambda} V$$

$$= \frac{U_{\text{Rb-88}}}{\lambda} FT_S$$

F = sampling flow rate = 30 liters/min
FR = filter retention = 1.0

Utilizing these relationships, the activity concentration is determined to be

$$C_S = \frac{U_{\text{Rb-88}}}{\lambda} FT_S(\text{FR}) \frac{1 - \exp(-\lambda T_S)}{\lambda T_S} \exp(-\lambda T_T)$$

$$\times [1 - \exp(-\lambda T_C)]e$$

$$U_{\text{Rb-88}} = \frac{\lambda^2 C_S}{e(\text{FR})F[1 - \exp(-\lambda T_S)] \exp(-\lambda T_T)[1 - \exp(-\lambda T_C)]}$$

$$= \frac{(0.0392/\text{min})^2 (1.5 \times 10^5 \text{ counts})}{(0.1 \text{ counts/disintegration})(1)(30 \text{ liters/min} \times 1000 \text{ cm}^3/\text{liter})}$$

$$\times \frac{1}{[1 - \exp(-0.0392/\text{min} \times 30 \text{ min})] \exp(-0.0392/\text{min} \times 10 \text{ min})}$$

$$\times \frac{1}{[1 - \exp(-0.0392/\text{min} \times 30 \text{ min})]}$$

$$= (0.238 \text{ disintegration/min-cm}^3)(1 \text{ min}/60 \text{ sec})(1.0 \text{ }\mu\text{Ci}/3.7$$

$$\times 10^4 \text{ disintegration/sec})$$

$$= 1.07 \times 10^{-7} \text{ }\mu\text{Ci/cm}^3$$

Question 5.46

The airborne activity concentration of Kr-88 is the same as the Rb-88 activity concentration. This assumes that the physical removal rate of Rb-88 is negligible compared with its decay rate. An examination of the relative half-lives of Kr-88 and Rb-88 suggests that secular equilibrium conditions are met, which also supports the contention.

Kr-88: T_{parent} = 2.84 hr

Rb-88: T_{daughter} = 17.7 min × (1 hr/60 min) = 0.295 hr

$T_{\text{parent}} \gg T_{\text{daughter}}$ (which supports the secular equilibrium contention).

SOLUTIONS FOR CHAPTER 6

Scenario 64

Question 6.1

Instantaneous Grab Samples. These are one-time or momentary grab samples that provide the instantaneous concentration during the sampling time. Examples of techniques employing this method include: Kusnetz Method, Tsivoglou Method, sealed ionization chamber, scintillation (Lucas) cells, Thomas Method, Rolle Method, two-filter method, and alpha spectroscopy.

Integrated Samples. This is a cumulative method yielding a time-averaged concentration over the duration of the measurement. Examples include: track etch detectors, charcoal canisters, and electrets (E-perms).

Continuous Samples These are a series of grab measurements performed at prescribed time intervals that yield a chronological record of the instantaneous radon concentration. Examples include: continuous working level meters, flow-through scintillation cells, and flow-through pulse ionization chambers.

Question 6.2

Alpha Track Detectors Alpha or nuclear track detectors are composed of cellulose nitrate or polycarbonate strips set inside a filtered chamber. Rn-222 diffuses into the chamber via the filter and exposes the strip to the radon gas. After exposure, the strip is chemically etched to enlarge the alpha tracks. The tracks are optically counted, and the track density is calibrated to the average radon concentration over the exposure period.

Activated Charcoal. Radon gas is adsorbed onto an activated charcoal bed. The bed is counted using an appropriate counting method such as NaI gamma ray counting of the photons emitted by progeny ingrowth in the activated charcoal. The Bi-214 photons are the dominant photon contributor.

Electrets (E-Perms). An electret is a dielectric disk. Charged ions formed by the decay of radon and its progeny are collected inside a fixed chamber volume by the quasi-permanently charged electret. The voltage discharge of the disk is linearly proportional to the average radon concentration over the exposure period.

Question 6.3: c

The short-lived Rn-222 decay products include: Bi-214, Pb-214, Po-218, and Po-214. The radon series is summarized in Table 6.1.

Question 6.4: b

Indoor unattached radon decay products are not usually measured using cyclone precollectors.

Question 6.5: c

The fraction of unattached radon decay products in the air does not depend on the radon concentration.

Question 6.6

$$\text{NWLM} = \frac{C_{\text{WL}} F_{\text{OCC}}}{k}$$

$$\text{NWLM} = \frac{(0.09 \text{ WL})(24 \text{ hr/day})(365 \text{ day/year})(0.75)}{(170 \text{ WL-hr/WLM})}$$

$$= 3.5 \text{ WLM/year}$$

Question 6.7

1. *Breathing Rate (BR)*
    a. NCRP 78: The curve is a relatively flat straight line because the increase in deposition caused by the larger breathing rate is slightly offset by the decreased tidal volume.
    b. BEIR IV: As the breathing rate increases, the total quantity of daughter deposition will increase, but only slightly. The fractional deposi-

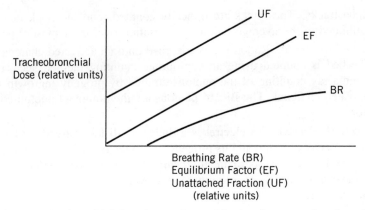

Fig. S6.1 Tracheobronchial dose from radon and its daughters as a function of breathing rate, equilibrium factor, and the unattached fraction.

tion in the upper bronchial region decreases and the total deposition increases less rapidly than the increased alpha energy inhaled. These factors results in a dose versus breathing rate curve that resembles a square root function because of the square root variation of mean bronchial dose with minute volume. Minute volume is the product of the tidal volume and the breathing rate.

2. *Equilibrium Factor (EF)*. As the equilibrium factor increases from zero to unity, the dose will increase linearly with a slope of one.
3. *Unattached Factor (UF)*. As the unattached fraction increases, the tracheobronchial dose increases linearly from some constant positive y-intercept value.

The qualitative variation in the tracheobronchial dose as a function of the breathing rate, equilibrium factor, and unattached fraction is summarized in Fig. S6.1.

Scenario 65

Question 6.8

Mechanisms by which airborne concentrations of radioactive materials are reduced during atmospheric transport include:

1. Radioactive decay
2. Diffusion or dispersion
3. Precipitation removal (rainout or washout)
4. Gravitational settling
5. Ground contact

Question 6.9

The dose rate for a constant intake of radioactivity, as the activity is building up, is given by

$$\dot{D}(t) = \dot{D}_{max}[1 - \exp(-0.693t/T_{1/2})]$$
$$= \dot{D}_{max} F_{BU}$$

Once the intake stops, the dose rate decreases as expected from the following relationship:

$$\dot{D}(t) = \dot{D}_{max} F_{BU} \exp(-0.693t/T_{1/2})$$
$$= \dot{D}_{max} F_D F_{BU}$$
$$T_{1/2} = 1 \text{ day}$$

The graph is plotted using the factors

$$F = F_{BU} F_D$$

For this problem, intake occurs only during the first 6 days. During the subsequent 6 days, decay occurs. Using this information, the points defining the desired graph can be determined:

Day (t)	Buildup F_{BU}	Decay F_D	F
1	0.5	1.0	0.5
2	0.75	1.0	0.75
3	0.87	1.0	0.87
4	0.94	1.0	0.94
5	0.97	1.0	0.97
6	0.98	1.0	0.98
7[a]	0.98	0.5	0.49
8	0.98	0.25	0.25
9	0.98	0.13	0.13
10	0.98	0.06	0.06
11	0.98	0.03	0.03
12	0.98	0.02	0.02

[a]The intake stops after day 6. No further buildup occurs in days 7–12.

Question 6.10

The dose rate for a constant intake of radioactivity, as the activity is building up, is given by

$$\dot{D}(t) = \dot{D}_{max}[1 - \exp(-0.693t/T_{1/2})]$$
$$= \dot{D}_{max}F_{BU}$$

Curve 1 assumes an effective half-life of 1 day, and curve 2 assumes an effective half-life of 2 days.

Day (t)	Curve 1 Buildup F_{BU}	Curve 2 Buildup F_{BU}
1	0.5	0.29
2	0.75	0.5
3	0.87	0.65
4	0.94	0.75
5	0.97	0.82
6	0.98	0.87
7	0.99	0.91
8	1.00	0.94
9	1.00	0.96
10	1.00	0.97
11	1.00	0.98
12	1.00	0.98

Question 6.11

The input rate to the pond's surface is given by

$$r_d = (V)(Q)(X/Q)$$
$$= (1.0 \times 10^{-2} \text{ m/sec})(1.0 \times 10^8 \text{ Bq/sec})(1.8 \times 10^{-7} \text{ sec/m}^3)$$
$$= 0.18 \text{ Bq/m}^2\text{-sec}$$

Question 6.12

The maximum steady-state concentration in the pond is

$$C_{eq} = (r_d)(S)/(\lambda_e)(V)$$

$$\lambda_e = \frac{0.693}{8 \text{ days}} + \frac{0.693}{15 \text{ days}} = 0.13/\text{day}$$

$$C_{eq} = \frac{(0.5 \text{ Bq/m}^2\text{-day})(100 \text{ m} \times 10 \text{ m})}{(0.13/\text{day})(1 \text{ m} \times 10 \text{ m} \times 100 \text{ m})} = 3.76 \text{ Bq/m}^3$$

Question 6.13

$$C_f = (C_{eq})(I)/(\lambda_e)$$

$$\lambda_e = \frac{0.693}{8 \text{ days}} + \frac{0.693}{21 \text{ days}} = 0.12/\text{day}$$

$$C_f = \frac{(2\text{Bq/m}^3)(8 \times 10^{-5} \text{ m}^3/\text{kg-day})}{(0.12/\text{day})}$$

$$= 1.34 \times 10^{-3} \text{ Bq/kg}$$

Question 6.14

The question 6.13 model inaccuracies include:

1. Water evaporation which increases the I-131 concentration.
2. Decreased concentration due to rain and snow.
3. Decreased concentration due to settling.
4. The previous model assumes that all I-131 incorporated into the fish stays there. There will be an excretion function for the fish that reduces the activity.
5. The I-131 removal by other food-chain members is ignored.
6. Physical and chemical removal of I-131 by the pond.
7. Bottom feeding by fish may concentrate I-131.
8. Changes in pond pH may increase the evolution of I-131 from the pond.

Scenario 66

Question 6.15

Type of Sample	Preoperational Environmental Monitoring		Sample Analysis
	Location	Method/Frequency	
Air			
Particulates	Site boundary	Continuous/weekly filter change or High volume/24 hr every 7 days	1
	Nearest residence	Continuous/weekly filter change or High volume/24 hr every 7 days	1

Type of Sample	Preoperational Environmental Monitoring		Sample Analysis
	Location	Method/Frequency	
Air			
Particulates (continued)	Control location (background)	Continous/weekly filter change or High volume/24 hr every 7 days	1
Water			
Groundwater	Off-site existing wells (within 5 miles)	Grab sample/quarterly	2
	Site boundary (located hydrologically down-gradient)	Grab sample/quarterly	2
	Control location (located hydrologically up-gradient/ background)	Grab sample/quarterly	2
Surface water	Permanent on-site or off-site sources subject to direct drainage from contaminated site areas	Grab sample/quarterly	2
	Surface waters passing through the site or off-site sources subject to direct drainage from contaminated site areas	Grab sample/quarterly	2
Vegetation Food, and Fish			
Vegetation	Composite grab samples at the site boundary	Sample/spring and fall	3
Food	Composite grab samples within site boundary	Sample/spring and fall	4

Type of Sample	Preoperational Environmental Monitoring		Sample Analysis
	Location	Method/Frequency	
Vegetation Food, and Fish			
Food (continued)	Crops within a few miles of the site boundary	Grab sample/time of harvest	2
Fish	Collection of fish in the site environs that may be subject to site runoff or seepage from contaminated areas	Grab sample/spring and fall	4
Soil and Sediment			
Soil	Immediate disposal area (1 sample per acre)	Grab sample/once	5
	Within site boundary (1 sample per 10 acres)	Grab sample/once	5
Sediment	Upstream and downstream of surface waters	Grab sample/once	5
	On-site surface waters	Grab sample/once	5
Direct Gamma			
	Representative locations across the site	Environmental TLDs/once	6

Analysis:
1. Quarterly composite, gamma scan, gross alpha, and gross beta
2. Gamma scan, gross alpha, gross beta, tritium, and carbon-14
3. Gamma scan, gross alpha, gross beta, tritium, and carbon-14 (species composition based on biomass surrounding plant site)
4. Gamma scan, gross alpha, gross beta, tritium, and carbon-14 (composite of edible portions)
5. Gamma scan, gross alpha, and gross beta
6. Gamma exposure rates from TLD, pressurized ion chamber, or properly calibrated survey instruments

Question 6.16

Major exposure pathways expected during the institutional care period are as follows:

1. Direct gamma from ground contamination and/or dust clouds
2. Airborne contamination from dust or radioactive gas inhalation
3. Ground water from contamination migration to a well or stream
4. Contamination of surface waters through erosion or trench overflow

Question 6.17

The risks for the various facility options are given by the product of the risk coefficient and the population doses:

Below Ground (BG):

$$\text{ecd} = \text{excess cancer deaths}$$

$$r_{BG} = (1.00 \times 10^{-4} \text{ ecd/person-rem})(10 \text{ person-rem/year})$$
$$\times (400 \text{ years}) = 0.4 \text{ ecd}$$

Above/Below Ground (ABG):

$$r_{ABG} = (1.00 \times 10^{-4} \text{ ecd/person-rem})(9 \text{ person-rem/year})$$
$$\times (400 \text{ years}) = 0.4 \text{ ecd}$$

Above Ground (AG):

$$r_{AG} = (1.00 \times 10^{-4} \text{ ecd/person-rem})(30 \text{ person-rem/year})$$
$$\times (400 \text{ years}) = 1.2 \text{ ecd}$$

Question 6.18

The below-ground technology is the preferred technology. Because the site already meets the technical requirements for stability, an enhanced technology is not necessary to ensure the confinement of the radioactivity. Both above/below- and above-ground options would compromise the confinement that the site itself offers.

Worker doses and surface dose rates are higher with the above-ground disposal. Above-ground disposal does not take advantage of the site stability, and the radioactivity has more direct pathways to the environment.

Scenario 67

Question 6.19

$$dC/dt = S/V - \lambda C - IC$$
$$C = 25 \text{ pCi/liter}$$
$$V = 100 \text{ cm}^3$$

For steady-state conditions, $dC/dt = 0$ and

$$S = VC(\lambda + I)$$
$$S = (25 \text{ pCi/liter})(100 \text{ m}^3 \times 10^6 \text{ cm}^3/\text{m}^3 \times 1 \text{ liter}/1000 \text{ cm}^3)$$
$$\times (0.00755/\text{hr} + 0.2/\text{hr})$$
$$= (5.19 \times 10^5 \text{ pCi/hr})(1 \text{ }\mu\text{Ci}/1.0 \times 10^6 \text{ pCi})$$
$$= 0.519 \text{ }\mu\text{Ci/hr}$$

Question 6.20

$$S_w = NUC_w f_w$$
$$N = 5 \text{ members}$$
$$U = 200 \text{ liters/member-day}$$
$$C_w = 2000 \text{ pCi/liter}$$
$$f_w = 0.70$$
$$S = (5 \text{ members})(200 \text{ liters/member-day})(2000 \text{ pCi/liter})(0.7)$$
$$\times (1 \text{ }\mu\text{Ci}/1.0 \times 10^6 \text{ pCi})$$
$$= 1.4 \text{ }\mu\text{Ci/day} \times 1 \text{ day}/24 \text{ hr} = 0.058 \text{ }\mu\text{Ci/hr}$$

The source term from the water supply is approximately 10% of the total estimated radon source term for the home. Therefore, radon in the water supply alone cannot explain the source of the elevated radon level for the home.

Question 6.21

$$C_{\text{WL}} = C(\text{EF})k$$
$$C = 25 \text{ pCi/liter}$$
$$\text{EF} = 0.3$$

k = conversion factor (1 WL/100 pCi/liter)

C_{WL} = (25.0 pCi/liter)(0.3)(1 WL/100 pCi/liter)

= 0.075 WL

Question 6.22

To calculate the effective dose equivalent, the number of working level months per year should be calculated:

$AWLM = (OF)(C_{WL})k_1/k_2$

C_{WL} = 0.1 WL

OF = 0.7

k_1 = time conversion factor (365 days/year × 24 hours/day)

k_2 = 170 hr/month for WLM estimates

$AWLM$ = (365 days/year)(24 hr/day)(0.70)(0.10 WL)/170 hr/month

= 3.61 WLM/year

The dose equivalent rate to the tracheobronchial (TB) region of the lung is obtained from the AWLM:

$D_L = (DCF_{TB})(AWLM)(QF)$
D_L = committed dose equivalent rate to the TB region of the lung (rem/year)
DCF_{TB} = tracheobronchial dose conversion factor = 0.7 rad/WLM
QF = quality factor for alpha particles = 20 rem/rad
D_L = (3.61 WLM/year)(0.7 rad/WLM)(20 rem/rad) = 50.5 rem/year delivered to the TB region of the lung
$H = w_L D_L$
H = committed effective dose equivalent rate (rem/year)
w_L = weighting factor for the lung (0.06)
H = 50.5 rem/year × 0.06 = 3.03 rem/year

Question 6.23

Source. Because the water supply and building materials have been eliminated as the principal sources of radon, soil gas entry must be responsible for the home's radon levels.

Mitigation Methods

1. Ventilate the soil under the basement slab.
2. Seal cracks, sumps, and surfaces in the basement foundation.

3. Increase the home's ventilation with outside air.
4. Remove soil around the home's foundation and replace it with low-uranium-concentration material with low porosity.

Scenario 68

Question 6.24

K-40 and cosmic radiation are not significant factors in air monitoring. Both radon and thoron and their daughters can have a dramatic impact on air monitoring. When counting for alpha particles, it will be necessary to ensure that the natural products are properly taken into account.

Question 6.25

Cosmic radiation is a significant contributor to sample counting. It is a significant contributor to the background counting rate for low-level beta and gamma counting equipment even if the equipment is well-shielded from terrestrial radiation sources.

Question 6.26

In vivo counting must account for natural radiation sources. The human subjects contain significant quantities of K-40 and may also have radon/thoron daughter activity on their bodies. K-40 has a positive impact because it provides a qualitative check on the proper operation of the counting equipment.

The equipment in the counting chamber may contain K-40, uranium, and thorium. The air within the chamber may contain radon and thoron and their associated daughter products. Some cosmic radiation will also enter the counting chamber. All of these sources must be properly determined in order to obtain accurate *in vivo* results.

Question 6.27

Cosmic radiation, K-40, the thorium series, and the uranium series will all contribute to background measurements. These sources and their variation make it difficult to detect small man-made contributions to the radiation background.

Question 6.28

Cosmic radiation, K-40, the thorium series, and the uranium series will all contribute to background radiation levels. For this reason, one cannot obtain a zero background for calibrating low-level instruments.

Questions 6.29

Nearly all construction materials, soil, and masonry contain K-40 and trace uranium. Some also contain thorium. Fallout from atomic bomb tests also provide residual radiation levels when they are contained in the construction material.

Shielding materials are often contaminated with natural background or fallout sources. For this reason, care must be exercised in choosing shield materials. Pre-World War II steel is preferred if it is available.

Question 6.30

Radon and thoron daughters are often contaminants in low-level counting laboratories. K-40 and the various isotopes of the uranium and thorium series may be found in laboratory items such as glassware and ceramic materials. Moreover, some of the chemical reagents will contain K-40 and some of the isotopes of the uranium and thorium series.

Scenario 69

Question 6.31

Net count rate on the gas channel monitor is

$$R = R_{s+b} - R_b$$
$$= 1.0 \times 10^7 \text{ cpm} - 50 \text{ cpm}$$
$$= (1.0 \times 10^7 \text{ cpm})/(0.3 \text{ cpm/dpm})$$
$$= 3.33 \times 10^7 \text{ dpm}$$

The air concentration may be obtained from knowledge of the net count rate and the detector volume:

$$C = R/V$$
$$= \frac{(3.33 \times 10^7 \text{ dpm})}{50 \text{ cm}^3} \times \frac{1 \ \mu\text{Ci}}{2.22 \times 10^6 \text{ dpm}}$$
$$= 0.3 \ \mu\text{Ci/cm}^3$$

The stack release source term (Q) in Ci/min may be obtained from this information and the stack flow rate (F):

$$Q = FC$$
$$= (1.0 \times 10^5 \text{ ft}^3/\text{min})(0.3 \ \mu\text{Ci/cm}^3)(28.32 \text{ liters/ft}^3)$$
$$\times (1000 \text{ cm}^3/\text{liter})(1.0 \text{ Ci}/1 \times 10^6 \ \mu\text{Ci})$$
$$= 850 \text{ Ci/min}$$

Because the release is expected to last for 4 hr (t), the total quantity of Kr-85 released (q) is

SOLUTIONS FOR CHAPTER 6   349

$$q = Qt$$
$$= (850 \text{ Ci/min})(4 \text{ hr})(60 \text{ min/hr})$$
$$= 2.04 \times 10^5 \text{ Ci}$$

The dose is related to these quantities by the relationship

$$D_i = \frac{X\bar{u}}{Q}(1/\bar{u})q\text{DCF}_i$$

Total Body Dose:

$$D_i = \frac{(5.0 \times 10^{-4} \text{ m}^{-2})}{(2.0 \text{ m/sec})}(2.04 \times 10^5 \text{ Ci}) \, 4.7 \times 10^{-4} \frac{\text{rem-m}^3}{\text{Ci-sec}}$$
$$= 2.40 \times 10^{-2} \text{ rem}$$

Skin Dose:

$$D_i = \frac{(5.0 \times 10^{-4} \text{ m}^{-2})}{(2.0 \text{ m/sec})}(2.04 \times 10^5 \text{ Ci}) \, 6.0 \times 10^{-2} \frac{\text{rem-m}^3}{\text{Ci-sec}}$$
$$= 3.06 \text{ rem}$$

Question 6.32

Specific notifications would be specified in the facility's emergency plan and its associated implementing procedures. As a minimum, notifications to the host county emergency management coordinator, the state radiation protection branch, and Federal regulators (U.S. Nuclear Regulatory Commission or U.S. Department of Energy) would be recommended.

Because the whole-body exposures are low and the event is occurring and will end during the night, people will probably be indoors. This fact and the low doses suggest that no evacuation is warranted.

Question 6.33

Both on-site and off-site monitoring teams should be deployed. During the event, they should measure air concentrations of Kr-85 in order to verify the accuracy of the dose projections. Following the event, their measurements will confirm that the event has been terminated. In addition, ground deposition samples should be obtained for use by state and local officials to verify that none of the Kr-85 (10.7-year half-life) remains.

Question 6.34

Following the termination of the release, environmental monitoring TLDs should be collected in order to verify the population dose assessments.

Scenario 70

Question 6.35

Plant Source	Pathway
Gaseous effluents	Submersion exposure
	Inhalation
	Deposition on crops and subsequent ingestion by man
	Deposition on animal feed, ingestion of livestock meat
	Deposition on animal feed, ingestion of livestock milk
	Ground deposition, water contamination, and water immersion
	Ground deposition, water contamination, and water ingestion
	Ground deposition, water contamination, and aquatic food consumption
	Ground deposition to direct exposure from the ground
Liquid effluents	Immersion exposure
	Shoreline exposure
	Ingestion of water
	Uptake by aquatic food, ingestion of aquatic food
	Ground-water deposition, ingestion of livestock meat
	Ground-water deposition, ingestion of livestock milk
Direct radiation and shine from N-16 sources or other radionuclides in the turbine hall steam piping systems	Direct exposure

Question 6.36

The three potentially dominant pathways include:

1. Direct exposure from N-16. Any failed fuel would add to this direct exposure.
2. The submersion dose could be dominant, depending on the fuel performance and how well the turbine hall shields the N-16 gamma rays.

3. Without charcoal absorbers, the gaseous release of radionuclides could be significant.

The dominant pathway would be determined following the assessment of the plant's operating parameters and design characteristics. Calculations, including the information from question 6.37, would be needed to determine the dominant pathway.

Another significant pathway could be the cow's milk pathway. I-131 releases, subsequent ground deposition, and consumption by cattle would also require a more detailed assessment.

Question 6.37

To refine the pathway exposure estimates, additional factors including the following should be evaluated:

1. Population data by sector surrounding the plant.
2. Land use census.
3. Characterization of the population's behavior including its living habits, food preferences, and recreational preferences.
4. Meteorological data including seasonal averages and historical patterns.
5. Plant operating characteristics including the time between release and population exposure.
6. Biological census characterizing biota and their concentration factors for the nuclides that could be released from the facility.
7. Concentration factors for the various pathways.
8. Stream and aquifer volumes and associated flow rates. These water bodies will dilute any liquid effluent releases.
9. Facility release rates and source terms classified by their magnitude, particle size, and chemical form.

The results of a more detailed pathway analysis may suggest design changes to reduce the population dose from the higher-dose pathways. These changes could include a heavier (thicker) turbine casing to shield the N-16 gamma rays or the addition of charcoal absorbers to the gaseous discharge system.

SOLUTIONS FOR CHAPTER 7

Scenario 71

Question 7.1: d

The capture of a negative pion at the end of its range results in the transfer of about 100 MeV to the tumor. This energy is transferred via scattering or nuclear interactions to the atoms, molecules, neutrons, or protons at the tumor site. The resulting particles have high LET, and the charged particles have short ranges which selectively deposit the pion's energy at the tumor site. By comparison, Co-60 will irradiate all tissues including the tumor and healthy tissue.

Question 7.2: a

The dominant energy-loss mechanism for charged particles in matter is the collision of these particles with atomic electrons. These collisions do not appreciably deflect the path in tissue of a heavy charged particle like a pion, which is much more massive than the electron.

With an electron or positron beam, the beam particles have the same mass as the atomic electrons, and thus large-angle deflections and beam spreading occur.

Elastic nuclear collisions through the Coulomb force have a much smaller cross section, but these collisions involve multiple small-scattering angles which spread the beam. For heavy charged particles, multiple scattering is a significant effect.

For negative pions, multiple scattering by atomic nuclei is the dominant effect (response "a"). Response "e" would apply for electron or positron beams, but it is not the best general answer.

Question 7.3: c

Neutrons, photons, and muons are the penetrating radiations generated with the accelerator running. Residual photons are present from induced radioactivity and can also penetrate thick shielding.

Question 7.4: a

Induced beta and gamma activity will be present in the treatment area.

Question 7.5: b

This energy difference is required to overcome the nuclear binding energy in order to fragment the oxygen, carbon, or nitrogen nucleus.

Question 7.6: a

A negative pion decays into a negative muon, which then decays into an electron. Some neutron contamination is also expected.

Question 7.7

The distance traveled by the pion is expressed in terms of the standard particle range relationship:

$$R = \text{range in g/cm}^2$$
$$t = \text{physical distance traveled (cm)}$$
$$p = \text{density of the material in g/cm}^3$$
$$R = tp$$
$$t = R/p = (9.1 \text{ g/cm}^2)/(0.95 \text{ g/cm}^3) = 9.6 \text{ cm}$$

Question 7.8

Assumptions

1. Protons and heavier fragments have ranges less than 1 cm and stop in the sphere.
2. Neutrons and gamma photons lose negligible energy in the sphere when compared to protons and heavy fragments.

Based upon these assumptions, the energy deposited into the 1.0-cm sphere per stopped negative pion is:

Particle	Energy Deposited (MeV)
Protons	20
Heavier fragments	17
	37

354 SOLUTIONS FOR CHAPTER 7

The absorbed dose is defined in terms of energy deposited per unit mass. The mass of material contained within the 1.0-cm sphere is given by

$$M = \tfrac{4}{3}\pi r^3 p$$

where

M = mass of material (g) contained within a sphere of radius r
r = radius of sphere (cm) = 1.0 cm
p = material density (g/cm^3) → 1.0 g/cm^3 is a reasonable approximation for water.
$M = (\tfrac{4}{3})(3.14)(1.0 \text{ cm})^3 (1.0 \text{ g/cm}^3) = 4.19$ g

The dose D is obtained from its definition, energy deposited per unit mass, and appropriate conversion factors. It is given by

$$D = \frac{(37 \text{ MeV})(1.6 \times 10^{-6} \text{ erg/MeV})(1 \text{ rad}/100 \text{ erg/g})}{4.19 \text{ g}}$$

$$= 1.4 \times 10^{-7} \text{ rad}$$

Question 7.9

The assumption that charged particles stop in the 1.0-cm sphere is accurate. A 20-MeV proton has a range of about 0.4 cm. The following additional data would be needed to refine the dose estimate:

1. A more accurate neutron and gamma-ray energy spectrum is needed. In addition, the double differential cross section in terms of energy loss and angle is needed for hydrogen and oxygen.
2. Mean free paths for neutrons and gamma rays in water at the energies of interest.

Scenario 72

Question 7.10: d

The most likely source of the Al-28 decay is the neutron activation of Al-27 resulting from a gamma–neutron reaction with deuterium in the beam stop. The neutrons may also be derived from electron-induced reactions such as $(e, e'n)$.

Question 7.11: e

Equation (e) describes the Compton scattering of the electron beam.

Question 7.12: c

No radiation hazard would prevent the researcher from attending to his experiment.

SOLUTIONS FOR CHAPTER 7   355

Question 7.13

The following radiological controls would be appropriate:

1. Initial neutron surveys at low power
2. Activation surveys at high power
3. Determination of airborne contamination and the need for respiratory protection

Question 7.14

The following data would be needed to properly evaluate the shielding design:

1. Physical configuration of the beam and specification of occupied areas
2. Gamma-ray attenuation coefficients
3. Neutron cross sections
4. Probability and consequences of gamma–neutron reactions including relevant isotopes produced and their radiological characteristics

Question 7.15: e

The shielding requirements are essentially the same because the broad-beam TVL for lead for the two accelerators are nearly equal.

Question 7.16: a

To produce neutrons, sufficient energy to overcome the nuclear binding energy must be provided. Because the binding energy per nucleon exceeds 4 MeV, the 40-MeV machine produces significantly more neutrons.

Question 7.17: a

The bremsstrahlung production is sufficiently high for both 4- and 40-MeV electron LINACs to control the shielding design.

Scenario 73

Question 7.18

The following gaseous products are expected:

Radioactive	Toxic
C-11	NO_x
O-15	O_3
N-13	

Question 7.19

Target Z. High-Z targets increase bremsstrahlung production, which increases photonuclear and ozone production.

356   SOLUTIONS FOR CHAPTER 7

Beam Current. Both photonuclear and toxic gas production are directly related to beam current, so increasing current will increase their production rate.

Question 7.20

Essentially all neutrons produced between 35 and 100 MeV are produced within the giant resonance region. This region is sharply peaked in an energy region between the reaction threshold and about 35 MeV. The cross section is significantly reduced and relatively constant between 35 and 100 MeV. Other mechanisms are present but account for only a small percentage of the neutrons produced.

Question 7.21

If an electron accelerator is properly shielded for bremsstrahlung, it is automatically shielded sufficiently for neutrons produced if the following conditions are met:

1. The electron energy is less than the photo-pion threshold.
2. The shielding is concrete or another low-Z material

Note: Mazes/labyrinths are separate considerations.

Question 7.22: a

The reaction (production) rate R equals $N\sigma\phi$ and ϕ is proportional to the beam current. Therefore, the production rate is a linear function of the beam current.

Scenario 74

Question 7.23

The 10-min nuclide will require about two half-lives to reach the 2-Bq/cm^3 level. During this period, the shorter-lived nuclide will decay through about 10 half-lives. Therefore, the time can be estimated sufficiently accurately using only the longer-lived radionuclide.

$$C(t) = C(0) \exp[-(v/V + \lambda)t]$$

$$2 \text{ Bq/cm}^3 = (6.3 \times 10^4 \text{ Bq/cm}^3) \exp\{-[(4.0 \text{ m}^3/\text{sec}/560 \text{ m}^3)$$
$$+ (0.693/10 \text{ min} \times 60 \text{ sec/min})]t\}$$

$$t = -\ln(2/6.3 \times 10^4)/[(4/560) + (0.693/600)] \text{ sec}$$

$$= 1248.2 \text{ sec} = 20.8 \text{ min}$$

Question 7.24

The concentration of the toxic gas varies after shutdown.

$$Z(t) = Z(0) \exp[-(v/V + 1/T)t]$$

$$0.1 \text{ ppm} = 3.5 \text{ ppm} \exp\{-[(4 \text{ m}^3/\text{sec} \times 60 \text{ sec/min}/560 \text{ m}^3) + (1/25 \text{ min})]t\}$$

$$t = -\ln(0.1/3.5)/(4 \times 60/560 + 1/25) \text{ min}$$

$$t = 7.59 \text{ min}$$

Scenario 75

Question 7.25

Assuming that the distribution of neutrons is isotropic, the dose equivalent rate can be written as

$$H = IKk_1 P(\text{DCF})/(4\pi r^2)$$

where

I = proton beam current = 25.0×10^{-6} A
k = charge/proton = 1.602×10^{-19} coulomb/proton
K = $1/k$ = 6.24×10^{18} protons/A-sec
k_1 = time conversion factor (3600 sec/hr)
P = neutron production rate = 1.8×10^{-6} neutrons/proton
DCF = dose conversion factor = 3.5×10^{-8} rem-cm^2/neutron
r = distance from the target = 40 cm

$$H = (25 \times 10^{-6} \text{ A})(6.24 \times 10^{18} \text{ protons/A-sec})(3600 \text{ sec/hr})$$
$$\times (1.8 \times 10^{-6} \text{ neutrons/proton})(3.5 \times 10^{-8} \text{ rem-cm}^2/\text{neutron})/$$
$$(4\pi)(40 \text{ cm})^2 = 1.8 \text{ rem/hr}$$

Question 7.26: b

Assuming a point source approximation, exponential attenuation, and a buildup factor of unity, we obtain the following attenuation relationship:

$$\text{AF} = \exp(-ut_b)(r_a/r_b)^2$$

where

AF = attenuation factor
r_a = 40 cm = distance from question 7.25.
r_b = 400 cm = distance from question 7.26.
t_b = 50 cm = shield thickness
u = neutron removal cross section for concrete 0.08 cm^{-1}

$$AF = \exp(-0.08/cm \times 50 \text{ cm})(40 \text{ cm}/400 \text{ cm})^2$$
$$= 1.8 \times 10^{-4}$$

Question 7.27: a

A BF_3 proportional counter in a polyethylene moderator would have good sensitivity to neutrons while providing the best gamma discrimination.

Poorer Answers

b. GM tube at greater than 2 atmospheres in a polyethylene moderator: The bare GM tube has low neutron sensitivity and no gamma discrimination in a mixed field.
c. Silver-wrapped GM tube inserted in a polyethylene moderator: This detector is neutron-sensitive but has no gamma discrimination.
d. LiI(Eu) scintillator inserted in a polyethylene moderator: The LiI(Eu) system has some sensitivity to photons.
e. Cadmium-wrapped LiI(Eu) scintillator: There is no such detector.

Question 7.28: c

In the laboratory reference frame, both energy and fluence rate are peaked in the forward direction.

Scenario 76

Question 7.29

The current can be determined from the relationship

$$P = IV$$
$$P = \text{average beam power (watts)}$$
$$P = P_{peak}(DF)$$

where

P_{peak} = peak beam power = 5 MW
DF = duty factor = 0.01
$P = (5 \times 10^6 \text{ W})(0.01) = 5 \times 10^4 \text{ W}$
V = LINAC terminal voltage
V = 10 MeV = 1×10^7 eV = 1×10^7 volts $\times e$ for a proton beam of charge e
I = average beam current
$I = P/V = (5 \times 10^4 \text{ W})/(1 \times 10^7 \text{ V}) = 5 \times 10^{-3}$ A

The desired production rate can be found from the empirical ozone production relationship:

Production rate = $(600 \text{ eV cm}^{-4} \text{ A}^{-1} \text{ sec}^{-1}) GId$

$= (600 \text{ eV cm}^{-4} \text{ A}^{-1} \text{ sec}^{-1})(10.3 \text{ molecules}/100 \text{ eV})$

$\times (5 \times 10^{-3} \text{ A})(200 \text{ cm}) = 61.8 \text{ molecules cm}^{-3} \text{ sec}^{-1}$

Question 7.30

The steady-state concentration can be obtained by integrating the concentration rate expression:

$$\dot{C}(t) = \dot{C}(0) \exp[-(v/V + 1/T)t]$$

from $t = 0$ to time t. Because the production rate $\dot{C}(0)$ is a constant, the concentration at time t, $C(t)$, can be determined:

$$C(t) = \frac{\dot{C}(0)}{(v/V + 1/T)} \{1 - \exp[-(v/V + 1/T)t]\}$$

The steady-state concentration $C(\infty)$ occurs at large times—that is, $t \to \infty$:

$$C(\infty) = \frac{\dot{C}(0)}{v/V + 1/T}$$

$$= \frac{100 \text{ molecules cm}^{-3} \text{ sec}^{-1}}{[(5 \text{ m}^3 \text{ sec}^{-1}/75 \text{ m}^3) + (1/1800 \text{ sec})]}$$

$$= 1488 \text{ molecules/cm}^3$$

Question 7.31

After equilibrium was established, the cell concentration $C(0)$ was 10 ppm. After beam shutdown, this concentration $C(t)$ will decrease as a function of time:

$$C(t) = C(0) \exp[-(v/V + 1/T)t]$$

Solving for t yields the time to reach the desired concentration:

$t = \ln[C(t)/C(0)]/[-(v/V + 1/T)]$

$= \ln(0.1 \text{ ppm}/10 \text{ ppm})/\{-[(5 \text{ m}^3 \text{ sec}^{-1}/75 \text{ m}^3) + (1/2000 \text{ sec})]\}$

$= 68.6 \text{ sec}$

Scenario 77

Question 7.32

The reaction of interest is $p + \text{O-16} \rightarrow \text{C-11} + \text{Li-6}$. C-11, like most proton accelerator products, is a positron emitter. Consequently, the detected photons have an energy of 0.511 MeV with a yield of 2 gammas per disintegration of C-11.

The activity of C-11 present 1 hour after the exposure is given by

$$A = A_0 \exp(-\lambda t)$$

The initial activity A_0 may be written as

$$A_0 = N\sigma\phi\lambda$$

where

N = the number of O-16 atoms in the target tissue
= $(1 \text{ cm}^2)(10 \text{ cm})(1.0 \text{ g/cm}^3)$
   $\times (6.02 \times 10^{23} \text{ molecules/mole})$
   $\times (1 \text{ atom O-16/molecule water})/(18 \text{ g/mole})$
= 3.34×10^{22} atoms of O-16
σ = C-11 production cross section = 20 mb/atom of O-16
ϕ = proton fluence
$= \dfrac{(1.0 \times 10^{12} \text{ protons/pulse})(10 \text{ pulses/min})(1 \text{ min})}{1.0 \text{ cm}^2}$
= 1.0×10^{13} protons/cm^2
λ = C-11 half-life = 20.4 min

The expected count rate (R) in cpm is related to the activity by

$$R = eA$$

where

e = efficiency of detector = 0.1 counts/dis
$R = e\lambda N\sigma\phi \exp(-\lambda t)$
= (0.1 counts/dis)(0.693/20.4 min)
 $\times (3.34 \times 10^{23} \text{ atoms O-16})(20 \times 10^{-27} \text{ cm}^2/\text{O-16 atoms})$
 $\times (1.0 \times 10^{13} \text{ protons/cm}^2)(1 \text{ dis/proton})$
 $\times \exp[-(0.693)(60 \text{ min})/20.4 \text{ min}]$
= 2.96×10^7 cpm

Question 7.33

For simplicity, assume that the detector efficiency for the accelerator background gamma radiation is similar to that for the Ra-226 gamma rays.

The NaI detector yields 400 cps in an exposure rate of 10 μR/hr due to Ra-226 gamma rays. A 1-mR/hr gamma background would correspond to a count rate R of

$$R_B = (1 \text{ mR/hr})(400 \text{ cps}/[10 \text{ }\mu\text{R/hr}])(1000 \text{ }\mu\text{R/mR})$$
$$\times (60 \text{ sec/min})$$
$$= 2.4 \times 10^6 \text{ cpm}$$

From the previous question, the expected arm count rate R equals 2.96×10^7 cpm. The background is a fraction of this value:

$$R_B/R = 2.4 \times 10^6 \text{ cpm}/2.96 \times 10^7 \text{ cpm} = 0.08$$

Because the background is only about 8% of the expected count rate, the activation of the arm should be easily detected.

Question 7.34

It is unlikely that the physicist's recollection of the event will be accurate, and a careful evaluation is warranted to reconstruct the sequence of events. A videotaped simulation with the worker reenacting the event should be performed. The position of his arm and body relative to the proton beam needs to be accurately determined. Other actions to quantify the physicist's dose include: whole-body counting to assess the arm's isotopic activity, analysis of hair and blood for induced activity, recreating the event with a phantom and appropriate dosimetry, and medical monitoring of the physicist for acute radiation syndrome symptoms and for indications of tissue damage in the arm. The calculated dose will be an important parameter in determining future medical decisions.

As a bounding case, consider that the 6-GeV proton beam is totally absorbed by the arm. For a 1-cm^2 beam and 10-cm-thick arm, this would correspond to a volume (V) of

$$V = \pi r^2 h = (3.14)(0.5 \text{ cm})^2 (10.0 \text{ cm}) = 7.85 \text{ cm}^3$$

The dose (D) delivered to the arm is

$$D = \frac{kI(\text{PRF})Et}{Vp}$$

where

I = proton intensity = 1.0×10^{12} protons/pulse
PRF = pulse repetition rate = 10 pulses/min

p = arm density = 1.0 g/cm^3
E = proton beam energy = 6 GeV/proton
k = conversion factor
  = (1.6 × 10^{-12} erg/eV)(1 rad/100 erg g^{-1})
    × (1 × 10^9 eV/GeV)
t = exposure time = 1 min
D = [(1.0 × 10^{12} protons/pulse)(10 pulses/min)/(7.85 cm^3)(1.0 g/cm^3)]
    × (6 GeV/proton)(1 × 10^9 eV/GeV)(1.6 × 10^{-12} erg/eV)
    × (1 rad/100 erg g^{-1})(1.0 min)
  = 1.22 × 10^8 rad

This calculation is conservative for the following reasons:

1. The 6-GeV beam will penetrate the arm
2. The physicist probably did not hold his arm steady for the duration of the exposure

However, the dose is sufficiently large to be of medical concern. The extent of the tissue damage will probably be significant, and medical attention is warranted. The affected area should be frequently monitored for changes that may occur.

In addition to the dose to the arm, the whole-body dose should be assessed. The whole-body dose is due to secondary neutron and gamma radiation resulting from reactions within the exposed area. The video may suggest that other areas of the body were subjected to the direct beam. These will also require evaluation.

PART IV

APPENDICES

APPENDIX I

SERIAL DECAY RELATIONSHIPS

Many radioactive nuclides decay in a single transition to a stable nuclide. However, there are cases when the transition occurs to a system that is also unstable. These serial decays are encountered in neutron-induced fission products of U-235 and Pu-239 and in the natural decay series.

The decay relationships are usually derived on the assumption at time $t = 0$ only parent atoms exist and all daughter activity is zero. The parent nuclide is labeled by the subscript a, and its daughters are labeled with b, c, d, and so on. N is the number of atoms at time t, and λ is the decay constant.

The serial decay relationships can be derived for any number of daughters. We illustrate the decay relationships for the decay of the parent and first, second, and third daughters:

$$N_a = N_{a0} \exp(-\lambda_a t) \tag{I.1}$$

$$N_b = N_{a0}[a_1 \exp(-\lambda_a t) + a_2 \exp(-\lambda_b t)] \tag{I.2}$$

$$N_c = N_{a0}[a_3 \exp(-\lambda_a t) + a_4 \exp(-\lambda_b t) + a_5 \exp(-\lambda_c t)] \tag{I.3}$$

$$N_d = N_{a0}[a_6 \exp(-\lambda_a t) + a_7 \exp(-\lambda_b t) + a_8 \exp(-\lambda_c t) + a_9 \exp(-\lambda_d t)] \tag{I.4}$$

where

N_{a0} = number of parent atoms present at $t = 0$

$$a_1 = \lambda_a/(\lambda_b - \lambda_a) \tag{I.5}$$

$$a_2 = -\lambda_a/(\lambda_b - \lambda_a) \tag{I.6}$$

$$a_3 = \lambda_a\lambda_b/[(\lambda_b - \lambda_a)(\lambda_c - \lambda_a)] \tag{I.7}$$

$$a_4 = \lambda_a\lambda_b/[(\lambda_a - \lambda_b)(\lambda_c - \lambda_b)] \tag{I.8}$$

$$a_5 = \lambda_a\lambda_b/[(\lambda_a - \lambda_c)(\lambda_b - \lambda_c)] \tag{I.9}$$

$$a_6 = \lambda_a\lambda_b\lambda_c/[(\lambda_b - \lambda_a)(\lambda_c - \lambda_a)(\lambda_d - \lambda_a)] \tag{I.10}$$

$$a_7 = \lambda_a\lambda_b\lambda_c/[(\lambda_a - \lambda_b)(\lambda_c - \lambda_b)(\lambda_d - \lambda_b)] \tag{I.11}$$

$$a_8 = \lambda_a\lambda_b\lambda_c/[(\lambda_a - \lambda_c)(\lambda_b - \lambda_c)(\lambda_d - \lambda_c)] \tag{I.12}$$

$$a_9 = \lambda_a\lambda_b\lambda_c/[(\lambda_a - \lambda_d)(\lambda_b - \lambda_d)(\lambda_c - \lambda_d)] \tag{I.13}$$

The activity (A) is readily determined from these relationships because $A = \lambda N$.

In practical applications, the case involving a parent and two daughter nuclides frequently occurs. In general, the activity of the daughters is not always zero. The serial decay of parent nuclide a to a daughter b that subsequently decays to nuclide c is given by

$$a \xrightarrow{\lambda_a} b \xrightarrow{\lambda_b} c \tag{I.14}$$

$$N_b = \frac{\lambda_a N_{a0}}{\lambda_b - \lambda_a}[\exp(-\lambda_a t) - \exp(-\lambda_b t)]$$

$$+ N_{b0}\exp(-\lambda_b t) \tag{I.15}$$

where

N_b = number of b atoms at time t
N_{a0} = number of a atoms present at time $t = 0$
N_{b0} = number of b atoms present at time $t = 0$
t = time of interest

The activity follows from Eq. (I.15):

$$A_b = \frac{\lambda_b A_{a0}}{\lambda_b - \lambda_a}[\exp(-\lambda_a t) - \exp(-\lambda_b t)]$$

$$+ A_{b0}\exp(-\lambda_b t) \tag{I.16}$$

where A_{b0} is usually assumed to be zero for simplicity. For example, Eq. (I.16) is used in medical applications involving the milking or elution of a Mo-99 generator to obtain the desired quantities of Tc-99m. For cases such as a Mo-99 generator, both chemical and fractional yields must be considered and

Eq. (I.16) must be modified to reflect the physical limitations of the generator and leaching technology

$$A_b = CY \frac{\lambda_b A_{a0}}{\lambda_b - \lambda_a} [\exp(-\lambda_a t) - \exp(-\lambda_b t)]$$
$$+ A_{b0} \exp(-\lambda_b t) \qquad (I.17)$$

where C is the chemical yield and Y is the fractional yield.

APPENDIX II

BASIC SOURCE GEOMETRIES AND ATTENUATION RELATIONSHIPS

The evaluation of radiation levels from a source is a fundamental problem in health physics. Common source configurations include the point, line, disk, and slab geometries. Although numerous computer codes are available for the assessment of these and more complicated geometries, it is often advantageous to have the capability to make rapid assessments using hand calculations before performing more detailed calculations. Knowledge of these basic geometries, as well as capability to utilize them in dose estimates, will be valuable to the practicing health physicist.

SOURCE CONFIGURATIONS–NO SHIELDING

Point Source Geometry

The point source approximation is applicable whenever the dose is calculated at a distance that is at least three times the largest source dimension. The student should test this statement by performing sample calculations for various geometries. A comparison of point and disk geometries will be provided later in this appendix.

The exposure, absorbed dose, or dose rate $\dot{X}(r)$ at a location a distance r from a point source is given by

$$\dot{X}_0(r) = AG/r^2 \qquad (\text{II}.1)$$

where

$\dot{X}_0(r)$ = unattenuated exposure rate (R/hr), absorbed dose rate (rad/hr), or dose rate (rem/hr) at point r. Although $\dot{X}(r)$ is normally expressed in terms of the Roentgen, the conversion to either rem or rad units is straightforward and these units are used rather loosely in external dose assessments.
A = source activity (Ci)
G = gamma constant (R-m^2/hr-Ci)
r = distance from the point source(m)

If the value of G is not provided, $\dot{X}_0$ can be estimated from the properties of the decaying nuclide using the approximate relationship

$$\dot{X}_0(r) = 0.5CE/r^2 \tag{II.2}$$

where

C = activity of the nuclide (Ci)
E = total weighted photon energy of the nuclide
$\quad = \sum_i E_i Y_i$
E_i = energy of the ith gamma ray from the nuclide
Y_i = yield of the ith gamma ray from the nuclide
$G = 0.5E$

The generalization of this relationship to multiple source nuclides is achieved by summing over the various source nuclides.

The approximation of Eq. (II.2) is limited by how accurately the gamma constant is represented by the approximation $G = 0.5E$. In general, this approximation should be within 20% of the actual value of the gamma constant. For the case of Na-24, Co-60, and Cs-137, the approximation is within +12.0%, −5.3%, and −15.2%, respectively, of the actual gamma constant.

Equation (II.1) may be combined with the shielding relationships to account for attenuation and buildup through a shield. This will be explored in more detail later in this appendix.

For those situations in which the gamma constant is not available, or for which an accuracy greater than given by the gamma constant approximation is desired, the total gamma dose equivalent at a distance r from the point source can be written in terms of the gamma energy of the source, source strength, mass energy absorption coefficient, and assumptions regarding the emission of the source. If an isotropic emission is assumed, the gamma dose rate for a source with a single gamma ray can be written as

$$\dot{H}_0 = (S/4\pi r^2)(u_{en}/p) E_{gamma} \tag{II.3}$$

where

$\dot{H}_0$ = unattenuated gamma dose rate (rem/hr)
S = source strength (gammas/sec)
r = distance from the source (cm) (*Note:* This relationship can be generalized to line, disk, or slab sources by substituting the appropriate geometry factor.)
u_{en}/p = Mass energy absorption coefficient (cm^2/g)
p = density (g/cm^3)
E_{gamma} = weighted energy of the emitted gamma ray
= EY

For sources characterized by the emission of multiple gamma rays, the dose rate from a point source may be written in the form

$$\dot{H}_0 = \sum_i (S_i/4\pi r^2) \sum_j (u_{en}^j/p) E_{ij} Y_{ij} \quad \text{(II.4)}$$

where

$\dot{H}_0$ = dose equivalent rate (rem/hr)
S_i = activity of isotope i (μCi or dis/sec)
E_{ij} = energy of the jth gamma ray for the isotope i
Y_{ij} = yield of the jth gamma ray for the isotope i
u_{en}^j/p = mass attenuation coefficient (cm^2/g) at energy E_{ij}

The total unattenuated neutron dose equivalent rate for a point source can be similarly defined:

$$\dot{H}_{on} = (S/4\pi r^2)k \quad \text{(II.5)}$$

where

$\dot{H}_{on}$ = neutron dose rate (rem/hr)
S = source strength (neutrons/sec)
r = distance from the source (*Note:* This relationship can also be generalized to other source geometries.)
k = factor converting neutrons sec^{-1} cm^{-2} to rem/hr

Line Source Geometry

The unattenuated radiation field due to a line source of length L, illustrated in Fig. II.1, is given by

$$\dot{X}_0(Q) = GC_L\theta/w \quad \text{(II.6)}$$

SOURCE CONFIGURATIONS—NO SHIELDING

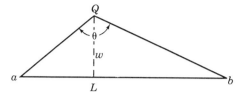

Fig. II.1 Geometry for computing the gamma-ray dose field at point Q from a line source (ab).

where

$\dot{X}_0(Q)$ = unattenuated exposure rate (R/hr) or dose rate (rem/hr) at point Q due to the line source ab.
w = perpendicular distance of the point Q to the line source
Q = point at which dose is to be determined
A = total source activity (Ci)
L = length of the line source
θ = angle subtended by the line source at the point of interest
C_L = source activity per unit length of the line source
  = A/L (Ci/m, Ci/ft, Ci/cm, etc.)
G = gamma constant (R-m²/hr-Ci)

This approximation is particularly useful for calculations involving piping, fluid lines, or resin columns containing radioactive material. For these calculations, the pipe wall thickness is often ignored without a significant loss in accuracy.

Disk Source Geometry

The unattenuated radiation field due to a thin disk source of radioactivity, illustrated in Fig. II.2, is given by

$$\dot{X}_0(Q) = \pi G C_a \ln \frac{r^2 + h^2}{h^2} \qquad (\text{II}.7)$$

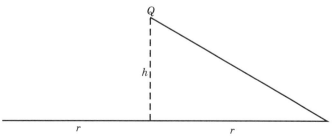

Fig. II.2 Geometry for computing the gamma-ray dose field at point Q from a thin disk source of radius r.

where

$\dot{X}_0(Q)$ = unattenuated exposure rate (R/hr) or dose rate (rem/hr) at point Q lying on the axis of the disk
Q = lies a perpendicular distance h from the disk source
A = Total source activity (Ci)
a = area of the disk source
C_a = Source activity per unit area of the disk source
  = A/a (Ci/m^2, Ci/ft^2, Ci/cm^2, etc.)
G = gamma constant (R-m^2/hr-Ci)
r = radius of the disk source

Disk relationships are useful to evaluate the radiation field associated with either (a) a spill of radioactive liquid, (b) the field above a resin bed, or (c) surface sources such as a radioactive pool.

A comparison of exposure rates from point and thin disk source geometries illustrates when a source may be reasonably approximated by a point source. This is illustrated by considering Scenario 60, question 5.26, which evaluated the exposure rate a distance h from a 10-m-radius thin disk source containing 3.0 Ci of Co-60.

Table II.1 summarizes the exposure rates from 3.0-Ci point and disk sources at a point h meters from the source. This table suggests that point and disk dose rates are within 1% of each other when the distance from the source is at least three times the largest source dimension. For the disk source, this occurs at a distance three times the disk diameter—that is, $h/d = 3$.

Table II.1 Comparison of Disk and Point Source Approximations for a 10-Meter-Radius Thin Disk Source

	Exposure Rate (mR/hr)		
h (m)	Disk	Point	h/d
2	127	975	0.1
10	27	39	0.5
20	8.70	9.75	1.0
30	4.11	4.33	1.5
40	2.36	2.44	2.0
50	1.53	1.56	2.5
60	1.07	1.08	3.0
70	0.788	0.796	3.5
80	0.605	0.609	4.0
90	0.479	0.481	4.5
100	0.388	0.390	5.0

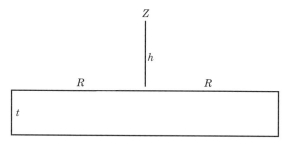

Fig. II.3 Geometry for computing the gamma-ray dose field at point Z from a slab source of radius R.

Slab Source Geometry

Consider a slab of uniformly distributed radioactive material characterized by a linear absorption coefficient u. The slab source is illustrated in Fig. II.3. If the slab is defined to be of cylindrical shape with a radius R and thickness t, the dose rate at point Z a distance h above the centerline of the slab is

$$\dot{X}_0(h) = \pi G(C_v/u)(1 - e^{-ut}) \ln \frac{R^2 + h^2}{h^2} \qquad (II.8)$$

where

$\dot{X}_0(h)$ = exposure rate (R/hr) or dose rate (rem/hr) at point Z on the axis of the slab at a distance h from the surface
G = gamma constant (R-m²/hr-Ci)
C_v = activity concentration (Ci/cm³)
  = A/v
A = total activity of the slab
v = volume of the source (cm³)
  = $\pi R^2 t$

This relationship is applicable to assessments of dose rates from distributed bodies such as spent-fuel pools, contaminated slabs, or tanks of contaminated fluids which contain uniformly distributed activity. The choice of using a disk or slab geometry is governed by a number of factors including the accuracy desired, knowledge of the radionuclide concentration, distribution, content, and distance of the receptor from the source.

ATTENUATION BY A SHIELD WITHOUT BUILDUP

The attenuation of gamma rays by a shielding medium can be approximated by considering shield attenuation. For thin shields with inherently small buildup factors of nearly unity, the attenuation may be represented by

$$I(x) = I_0 \exp(-ux) \qquad (\text{II.9})$$

where

$I(x)$ = attenuated quantity (dose rate, fluence rate, flux, or activity) due to a thickness x of shielding material
I_0 = unattenuated initial value of the quantity
u = attenuation coefficient (cm^{-1})
 = 1/relaxation length
x = shield thickness (cm)

ATTENUATION BY A SHIELD WITH BUILDUP

For thicker shields, where the buildup factor is greater than unity, a better approximation to the attenuation of gamma rays by a shielding medium is given by

$$I(x) = I_0 B(E, Z, ux) e^{-ux} \qquad (\text{II.10})$$

where the buildup factor (B) is a function of the gamma-ray energy (E), the atomic number (Z) of the shield material, the attenuation coefficient (u), and the thickness (x) of the shield material. This relationship also applies to the various geometries considered in this chapter.

The ratio $I(x)/I_0$ defines the transmission factor (T):

$$T = I(x)/I_0 = B \exp(-ux) \qquad (\text{II.11})$$

The reader should note that u may be expressed in a variety of ways such as the mass attenuation coefficient (cm^2/g) times the material density (g/cm^3) or as an attenuation coefficient with units of reciprocal length.

A common problem involves the determination of the shield thickness required to reduce the radiation level to a desired value. The desired thickness may be obtained by solving Eq. (II.10) for the required thickness:

$$x = (-1/u) \ln[I(x)/I_0 B] \qquad (\text{II.12})$$

If the source emits multiple gamma rays during its decay, the photon fluence reduction due to a shield $I(x)$ is given by

$$I(x) = I_0 \sum_i B_i f_i \exp(-u_i x) \qquad (\text{II.13})$$

where

B_i = buildup factor for energy E_i
u_i = attenuation coefficient of the shield for energy E_i
x = thickness of the shield
E_i = energy the ith gamma ray
f_i = fraction of I_0 contributed by E_i or the partial yield for the ith gamma ray
    = $Y_i / \sum_i Y_i$
Y_i = yield of the ith gamma ray

APPENDIX III

NEUTRON-INDUCED GAMMA RADIATION SOURCES

Neutron sources or beams, encountered in accelerator or power reactor environments, generate additional radiation sources as they penetrate matter. The neutron interactions produce activated material that remains a hazard long after the neutron source is terminated. The neutron-induced interactions include activation, capture, and fission reactions. Each of these mechanisms present a significant radiation hazard.

ACTIVATION SOURCES

The activation of a material as a function of time depends upon the material being activated and the beam activating the material. The activity buildup consists of both production and decay terms and may be written as

$$A = N\sigma\phi[1 - \exp(-\lambda t_{\text{irrad}})] \exp(-\lambda t_{\text{decay}}) \qquad (\text{III}.1)$$

where

- A = activity of the sample as a function of time
- N = number of atoms in the sample that are activated $N = m\overline{A}/\text{GAW}$
- m = mass of the sample (If multiple isotopes are activated, the mass of each constituent must be considered.)
- $\overline{A}$ = 6.023 × 10²³ atoms/GAW (Avogadro's number)
- GAW = gram atomic weight or mass of a mole of material (g)
- σ = cross section for the reaction induced by the flux ϕ (barns/atom)
- ϕ = fluence rate or flux (neutrons/cm²-sec)
- λ = decay constant of the activated material
- t_{irrad} = time the sample was irradiated or exposed to the flux
- t_{decay} = decay time or time the sample was removed from the flux

For material that is activated for a long time relative to its half-life, the activity reaches a constant value or saturates. Saturation occurs as the irradiation time becomes much larger than the decay half-life, and the decay time is short relative to the decay half-life. Under these circumstances, the activity approaches A_{sat}, which is the saturation activity

$$A_{sat} = N\sigma\phi \tag{III.2}$$

Determining the saturation activity is an important exercise because it represents a bounding case for dose rate assessments. Design work often utilizes the saturation activity to ensure that the design will bound any operating condition.

Many common activation gamma sources involve the absorption of a neutron with the emission of a gamma ray. Usually this process involves thermal neutrons. Other reactions involve high-energy or fast neutrons and produce high-LET protons via (n, p) reactions. Examples of activation sources are contained in Table III.1. These sources are frequently produced in reactor or accelerator environments.

Table III.1 Activation Gamma Sources

Reaction	Activation Cross Section (barns)	Half-life	Energy (MeV)	Yield Gammas/ Decay
^{23}Na$(n, \gamma)^{24}$Na	0.534	14.96 hr	1.369	1.00
			2.754	1.00
^{54}Fe$(n, p)^{54}$Mn	1.0	314 days	0.835	1.00
^{55}Mn$(n, \gamma)^{56}$Mn	13.3	2.576 hr	0.847	0.99
			1.811	0.29
			2.11	0.15
^{59}Co$(n, \gamma)^{60}$Co	37.2	5.263 years	1.173	1.00
			1.332	1.00
^{58}Fe$(n, \gamma)^{59}$Fe	1.2	45.6 days	1.095	0.56
			1.292	0.44
^{58}Ni$(n, p)^{58}$Co	1.0	71.3 days	0.51	0.30
			0.81	0.99
			0.865	0.014
			1.67	0.006
^{94}Zr$(n, \gamma)^{95}$Zr	0.075	65.5 days	0.724	0.49
			0.756	0.49
			0.765	1.00

Source: O. J. Wallace, WAPD-TM-1453.

During activation, the absorbed neutron produces a radioactive nuclide that decays with a characteristic half-life. This may be contrasted with capture reactions which involve nearly an instantaneous gamma-ray emission following the capture.

CAPTURE GAMMA SOURCES

Capture gamma sources involve the absorption of a neutron with the nearly instantaneous emission of a gamma ray. Capture reactions are often defined in terms of bulk materials and play an important role in design calculations. Calculations of nuclear radiation heat generation and primary or biological shield design must properly account for capture sources. Examples of capture gamma sources are provided in Table III.2.

FISSION GAMMA SOURCES

Fission gamma sources are produced by the neutron spectrum of a reactor. Table III.3 summarizes the characteristics of two common fission gamma

Table III.2   Capture Gamma Sources

Reaction	Absorption Cross Section (barns)	Density (g/cm^3)	Energy (MeV)	Yield Gammas/ Capture
Hydrogen capture	0.33	8.988×10^{-4}	2.2	1.0
Iron capture	2.53	7.874	1.0	1.85
			7.0	0.86
Zirconium capture	0.180	6.53	1.0	0.7
			5.0	1.42
Uranium capture	7.68	18.95	0.7	3.63
			2.5	1.2
Water capture	0.66	0.998	2.2	1.0
Hafnium capture	105.0	13.29	1.0	0.75
			3.0	2.366
Stainless steel capture	0.0332	7.9	1.0	1.69
			7.0	0.93
Inconel capture	0.0462	8.51	0.5	2.0
			5.5	1.51

Source: O. J. Wallace, WAPD-TM-1453.

Table III.3 Fission Gamma Sources

Source	Energy (MeV)	Yield Gammas/ Fission
^{16}N gammas	2.74	0.01
	6.13	0.69
	7.12	0.05
Prompt and delayed fission gammas	0.80	7.22
	2.0	2.26
	4.0	1.02

Source: O. J. Wallace, WAPD-TM-1453.

sources. The N-16 source is derived from the O-16 (n, p) reaction with the subsequent decay of the short-lived N-16 system. The prompt and delayed fission gamma source represents a collective summary of the uranium fission gamma spectrum. These sources tend to have more varied energies and yields when compared with either the activation or capture sources for specific isotopes.

APPENDIX IV

SELECTED TOPICS IN INTERNAL DOSIMETRY

As a health physicist, you may be required to evaluate the dose to internal organs and to the whole body as a result of the intake of radioactive material. This intake may occur, either intentionally or as the result of an unplanned activity, in a variety of ways including inhalation, injection, and ingestion. Internal dosimetry involves the use of numerous calculational models to evaluate the consequences of the intake. This appendix will present a variety of approaches that introduce the essential elements of internal dosimetry and its radiological basis.

The calculation of internal dose by any model has a number of common elements. The dose depends upon the activity deposited in the organ (q), the energy of the emitted radiation (E) absorbed by the organ, and the mass of the organ (m) at risk. Because dose (D) is defined in terms of energy deposition per unit mass, internal dose rate ($\dot{D}$) formulas contain factors of the form

$$\dot{D} = \frac{kqE}{m} \tag{IV.1}$$

where k is a conversion factor which expresses the dose rate in the desired units.

Internal dose is dominated by alpha and beta radiation because these radiations deposit their energy in distances shorter than typical organ sizes. Gamma rays contribute to the internal dose, but to a lesser extent than either beta or alpha radiation.

INTERNAL DOSE ASSESSMENT MODELS

An anatomical model, used to approximate the human body, must be formulated in order to calculate internal doses. The model will consist of mathematical equations that define the sizes and shapes of organs and tissue that are assembled together to form a representation of the human body. The degree of sophistication of the internal dosimetry models varies from simplistic single-compartment models to the more complex lung model currently under development. The current lung model has a strong resemblance to the actual body tissue and is considerably more complex than the simplistic geometric forms used in the initial internal dosimetry models.

INTERNAL DOSIMETRY DEFINITIONS

Before considering specific models, several terms will be defined to ensure that the subsequent discussion is clear. These terms include intake, uptake, deposition, and content.

Intake is the quantity of a radionuclide taken into the body by inhalation, ingestion, or injection. An uptake is that quantity of a radionuclide that is absorbed into systemic circulation by injection directly into the blood or by absorption via the respiratory or gastrointestinal (GI) tract. Deposition is that quantity of a radionuclide that is deposited into an organ. Finally, content refers to the quantity of a radionuclide that resides in the compartment of interest.

In the subsequent discussion, the focus will be upon the more common internal dosimetry models. These models include the ICRP-2/10, MIRD, ICRP-26/30, and ICRP-60/61 internal dosimetry formulations.

ICRP-2/10 METHODOLOGY

Internal Dose Assessment—Single-Compartment Model

An early internal dosimetry model was described in the 1959 ICRP Committee 2 Report, which utilized the concept of a body and organs characterized by effective radii. This approximation made it possible to calculate the dose to an organ from the activity residing within that organ. The concept of a single-compartment model was developed to address the absorption and elimination of radioactive material from that organ.

Although simplistic by today's standards, the single-compartment model has considerable utility and will be addressed in terms of its key properties and dosimetry relationships. Moreover, it will serve as a basis for formulating more complex models describing both chronic and acute internal exposures.

The following methodology is based upon the single-compartment model in which the radionuclide is uniformly distributed in the organ of interest. Both chronic and acute exposures will be considered, and dose rate and integrated dose relationships will be provided.

Single Uptake—Single-Compartment Model

An initial uptake of activity (q_{i0}) of radionuclide i results in a deposition ($q_{i0}f_2$) in the organ of interest. The activity $q_i(t)f_2$ post uptake is given by

$$q_i(t)f_2 = q_{i0}f_2 \exp(-\lambda_e t) \tag{IV.2}$$

where λ_e is the effective decay constant and f_2 is the fraction of the material entering the organ of interest.

The initial dose equivalent rate ($\dot{D}_0$) to the organ of interest due to an initial deposition of $q_0 f_2$ is

$$\dot{D}_0 = q_0 f_2 E/m \tag{IV.3}$$

where E is the effective energy deposition and m is the mass of the organ. The effective energy, for alpha and beta emitters, may be written as

$$E = E_{\text{alpha}} \quad \text{(alpha emitters)}$$

$$E = E_{\max} f/3 \quad \text{(beta emitters)} \tag{IV.4}$$

where E_{alpha} is the energy of the emitted alpha particle, $E_{\max}$ is the maximum beta energy, and f is the probability that a beta is emitted. If q_0 is expressed in μCi, E in MeV, and m in grams, Eq. (IV.3) takes the form for the initial dose rate in rad/hr:

$$\dot{D}_0 = [q_0(\mu\text{Ci})f_2 E(\text{MeV})/m(\text{g})](3.7 \times 10^4 \text{ dis}/\mu\text{Ci-sec})$$
$$\times (1.6 \times 10^{-6} \text{ erg/MeV})(1 \text{ rad}/100 \text{ erg/g})(3600 \text{ sec/hr}) \tag{IV.5}$$

$$\dot{D}_0 = 2.13 q_0 f_2 E/m \tag{IV.6}$$

The dose equivalent rate $\dot{D}(t)$ at a time t post uptake to the organ of interest containing a uniform deposition of $q_0 f_2$ is

$$\dot{D}(t) = \dot{D}_0 \exp(-\lambda_e t) \tag{IV.7}$$

The total dose equivalent accumulated in the organ of interest from the single intake at $t = 0$ to any time t post uptake is obtained by integrating Eq. (IV.7) from $t = 0$ to time t:

$$D = (\dot{D}_0/\lambda_e)[1 - \exp(-\lambda_e t)] \tag{IV.8}$$

or

$$D = 2.13[A_0 E/m\lambda_e][1 - \exp(-\lambda_e t)] \tag{IV.9}$$

where $A_0 = f_2 q_0$.

For the case of long times after deposition, $t \to \infty$, the total dose due to the intake is

$$D = (\dot{D}_0/\lambda_e) \tag{IV.10}$$

or

$$D = 2.13 A_0 E/m\lambda_e \tag{IV.11}$$

The form of this equation assumes the deposition of a single radionuclide. The deposition of multiple radionuclides may be assessed by performing a calculation for each isotope.

The calculation of the activity of a nuclide distributed throughout the body or organ can be estimated from the definition of activity:

$$A = \lambda N = (0.693/T_{1/2})N \tag{IV.12}$$

where

A = activity in the organ
λ = disintegration constant
$T_{1/2}$ = half-life
N = number of atoms of the isotope
  = mG/M
m = mass of the isotope in the organ
M = gram atomic weight (GAW) of the radionuclide in the organ
G = Avogadro's number = 6.02×10^{23} atoms/GAW

Constant Rate of Uptake—Single-Compartment Model

A constant rate of uptake (P) of a particular radionuclide by the organ of interest (μCi/hr) results in a burden of $q(t)f_2$ which increases with time t:

$$q(t)f_2 = (P/\lambda_e)[1 - \exp(-\lambda_e t)] \tag{IV.13}$$

where P has the units of activity per unit time such as μCi/day. An approach similar to that for a single uptake leads to the following relationship for the dose rate delivered to the organ of interest from a continuous uptake P:

$$\dot{D}(t) = [(P/\lambda_e)E/m][1 - \exp(-\lambda_e t)] \tag{IV.14}$$

It also leads to the relationship for the dose equivalent delivered to the organ of interest during the time interval from $t = 0$ to the time t:

$$D = \frac{(P/\lambda_e)Et}{m}\left[1 - \frac{1 - \exp(-\lambda_e t)}{\lambda_e t}\right] \quad \text{(IV.15)}$$

Variation of qf_2 After the Cessation of the Uptake

The deposition (qf_2) in an organ of interest at the end of a continuous uptake interval may be considered an initial burden at that instant. This activity burden will subsequently decrease in the same manner as the initial deposition from a single uptake. If t_1 defines the interval of continuous uptake and t_2 defines the time interval post uptake, then the deposition's variation is given by

$$q(t_2)f_2 = (P/\lambda_e)[1 - \exp(-\lambda_e t_1)]\exp(-\lambda_e t_2) \quad \text{(IV.16)}$$

The dose equivalent rate at the end of the uptake interval is

$$\dot{D}(t_1) = (P/\lambda_e)(E/m)[1 - \exp(-\lambda_e t_1)] \quad \text{(IV.17)}$$

and the dose equivalent rate during the decay period post intake is

$$\dot{D}(t_2) = \dot{D}(t_1)\exp(-\lambda_e t_2) \quad \text{(IV.18)}$$

The dose equivalent accumulated by an organ of interest during the uptake interval from 0 to t_1 is obtained by integrating Eq. (IV.14) from $t = 0$ to t_1:

$$D(0 \text{ to } t_1) = \frac{(P/\lambda_e)Et_1}{m}\left[1 - \frac{1 - \exp(-\lambda_e t_1)}{\lambda_e t_1}\right] \quad \text{(IV.19)}$$

Similarly, the dose equivalent accumulated by an organ of interest during the time post uptake from $t_2 = t_1$ to infinity is derived from Eq. (IV.18):

$$D(t_1 \text{ to } \infty) = \frac{(P/\lambda_e)Et_1}{m}\left[\frac{1 - \exp(-\lambda_e t_1)}{\lambda_e t_1}\right] \quad \text{(IV.20)}$$

Adding Eqs. (IV.19) and (IV.20) leads to the total dose equivalent accumulated in the organ from intake to complete decay:

$$D(0 \text{ to } \infty) = \frac{(P/\lambda_e)Et_1}{m} \quad \text{(IV.21)}$$

MIRD THEORY

The Committee on Medical Internal Radiation Dose (MIRD) of the Society of Nuclear Medicine developed a methodology to perform radiation absorbed dose calculations. These calculations are performed to assess the risks of the application of radiopharmaceuticals to medical studies involved in imaging, therapy, or metabolic applications.

The MIRD technique (schema) is a computational framework that facilitates these absorbed dose calculations to specified target organs from radioactive decays that occur within source organs. The source organs contain the radioactive isotope, and the target is the organ in which the dose is computed. The target and source organ may be the same tissue.

Simplified MIRD Equation

To introduce the methodology, a simplified MIRD dose equation will be derived. Initially, a single radiopharmaceutical, single radiation type, and single source organ will be considered.

To define the MIRD schema, it is necessary to define several terms. The mean energy emitted per transition (Δ), in Gy-kg/Bq-sec or rad-g/μCi-hr, is given by the product of the mean particle energy (E) (in MeV or joules) and the number of particles emitted per nuclear transition (n):

$$\Delta = KEn \qquad (IV.22)$$

where K is a conversion factor. Particles are defined to be either photons, beta particles, or positrons within the simplified MIRD model.

The cumulated activity or the total number of nuclear transitions occurring within the source organ ($\tilde{A}$) from $t = 0$ to some time t is given by

$$\tilde{A} = \int_0^t A(t)\, dt \qquad (IV.23)$$

The activity as a function of time is

$$A(t) = A(0)\exp(-\lambda_{\text{eff}}t) \qquad (IV.24)$$

Using Eq. (IV.24), the expression for $\tilde{A}$ takes a closed form if the integration interval is taken from $t = 0$ to infinity. For this case, $\tilde{A}$ takes the form

$$\tilde{A} = A(0)T_{\text{eff}}/\ln(2) = 1.44 T_{\text{eff}} A(0) \qquad (IV.25)$$

where

$$T_{\text{eff}} = (T_{\text{physical}} \times T_{\text{biological}})/(T_{\text{physical}} + T_{\text{biological}}) \qquad (IV.26)$$

The reader should note that the initial activity in the organ [$A(0)$] is not the intake activity but is only a fraction (f_2) of the total activity of radionuclide i in the body (q):

$$A(0) = qf_2 \qquad (IV.27)$$

where f_2 is the fraction of the intake activity reaching the organ of interest.

The total energy emitted by the source organ is the product of Δ and the cumulated activity. However, only a fraction (f) of this energy will be deposited in the target organ, which is the location of interest in the dose assessment.

With these quantities and a knowledge of the mass of the target organ (m_T), the mean absorbed dose ($\overline{D}$) can be defined:

$$\overline{D} = \tilde{A}\Delta f/m_T \qquad (IV.28)$$

The MIRD methodology also defines the specific absorbed fraction (F):

$$F = f/m_T \qquad (IV.29)$$

where

$$f = \frac{\text{energy absorbed by the target}}{\text{energy emitted by the source}} \qquad (IV.30)$$

and m_T is the target mass. The specific absorbed fraction represents the mean target dose per unit energy emitted by the source. Therefore, the mean absorbed dose can be written as

$$\overline{D} = \tilde{A}\Delta F \qquad (IV.31)$$

The MIRD schema defines the mean dose to the target per unit cumulated activity (S):

$$S(T \leftarrow S) = \Delta f/m_T = \Delta F \qquad (IV.32)$$

which permits another mean absorbed dose relationship to be written:

$$\overline{D} = \tilde{A}S(T \leftarrow S) \qquad (IV.33)$$

where

S = mean absorbed dose per unit cumulated activity in mGy/MBq-sec or rad/μCi-hr

In Eq. (IV.33), most of the metabolic factors are contained in the $\tilde{A}$ term, which depends on uptake by the source organ and biological elimination of the radiopharmaceutical by the source organ. The S factor represents much of the physical decay process including the decay characteristics of the radionuclide, the range of the emitted radiations, and the organ size and configuration. If a standard anatomy is utilized, S can be calculated and tabulated for a variety of radionuclides and source–target combinations. MIRD Publication 11 provides a tabulation of these S values.

Alternate MIRD Equation

The basic MIRD equation can be rewritten in terms of the activity administered to the patient (A_0):

$$\overline{D}/A_0 = (\tilde{A}/A_0) S \tag{IV.34}$$

where both sides of Eq. (IV.33) have been divided by A_0. The residence time (T) of a radiopharmaceutical in the source organ is defined as

$$T = \tilde{A}/A_0 \tag{IV.35}$$

The mean dose per unit administered activity ($\overline{D}/A_0$) is expressed as

$$\overline{D}/A_0 = TS \tag{IV.36}$$

MIRD Equation

The MIRD schema can be expanded to include a single radiopharmaceutical which leads to multiple source organs and multiple radiation types. In order to define the more general MIRD equation, a number of terms need to be defined.

For radiation type i, the mean energy emitted per nuclear transition Δ_i is defined as

$$\Delta_i = KE_i n_i \tag{IV.37}$$

where E and n were defined previously. The total mean energy emitted per nuclear transition from all radiation types is expressed as

$$\Delta = \sum_i \Delta_i \tag{IV.38}$$

The absorbed fraction f_i is defined as

$$f_i(r_T \leftarrow r_S) = b/B \tag{IV.39}$$

where b is the ith-type radiation energy emitted in the source r_S and absorbed in target r_T. The label r defines an anatomical region, with r_S defining the source organ and r_T defining the target organ. B is the ith-type radiation emitted in the source. MIRD Pamphlet No. 3 defines the f_i values as summarized in Table IV.1.

In a similar fashion, the specific absorbed fractions F_i and the mean dose per unit cumulated activity $S(r_T \leftarrow r_S)$ are defined as

$$F_i(r_T \leftarrow r_S) = \frac{f_i(r_T \leftarrow r_S)}{m_T} \qquad (IV.40)$$

$$S(r_T \leftarrow r_S) = \sum_i \Delta_i F_i(r_T \leftarrow r_S) \qquad (IV.41)$$

Equations (IV.40) and (IV.41) are defined analogously to the single target equations.

These definitions permit the general MIRD equation describing the mean absorbed dose to the target organ $\overline{D}_T$ to be defined as a sum of dose contributions from all source organs (S):

$$\overline{D}_T = \sum_S \tilde{A}_S \sum_i KE_i n_i f_i(r_T \leftarrow r_S)/m_T \qquad (IV.42)$$

Equation (IV.42) may be rewritten in a variety of ways to suit the particular dose calculation:

$$\overline{D}_T = \sum_S \tilde{A}_S \sum_i \Delta_i F_i(r_T \leftarrow r_S) \qquad (IV.43)$$

$$\overline{D}_T = \sum_S \tilde{A}_S S(r_T \leftarrow r_S) \qquad (IV.44)$$

and

$$\frac{\overline{D}_T}{A_0} = \sum_S T_S S(r_T \leftarrow r_S) \qquad (IV.45)$$

Table IV.1 f_i **Values**

Radiation Type	$f_i(r_T \leftarrow r_S)$
Penetrating radiations (photons with energy > 10–20 keV)	0 to 1
Nonpenetrating radiations (beta particles, positrons, alpha particles, and photons with energies < 10–20 keV):	
When $S = T$	1
When $S \neq T$	0

STOCHASTIC AND NONSTOCHASTIC EFFECTS

Stochastic effects are those in which the probability of the effect increases with increasing dose without threshold. That is, any dose has a probability of causing the effect. Stochastic effects may result from injury to a small number of cells.

Cancer and hereditary effects are examples of stochastic effects. Once these types of effects are induced, their severity is already determined. The severity of cancer or hereditary effects does not increase with dose. Only the probability of their incidence increases with dose.

Nonstochastic effects are those in which the severity of the effect varies with the dose. That is, if the dose is kept below a given threshold, no effect will be observed. Nonstochastic effects result from the collective damage to a large number of cells in an organ or tissue. Examples of nonstochastic effects include cataracts, impairment of fertility, and depletion of blood-forming cells in bone marrow.

ICRP-26/30 DOSE LIMITS

In the late 1970s, ICRP-26/30 established a dose limitation system designed to prevent nonstochastic effects and to limit the stochastic effects of radiation. The system of dose control and limitation recommended in 1977 was intended to be applied to controllable activities encountered during normal facility operations.

The ICRP dose limit recommendations for occupational exposure are formulated to prevent stochastic and nonstochastic effects. To prevent nonstochastic effects, the ICRP recommends that exposures be limited to 50 rem/year (0.5 Sv/year) to all tissues except the lens of the eye and that the lens of the eye be limited to 15 rem/year (0.15 Sv/year). To limit stochastic effects, 5 rem (50 mSv) in 1 year for uniform whole-body irradiation is recommended.

The nonstochastic limits are based on the 50-year organ doses ($H_{50,T}$), and the stochastic limits depend upon the effects of the organ doses upon the whole body:

$$H_E = \sum_T w_T H_{50,T} \qquad \text{(IV.46)}$$

where

H_E = effective dose equivalent (rem)
T = tissue
w_T = weighting factor for tissue (T)

The tissue weighting factors, derived on the basis of stochastic risk, are defined as follows:

Tissue	w_T
Gonads	0.25
Breast	0.15
Red bone marrow	0.12
Lung	0.12
Thyroid	0.03
Bone surfaces	0.03
Remainder (five highest other organs)	0.30

As in ICRP-2, the period of integration of the dose equivalent is the entire working lifetime, which is selected to be 50 years. The committed dose equivalent (CDE) to an organ ($H_{50,T}$) is defined as the total dose equivalent in an organ after the intake of a radionuclide into the body.

In order to not exceed either the stochastic or nonstochastic limits for internal exposure, the concepts of the annual limit on intake (ALI) and derived air concentration are introduced. Following ICRP-30, the ALI is defined to be the value of I which satisfies both of the following inequalities:

$$I \sum_T w_T H_{50,T} \leq 0.05 \text{ Sv} \quad \text{for stochastic effects} \quad \text{(IV.47)}$$

$$IH_{50,T} \leq 0.5 \text{ Sv} \quad \text{for nonstochastic effects} \quad \text{(IV.48)}$$

It should be noted that $H_{50,T}$ is specified as the dose per unit intake (Sv/Bq) which yields the correct units for I.

The derived air concentration (DAC) in Bq/m^3 for any radionuclide is that concentration in air which, if breathed by reference man for a working year under conditions of light activity, would result in the uptake of one ALI by inhalation:

$$\text{DAC} = \text{ALI}/2000 \text{ hr/year} \times 1.2 \text{ m}^3/\text{hr}$$

$$= \text{ALI}/2400 \text{ m}^3 \quad \text{(IV.49)}$$

The factor of 2400 m³ is obtained from the assumption of breathing 1.2 m³ per hour for 2000 working hours per year.

ICRP-30 METHODOLOGY

Metabolic Models

The metabolic models summarized within the ICRP-30 framework have a number of common characteristics. These characteristics include the following:

1. Simple, linear first-order equations adequately describe the translocation of material.

2. Contents mix in an instantaneous, uniform manner.
3. Organ deposition from the transfer compartments occurs rapidly. Removal is governed by specific retention times.
4. Direct excretion occurs from the organs and tissues after deposition.
5. Metabolic properties of a nuclide are described by a set of rate constants that are fixed for the 50-year deposition period.
6. Daughter atoms born in the body following deposition of the parent are usually metabolized with the properties of the parent.

These properties form the basis of the ICRP internal dosimetry models.

ICRP Lung Model

The mathematical model used to describe clearance from the respiratory tract consists of a nasopharyngeal (NP) region, a tracheobronchial (TB) region, a pulmonary region (P), and the lymph (L) nodes. Deposition is governed by either the activity median aerodynamic diameter (AMAD) or the mass median aerodynamic diameter (MMAD) of the inhaled aerosol.

The model describing the deposition of particles into the various respiratory tract regions is summarized in ICRP-30. The percentage of activity or mass of an aerosol which is deposited in the NP, TB, and P regions is given as a function of the AMAD of the aerosol distribution.

The deposition model is applicable to aerosols having an AMAD between 0.2 and 10 μm and with geometric standard deviations of 4.5 or less. The model does not apply to aerosol distributions with an AMAD of 0.1 μm or less. For distributions having an AMAD greater than 20 μm, complete NP deposition can be assumed.

The retention model describes the clearance from the respiratory region. Table IV.2 summarizes the key parameters of this model. The values for the removal half-times (T) and compartmental fractions (F) are tabulated for each of the three classes of retained material.

The respiratory tract clearance model assigns particles to one of three classes: D, W, and Y. The chemical composition of the aerosol will govern the class assignment which is similar to the ICRP-2 assignment of material in terms of its solubility characteristics. The ICRP-30 classes describe clearance half-times for material deposited in the lung which are on the order of days (D), weeks (W), and years (Y). Class D, W, and Y aerosols have clearance half-times less than 10 days, between 10 and 100 days, and greater than 100 days, respectively.

The values given for D_{NP}, D_{TB}, and D_P (left column), given in Table IV.2, are the regional depositions for an aerosol with an AMAD of 1 μm. ICRP 30 presents a method for correcting the dose calculations when the AMAD is not 1 μm.

Values of $H_{50,T}$ per unit intake, ALI, and DAC values given in ICRP-30 for individual radionuclides are given for an AMAD of 1 μm. ICRP-30 also

Table IV.2 ICRP-30 Respiratory Model's Compartment Fractions and Clearance Half-Times

Region	Compartment	Class D T (day)	D F	W T (day)	W F	Y T (day)	Y F
NP (D_{NP} = 0.30)	a	0.01	0.5	0.01	0.1	0.01	0.01
	b	0.01	0.5	0.40	0.9	0.40	0.99
TB (D_{TB} = 0.08)	c	0.01	0.95	0.01	0.5	0.01	0.01
	d	0.2	0.05	0.2	0.5	0.2	0.99
P (D_P = 0.25)	e	0.5	0.8	50	0.15	500	0.05
	f	NA^a	NA	1.0	0.4	1.0	0.4
	g	NA	NA	50	0.4	500	0.4
	h	0.5	0.2	50	0.05	500	0.15
L	i	0.5	1.0	50	1.0	1000	0.9
	j	NA	NA	NA	NA	∞	0.1

Source: ICRP-30, Part 1.
^{a}NA, not applicable.

provides the fractions (f_{NP}, f_{TB}, and f_P) of the committed dose equivalents in the reference tissue resulting from depositions in the NP, TB, and P regions, respectively. The $H_{50,T}$ value for aerosols other than 1 μm are determined from

$$H_{50,T}(x) = H_{50,T}(1\ \mu m) \sum_i f_i \frac{D_i(x)}{D_i(1\ \mu m)} \qquad (IV.50)$$

where i = 1, 2, and 3 refer to the NP, TB, and P lung regions, respectively, D_i are the deposition probabilities in the respiratory region for a given AMAD, and x is the aerosol size for which the dose is required. ICRP-30 recommends that if the AMAD is unknown, then data for 1 μm should be used.

The schematic drawing (Fig. IV.1) and Table IV.2 identify the various clearance pathways from compartments a–j in the four respiratory regions (NP, TB, P, and L).

The various absorption and translocation processes associated with the compartments are designated by the labels a through j. These processes describe the movement of material from the major respiratory tract regions to the gastrointestinal tract, systemic blood, and the lymph nodes. A brief summary of the pathways is provided in Table IV.3.

ICRP Ingestion Model

The ingestion or inhalation of radioactive material leads to the transport of this material into the GI tract. The GI tract is divided into the four regions illustrated in Fig. IV.2 and characterized in Table IV.4. The tract is divided into the

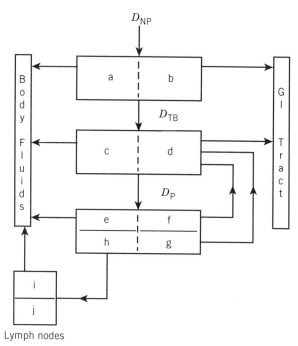

Fig. IV.1 ICRP-30 respiratory tract clearance model for dose calculations resulting from an intake of radioactive material. The schematic drawing identifies the model's clearance pathways. (From ICRP-30, Part 1.)

Table IV.3 ICRP-30 Absorption and Translocation Processes

Pathway	Initial Location	Final Destination	Description
a	NP	Systemic blood	Rapid uptake from NP directly to the systemic blood.
b	NP	GI tract	Rapid clearance by ciliary-mucous transport.
c	TB	Systemic blood	Rapid absorption from TB directly to the systemic blood.
d	TB	GI tract	Analogous to pathway b.
e	P	Systemic blood	Direct translocation from P to the systemic blood.
f	P	TB and then GI tract via pathway d	Relatively rapid clearance of pulmonary region to TB followed by a ciliary-mucous transport to the GI tract.
g	P	TB and then GI tract via pathway d	Slower but analogous to pathway f. The P translocation is governed by the material characteristics.
h	P	Lymph nodes	Slow removal from P via lymph nodes.
i	Lymph	Systemic blood	Partial or complete particle dissolution impacts the transport to the systemic blood.
j	Lymph	Lymph	Insoluble material is retained indefinitely.

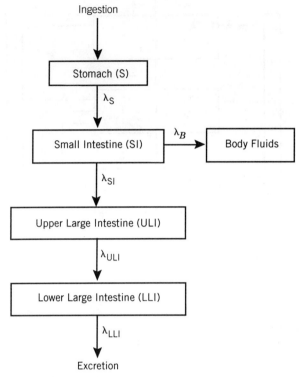

Fig. IV.2 ICRP-30 gastrointestinal tract model for dose calculations resulting from an intake of radioactive material. The schematic drawing identifies the model's clearance pathways. (From ICRP-30, Part 1.)

stomach (S), small intestine (SI), the upper large intestine (ULI), and the lower large intestine (LLI). Absorbed doses are modeled as averages over the particular tract section.

Table IV.4 specifies the rate constant for each of the sections of the GI tract.

Table IV.4 Gastrointestinal Tract Model for Reference Man

Section of GI Tract	Mass of Walls (g)	Mass of Contents (g)	Mean Residence Time (day)	λ (1/day)
Stomach (S)	150	250	1/24	24
Small intestine (SI)	640	400	4/24	6
Upper large intestine (ULI)	210	220	13/24	1.8
Lower large intestine (LLI)	160	135	24/24	1

Source: ICRP-30, Part 1.

The rate constant for removal of material from the small intestine into the body fluids (λ_B) is defined in ICRP-30 by

$$\lambda_B = \frac{f_1 \lambda_{SI}}{1 - f_1} \qquad \text{(IV.51)}$$

where f_1 is the fraction of the stable element reaching the body fluids after ingestion.

Table IV.4 provides the model parameters for each section of the GI tract. These parameters include the mass of the tissue walls, the mass of the contents (food) residing in the tissue, and the mean residence time of the food in each tract section.

Each of the sections in the GI tract is a single compartment, and the movement of material from one compartment to the next is represented by first-order kinetics and their associated linear differential equations. These equations are provided in ICRP-30.

The ICRP model for dosimetry in the GI tract assumes that material entering each segment of the tract instantaneously mixes uniformly with the contents and is removed from the segment at a constant instantaneous fractional rate. An insoluble radionuclide's total removal rate constant for a particular segment is given simply by the sum of the radionuclide's decay constant and the transfer rate constant for the segment determined from the mean residence time for the segment:

$$k^{(i)} = \lambda + \frac{1}{T^{(i)}} \qquad \text{(IV.52)}$$

where λ is the decay constant of the radionuclide and $T^{(i)}$ is the mean residence time in segment i. For specificity $i = 1, 2, 3,$ and 4, refer to the stomach, small intestine, upper large intestine, and lower large intestine, respectively. The dose equivalent delivered to the walls of a segment of the GI tract is calculated from one-half the energy spatial equilibrium dose delivered to the contents by particulate radiation emitted by the radionuclide in the segment.

Calculation of Doses

Following a methodology similar to that used in the development of the MIRD dose equation, the 50-year committed dose equivalent to a target organ (T) is calculated. Estimates are made of committed dose equivalents in a number of target organs from the activity in a given source organ. For each type of radiation i, $H_{50, T}$ in target organ T resulting from radionuclide j in source organ S is the product of two factors. The first factor is the total number of transformations of radionuclide j in S over a period of 50 years after intake. This factor is analogous to the MIRD $\tilde{A}$ factor. The second factor is the energy absorbed per gram in T, suitably modified with a quality factor, from radiation of type

i per transformation of radionuclide j in S. This factor is similar to the MIRD S-factor. For each radiation type i from radionuclide j, the committed dose equivalent is

$$H_{50,T}(T \leftarrow S)_i = (1.6 \times 10^{-10}) U_S \, \text{SEE}(T \leftarrow S)_i \quad \text{(IV.53)}$$

where

$H_{50,T}(T \leftarrow S)_i$ = 50-year committed dose equivalent for radiation type i of radionuclide j (Sv)

U_S = the number of transformations of radionuclide j in S over the 50 years following intake of the radionuclide

$\text{SEE}(T \leftarrow S)_i$ = the specific effective energy (MeV/g-transformation) for radiation type i, suitably modified by a quality factor, absorbed in T from each transformation in S.

Equation (IV.53) may be generalized to include all types of radiation (i) emitted by radionuclide j:

$$H_{50,T}(T \leftarrow S)_j = \left[\sum_i H_{50,T}(T \leftarrow S)_i \right]_j \quad \text{(IV.54)}$$

$$H_{50,T}(T \leftarrow S)_j = (1.6 \times 10^{-10}) \left[U_S \sum_i \text{SEE}(T \leftarrow S)_i \right]_j \quad \text{(IV.55)}$$

If radionuclide j has a radioactive daughter j', then we obtain

$$H_{50,T}(T \leftarrow S)_{j+j'} = 1.6 \times 10^{-10} \left\{ \left[U_S \sum_i \text{SEE}(T \leftarrow S)_i \right]_j \right.$$

$$\left. + \left[U_S \sum_i \text{SEE}(T \leftarrow S)_i \right]_{j'} \right\} \quad \text{(IV.56)}$$

For the intake of any mixture of radionuclides, the dose $H_{50,T}$ in the target from activity in the source S is given by

$$H_{50,T} = \sum_j H_{50,T}(T \leftarrow S)_j \quad \text{(IV.57)}$$

$$= 1.6 \times 10^{-10} \sum_j \left[U_S \sum_i \text{SEE}(T \leftarrow S)_i \right]_j \quad \text{(IV.58)}$$

where the sum over j is over all the radionuclides involved.

For the general case in which target T is irradiated by multiple radiation sources, Eq. (IV.58) becomes

$$H_{50,T} = 1.6 \times 10^{-10} \sum_S \sum_j \left[U_S \sum_i \text{SEE}(T \leftarrow S)_i \right]_j \qquad \text{(IV.59)}$$

where i represents the sum over radiation type, j labels the radionuclide, and S labels the source organ.

ICRP-60/61 METHODOLOGY

The Main Commission of the ICRP finished its most recent recommendations regarding the biological risks of exposure to ionizing radiation during its November 1990 meeting. These recommendations appeared in the Annals of the ICRP in 1991 as ICRP-60 and -61. ICRP-60/61 intends to advance the ICRP-26/30 recommendations that appeared in 1977.

Since the publication of its 1977 recommendations, the ICRP has periodically reviewed the basis of its recommendations in terms of biological risk. In parallel, the methodology utilized to define the risk to ionizing radiation was also assessed. Based upon these series of reviews, the ICRP now concludes that the risks of radiation-induced cancers are a factor of 3–4 greater than those published in 1977.

The ICRP-60 recommendations include situations where there is a probability of exposure, including accident situations and the disposal of solid radioactive waste. The recommendations and ongoing Task Group efforts also include those situations in which the source is not under control, including radon intrusion into homes. The new recommendations and Task Group efforts should be applicable to nearly all situations encountered by the health physics professional.

Detriment

There is considerable uncertainty regarding the best manner to characterize the consequences of the effects of exposure to low levels of ionizing radiation. These effects include the various forms of cancer and hereditary or stochastic effects resulting from exposures of a few tens of mSv/year (few rem/year). The detrimental effects include severe hereditary diseases or stochastic effects and nonstochastic cancers.

The ICRP considers health detriment effects resulting from low exposures to ionizing radiation to the organs and tissues of the body. The first consideration is the probability of a radiation-induced cancer death. The weighted probability of a nonfatal radiation-induced cancer or of a severe hereditary disease occurring in the children of irradiated individuals was also considered by the

Table IV.5 ICRP-60 Risk Factors from Exposure to Low Doses of Ionizing Radiation

Detrimental Effect	Risk Coefficient (Detriment) (1.0×10^{-4} effects/rem)	
	Public	Radiation Workers
Fatal cancers	5	4
Nonfatal cancers (weighted)	1.0	0.8
Hereditary effects	1.3	0.8
Total detriment	7.3	5.6

Source: ICRP-60.

ICRP. Finally, the ICRP considered the time of the appearance of these detrimental effects following the exposure to ionizing radiation. The contributions of these factors to the risk is summarized in Table IV.5.

The total health detriment following exposure to low doses of radiation is 7.3×10^{-4}/rem for a nominal population. The total detriment is slightly less for a working-age population of radiation workers whose ages range from 18 to 64 years. The differences between the public and the radiation worker risk coefficients are influenced by the omission of younger, more radiosensitive persons and the shorter period of potential reproduction during occupational exposure for persons commencing work at age 18.

Terminology

Although the exact nature of radiation and tissue interactions is not fully understood, ICRP-60 focuses on doses averaged over tissue rather than the dose at a point. The average absorbed dose in the tissue or organ (T) of interest, due to all radiations (R), weighted by a radiation weighting factor w_R is defined to be the equivalent dose (H_T):

$$H_T = \sum_R w_R D_{T,R} \qquad (\text{IV.60})$$

where $D_{T,R}$ is the average absorbed dose in tissue T due to radiation of type R. The radiation weighting factor replaces the quality factor which is related to the physical effects of the radiation at a point. Values of the radiation weighting factor as a function of the radiation type and energy are defined in Table IV.6. The equivalent dose is significantly different than the ICRP-26/30 dose equivalent, which is a point quantity. The effective dose (E), analogous to the effective dose equivalent, is also defined in ICRP-60. The effective dose expresses the relative detriment associated with each irradiated tissue or organ, and its response is expressed as if the whole body were irradiated. The tissue weighting factor (w_T) is obtained by expressing the detriment of each tissue-

Table IV.6 Radiation Weighting Factors[a]

Type and Energy Range[b]	Radiation Weighting Factor
Photons (all energies)	1
Electrons and muons (all energies)[c]	1
Neutrons	
< 10 keV	5
10 keV to 100 keV	10
> 100 keV to 2 MeV	20
> 2 MeV to 20 MeV	10
> 20 MeV	5
Protons, other than recoil protons (> 2 MeV)	5
Alpha particles, fission fragments, and heavy nuclei	20

[a] All values relate to the radiation incident on the body or, for internal sources, emitted from the source.
[b] The choice of values for other radiations is discussed in Annex A, ICRP-60.
[c] Excluding Auger electrons emitted from nuclei bound to DNA.

specific cancer or hereditary disease relative to the total aggregated detriment. With these definitions, the effective dose is defined as

$$E = \sum_T w_T H_T \tag{IV.61}$$

which can be written in terms of the double sum of the average absorbed dose in tissue T due to the radiation R:

$$E = \sum_R w_R \sum_T w_T D_{T,R}$$

or

$$E = \sum_T w_T \sum_R w_R D_{T,R} \tag{IV.62}$$

A comparison of the weighting factors for ICRP-60 and those from ICRP-26 and UNSCEAR 88 is provided in Table IV.7. The reader should note the wide variation in the selection of both organs and values of the weighting factors. Clearly, the selection of organs and weighting factors will continue to evolve as better data become available. The 10 remainder tissues for ICRP-60 include the following organs: adrenals, brain, small intestine, spleen, kidneys, muscle, pancreas, upper large intestine, thymus, and uterus. In the exceptional case in which one of the remainder tissues or organs receives an equivalent dose in excess of the highest dose in any of the 12 organs for which a weighting

Table IV.7 Weighting Factors for Various Models

Tissue	ICRP-26	UNSCEAR (1988)	ICRP-60
Gonads	0.25	—	0.20
Breast	0.15	0.05	0.05
Bone marrow (red)	0.12	0.17	0.12
Lung	0.12	0.17	0.12
Thyroid	0.03	—	0.05
Bone surfaces	0.03	—	0.01
Stomach	—	0.18	0.12
Colon	—	0.09	0.12
Esophagus	—	0.04	0.05
Bladder	—	0.05	0.05
Ovary	—	0.03	—
Skin	—	—	0.01
Liver	—	—	0.05
Multiple myeloma	—	0.03	—
Remainder	0.30	0.19	0.05

factor is assigned, a weighting factor of 0.025 should be applied to that remainder organ or tissue. A weighting factor of 0.025 should also be assigned to the average dose in the rest of the remainder.

ICRP-60 defines subsidiary dosimetric quantities including committed effective dose $E(t)$ and the committed equivalent dose $H(t)$. For intakes of radionuclides, the ICRP defines the committed equivalent dose as the time integral of the equivalent dose rate:

$$H_T(d) = \int_{t_0}^{t_0+d} \dot{H}(t)\, dt \qquad (IV.63)$$

where t_0 is the time that the intake occurs and d is the integration time following the intake. For occupational dose, d is taken to be 50 years; for nonoccupational exposures, 70 years is selected.

For intakes of radionuclides, the ICRP defines the committed effective dose $E(t)$ as the time integral of the effective dose rate. The integration times are the same as defined for the committed equivalent dose.

The committed effective dose $E(50)$ for the worker is explicitly written as

$$E(50) = \sum_{T=1}^{12} w_T H_T(50) + w_{\text{remainder}} \frac{\sum_{T=13}^{22} m_T H_T(50)}{\sum_{T=13}^{22} m_T} \qquad (IV.64)$$

where $H_T(50)$ is the committed equivalent dose, m_T is the mass of the remainder tissues, and $w_{\text{remainder}} = 0.05$. In Eq. (IV.64), the first sum is over the 12 organs/tissues with assigned weighting factors and the second sum is over the 10 remainder organs/tissues.

For the exceptional case, noted above for the remainder organs/tissues, the committed effective dose $E(50)$ for the worker is computed as

$$E(50) = \sum_{T=1}^{12} w_T H_T(50) + 0.025\, H_{T'}(50)$$
$$+ 0.025\, \frac{\sum_{T=13}^{22} m_T H_T(50) - m_{T'} H_{T'}(50)}{\sum_{T=13}^{22} m_T - m_{T'}} \quad \text{(IV.65)}$$

where $m_{T'}$ is the mass of the remainder tissue or organ in which the committed equivalent dose is calculated to be higher than that in any of the 12 specified tissues/organs with assigned weighting factors and $H_{T'}(50)$ is the committed equivalent dose in that remainder tissue/organ.

Framework

The recommendations of the ICRP are based upon the prevention of deterministic effects by keeping equivalent or effective dose limits below the relevant threshold for these effects. Implicit is the demand that all reasonable actions are taken to reduce the incidence of stochastic effects to acceptable levels.

Exposure to a radiation source involves the transit of radiation from within the source, through a series of media (air, shielding, worker clothing, etc.) and into the tissues of the irradiated person. The radiation source may originate in a variety of locations including the environment, industrial facility, medical establishment, public building, or individual residence. Although radiation may be controlled at any of these locations, the most effective way of controlling radiation exposure is at the source.

The ICRP defines activities related to radiation exposure and control. Activities that increase radiation exposures or risks are defined as "practices," and those activities that reduce exposure are "interventions." For both practices and interventions, some radiation exposure will occur. The magnitude of the exposure is predictable with a reasonably defined error. The possibility also exists that although there is the potential for an exposure, the exposure may not occur. The ICRP calls such exposures "potential exposures." Within the framework of these new terms, the ICRP formulates a system of radiation protection whose requirements are as follows:

1. Any practice should do more good than harm, and the practice should produce a net benefit to the exposed individual or to society.

2. Doses or the likelihood of being exposed should be as low as reasonably achievable. Exposures and risks should be optimized.
3. Individuals should be subject to dose limits or to the control of risk in the case of potential exposures.

Any radiation protection program should be judged in terms of its effectiveness. Assessments should be based upon the distribution of measured doses in the exposed population and on the effectiveness of actions taken to limit the probability of potential exposures.

Practices

For practices encountered during a facility's normal operation, the ICRP radiation protection system includes the requirement to optimize the radiation protection program elements to maximize the benefit of exposure to ionizing radiation. This requirement involves professional judgment, supported by quantitative procedures such as a cost benefit analysis, to determine the point at which any further improvements can be obtained only through the use of excessive resources. Optimization is a source-related process, and all sources of radiation are to be considered in applying the ICRP's limits.

Constraint

The ICRP-60 recommendations also include the concept of a "constraint" which indicates a restriction to be applied during optimization of individual doses resulting from a single source of exposure. The constraint represents an upper bound that must not be exceeded. Any option causing the constraint to be exceeded would be excluded from further consideration.

A constraint is not a dose limit which relates to the total dose received from all sources. However, the ICRP views the constraint as a regulatory requirement and not a design, target, or investigation level. Accordingly, it should be established by the regulatory agency based on industry judgment or experience at a well-managed operation performing the task under review.

The concept of a constraint differs from the ICRP's 1977 recommendations which used the dose limit to restrict the completion of a practice. The use of a constraint provides more control for those practices where a dose limit would be too lenient and therefore could lead to more worker exposure than warranted.

Intervention

An intervention is intended to improve the existing radiological conditions and, therefore, lower the collective exposures. Each intervention should be justified to ensure the proper balance of risk and benefit.

Two major examples of intervention are the actions taken following the discovery of elevated levels of radon in homes and following a potential or actual accidental release of radioactive material to the environment. In neither of these cases does the ICRP provide a recommendation about numerical values. For both subjects, task groups have been established. Until these task groups complete their work, ICRP Publications 39 and 40 remain applicable.

Dose Limits

It is unlikely that the dose limit will be the limiting factor during the conduct of operational activities. However, these limits are needed to ensure control of the total exposure from facility practices, to restrict the choice of dose constraints, and to protect against judgment errors during optimization.

For occupational exposures, an effective does limit of 20 mSv/year (2 rem/year) to be averaged over a period of 5 years is established. No more than 50 mSv (5 rem) should be received in a single year. Public effective dose limits should be limited to 1 mSv/year (100 mrem/year) averaged over 5 years. The ICRP does permit a higher public limit in a single year, provided that the average over 5 years does not exceed 1 mSv/year.

For the skin, the recommended annual occupational organ equivalent limit is 500 mSv (50 rem) averaged over 1 cm^2, regardless of the area exposed. The nominal depth for the dose evaluation is 7 mg/cm^2. The annual occupational dose limit for the hands and feet is also 500 mSv. The public dose limit for the skin, hands, and feet is 50 mSv. The annual limit for the lens of the eye is 150 mSv and 15 mSv for occupational and public exposure limits, respectively.

The basis for control of occupational exposure of women who are not pregnant is the same as that for men. However, if a woman is or may be pregnant, additional controls are required to protect the unborn child. Once pregnancy has been declared, the embryo/fetus should be protected by applying a supplementary equivalent dose limit of 2 mSv to the surface of the woman's abdomen for the remainder of the pregnancy, and internal exposure is restricted to 1/20 of an ALI.

ICRP-60 also recommends criteria for exemptions from regulatory control. Exemptions are recommended if individual doses are unlikely to exceed 10 μSv/year (1 mrem/yr), and the collective dose is not more than 1 person-Sv per year of practice.

CALCULATION OF ORGAN DOSES FROM INTAKES OF RADIOACTIVE MATERIAL

A common practical problem in health physics is the estimate of a worker's intake. This is often required for conditions in which an unanticipated intake occurred. The following relationships are not tied to specific ICRP recommendations, but are useful methods to assess intakes resulting from abnormal situations.

404 SELECTED TOPICS IN INTERNAL DOSIMETRY

Normally, organ doses are assessed using bioassay or whole-body counting. However, there are times when a quick assessment is required and only air sample data are available. When bioassay or whole-body counting can't be performed in a timely manner, an estimate of the intake can be obtained by knowing the average air concentration to which a worker was exposed. Knowledge of the intake is needed for the assessment of organ dose. The intake can be determined once the exposure time, air concentration, and worker breathing rate are determined:

$$I = kt(\text{BR})C \qquad (\text{IV.66})$$

where

I = intake (μCi)
t = exposure time (sec)
BR = breathing rate of individual exposed (liters/min)
C = average air concentration (μCi/ml)
k = (1 min/60 sec)(1000 ml/liter)

An alternate method of calculating the intake is based upon the use of lapel air samplers to determine the air concentration in the breathing zone of the worker. The activity on the air sampler's filter paper (A) may be obtained from the filter paper count rate (CR) and counter efficiency (e):

$$A = (\text{CR})/ek \qquad (\text{IV.67})$$

where

A = activity on the filter (μCi)
CR = count rate from the lapel air sampler filter (cpm)
e = counting efficiency (cpm/dpm)
k = conversion factor (2.22 × 10^6 dpm/μCi)

The average concentration inhaled by the worker ($\overline{C}$) is determined from the breathing zone filter's activity:

$$\overline{C} = A/ft \qquad (\text{IV.68})$$

where

$\overline{C}$ = average concentration inhaled (μCi/liter)
f = lapel air sampler flow rate (liters/min)
t = exposure duration (min)

The intake may be defined in terms of these quantities

$$I = (CR)(BR)/ekf(DF) \qquad (IV.69)$$
$$= \overline{C}t(BR)/DF \qquad (IV.70)$$

where

I = intake (μCi)
BR = worker's breathing rate (liters/min)
DF = protection factor of the worker's respirator

The calculation of organ doses follows from the intake:

$$D = I \sum_i f_i(DCF_i) \qquad (IV.71)$$

where

D = organ dose equivalent (rem)
i = number of isotopes in the air mixture
I = intake (μCi)
f_i = activity fraction of nuclide i in the inhaled air which can contain multiple radionuclides
DCF_i = dose conversion factor (rem/μCi intake)

Specific examples of the use of these relationships are presented in Chapter 4.

RELEASES

Unanticipated uptakes also occur when radioactive material is released into an area which may be followed by a release to the environment. Releases to a room may occur in both ventilated and unventilated areas. Releases of this type include rupture of a glove box with the subsequent release of its contents to a laboratory or work area, ruptures of process lines or vessels, or failures of glove box or process ventilation systems. These types of events can occur in any area that contains radioactive material.

Release of Activity into a Room without Ventilation

Assuming an instantaneous, uniform distribution of the activity into the room's volume and no dilution by the ventilation system, the dose to an individual in the room is expressed as

$$D = \overline{C}(BR)t(DCF) \qquad (IV.72)$$

where

D = dose to the individual in the room (rem)
$\overline{C}$ = average concentration in the room (Ci/m³)
   = fA/V
V = room volume (m³)
A = total activity that could be released (Ci) (It is assumed that the half-life of the radionuclide is much longer than the individual's residence time in the room.)
f = fraction of activity that is released and can be inhaled by an individual in the room
BR = breathing rate of the worker (m³/sec)
t = exposure time
DCF = dose conversion factor (rem/Ci inhaled)

The maximum dose that could be delivered to a person outside the room can be calculated by assuming that all activity is released to the environment in a short time and that the person stays at a given location for the entire release. The location is assumed to be on the plume centerline. With these conditions, the dose delivered to the individual is expressed as

$$D = fA(X/Q)(\text{BR})(\text{DCF}) \qquad (\text{IV}.73)$$

where

D = dose to the individual outside the room (rem)
X/Q = atmospheric diffusion factor at the receptor (sec/m³)
A = total activity that could be released (Ci)
f = fraction of activity that is released to the environment
BR = breathing rate of the worker (m³/sec)
DCF = dose conversion factor (rem/Ci inhaled)

Release of Activity into a Room with Ventilation

A ventilation system's design criteria specifies the number of complete air changes per hour in the room. If the ventilation system works as designed, the activity in the room will be reduced as a result of the air changes.

This situation is similar to the previous example. The major difference is that the ventilation system reduces the concentration of the room air. Following the release, the worker's dose is still given by Eq. (IV.72).

Assuming a uniform distribution of the radionuclide and exponential removal by the ventilation system, the concentration of activity in the room's air $C(t)$ as a function of time is

$$C(t) = C_0 \exp(-rt) \qquad (\text{IV}.74)$$

where

C_0 = airborne concentration assuming uniform mixing with the room air
r = ventilation system removal rate (hr^{-1})

Again, the assumption is made that the half-life is much longer than the duration of the accident. If this were not the case, an effective half-life (air removal + physical decay) would be employed in the calculations. Chapter 7 illustrates the use of effective half-lives and clearance of radionuclides by room air changes.

The average concentration ($\overline{C}$) in the room during the time of exposure is obtained by integrating $C(t)$ from time = 0 to t and by dividing by the associated time interval:

$$\overline{C} = C_0[1 - \exp(-rt)]/rt \qquad \text{(IV.75)}$$

The dose is readily obtained from Eq. (IV.72) once the average air concentration is known.

APPENDIX V

RADIATION RISK AND RISK MODELS

Previous material focused upon the various approaches used to calculate dose. Although dose is an important quantity, it is becoming more important to relate the long-term effects of receiving a radiation exposure to its impact upon the body. These impacts usually include the probability of the radiation exposure inducing cancer or death. The concept of risk via risk coefficients has been introduced to assess these effects and to compare the risk of radiation exposure to risks in other commercial and industrial activities. This appendix will outline the concept of risk and the various risk models.

RISK

Radiation is one of the most thoroughly studied agents associated with biological impacts. These impacts are quantified in terms of stochastic and nonstochastic effects and their associated health risks. The risk is often quantified in terms of a risk coefficient expressing excess-radiation-induced risk per unit radiation dose. Accordingly, the risk of the radiation exposure is often determined from the relationship

$$\text{Risk} = (\text{Risk Coefficient})(\text{Dose}) \qquad (V.1)$$

where the dose is the radiation exposure received and under evaluation. The risk coefficient varies depending upon the data under evaluation and the underlying modeling assumptions. Risk estimates are also influenced by the radiation characteristics (dose, dose rate, fractionalization, and radiation quality), biological characteristics (age, sex, genetic background, and nature of the tissue or organ), and the approach to the analysis (dose response model, projection model, and risk model).

In view of these factors, it is not surprising that there is considerable variance in risk estimates. For example, the ICRP-26 risk coefficient is 2×10^{-4} excess cancer deaths/rem, while BEIR-V with its 8×10^{-4} coefficient yields a larger characterization of the risk. A summary of risk coefficients derived from major studies is provided in Table V.1.

Equation (V.I) is often applied carelessly. This equation is only valid for a large ensemble of subjects (10,000–100,000) who have each received at least 10 rem of acute radiation exposure.

The total risk coefficient (r) is the sum of the risk coefficients for the organs or tissues (T) composing the modeled human body:

$$r = \sum_T r_T \quad (V.2)$$

Table IV.7 summarizes the various organs that are assumed in the ICRP-26, UNSCEAR 88, and ICRP-60 formulations. The formulations do not contain the same organs or level of organ risk. This table also provides the values of the organ weighting factors (w_T) for these models:

$$w_T = r_T/r \quad (V.3)$$

where the weighting factor is a dimensionless number with a value between zero and unity.

The value of Tables V.1 and IV.7 illustrate the wide modeling variations encountered in the risk estimates. The more recent models contain more organs that tend to generally have smaller weighting factors. The assignment of organs of the remainder illustrates this point. The remainder weighting factor decreased from 0.30 in 1977 (ICRP-30) to 0.05 in 1991 (ICRP-60).

Table V.1 Ionizing Radiation Risk Coefficient Summary

Year	Report	Risk Coefficient ($\times 1.0 \times 10^{-4}$) (Excess Cancer Deaths/rem)
1972	BEIR I	1
1976	ICRP-26	2
1980	BEIR III	2
1985	EPA NESHAP	4
1988	NRC BRC Policy	5
1990	BEIR V	8
1991	ICRP-60	7

BASIC EPIDEMIOLOGY

Studies of radiation risk utilize epidemiological input which requires a sample size dependent on the magnitude of the radiation exposure. The sample size, required for statistically meaningful results, is 5×10^4, 5×10^8, and 5×10^{12} for acute exposure of 10 rad, 100 mrad, and 1 mrad, respectively. These values illustrate how the size of the required exposed group varies with the absorbed whole-body dose. Meaningful results are possible for larger exposures, but the population size required for typical occupational exposures is prohibitive.

The BEIR III Committee did not know whether dose rates of gamma or x-rays of about 100 mrad/year are detrimental to humans. Somatic effects at these doses would be masked by environmental or other factors that produce the same types of health effects as does the ionizing radiation. Clearly, assessments of the impact of doses on the order of magnitude of 100 mrad or less are not physically possible.

Epidemiological studies must also consider a number of factors including sex, age, time since exposure, and the age at exposure. They are also of long duration because of the time required to follow the exposed population and control group.

The number of cancers expected in a cohort (E) is given by the sum

$$E = \sum_x c(x) r(x) \qquad (V.4)$$

where $r(x)$ is the annual incidence (morbidity) per person at age x per year, and $c(x)$ is the sum of all years spent by cohort members at age x. Once the expected incidence is determined, the number of excess cancers (EC) is readily obtained:

$$EC = O - E \qquad (V.5)$$

where O is the observed cancer incidence in the risk population.

The excess cancers per population year per incident exposure (Z) is given by

$$Z = (O - E)/N \qquad (V.6)$$

The quantity N has the units of population year-Gy:

$$N = d_i y_i \qquad (V.7)$$

where d_i is the dose to the ith group and y_i is the number of years the ith group is observed. Therefore, Z is expressed in excess cancers per population year-Gy.

With these definitions, the following terms commonly utilized in epidemiology can be defined: the relative risk (RR), standard mortality ratio (SMR), and excess relative risk (ERR):

$$RR = O/E \qquad (V.8)$$

$$SMR = 100 \, RR \qquad (V.9)$$

and

$$ERR = RR - 1 \qquad (V.10)$$

DOSE RESPONSE RELATIONSHIPS

Dose response relationships describe how an effect varies with dose. Currently, the two most popular dose response relationships are the linear and linear quadratic models. In the linear model, the effect increases linearly with dose and there is no threshold below which effects do not occur:

$$f(d) = a_1 d \qquad (V.11)$$

where d is the dose, f is the effect (such as excess cancers) being observed, and a_1 is a constant. The BEIR V Report suggests that a linear relationship is applicable for all cancers except for leukemia.

BEIR V suggests that a linear quadratic model is most representative of the leukemia data:

$$f(d) = a_2 d + a_3 d^2 \qquad (V.12)$$

RISK MODELS

There are two general types of risk models: the absolute risk model and the relative risk model. The absolute risk model assumes that the cancer risk from radiation exposure adds an increment to the natural incidence. In relative risk models the cancer risk from radiation exposure increases in an amount proportional to the natural incidence.

Age-specific models for cancer risk can be explicitly written as

$$r(d) = r_0[1 + f(d)g(B)] \quad \text{(for relative risk)} \qquad (V.13)$$

or

$$r(d) = r_0 + f(d)g(B) \quad \text{(for absolute risk)} \qquad (V.14)$$

where

$$f(d) = a_1 d \quad \text{(linear)} \qquad (V.15)$$

or
$$f(d) = a_2 d + a_3 d^2 \quad \text{(linear-quadratic)} \quad \text{(V.16)}$$

and $g(B)$ is the excess risk function. Table V.2 summarizes the BEIR V models used for a variety of cancer types.

As an example, consider the BEIR V Committee's preferred risk model for leukemia. BEIR V supports a relative risk model that is dependent upon the following factors: dose, dose squared, age at exposure, time after exposure, and interaction effects. There is an observed 2-year latency period. According to BEIR V, the preferred leukemia model is

$$f(d) = a_2 d + a_3 d^2 \quad \text{(V.17)}$$

$$g(B) = \exp[B_1 I(T \le 15) + B_2 I(15 < T \le 25)]; \quad E \le 20 \quad \text{(V.18)}$$

$$g(B) = \exp[B_3 I(T \le 25) + B_4 I(25 < T \le 30)]; \quad E > 20 \quad \text{(V.19)}$$

where T is the number of years after the exposure and E is the age at the time of the exposure. The I factors have the following values:

$$I(T \le x) = \begin{cases} 0 & \text{for } T > x \\ 1 & \text{for } T \le x \end{cases} \quad \text{(V.20)}$$

For leukemia, $x = 15$ years.

Table V.2 BEIR V Preferred Relative Risk Model[a]

Cancer Type	Dose Response Model	Comments
Leukemia	Linear + quadratic	Minimum latency of 2 years
Breast	Linear	Highest risk in women under age 15 at the time of exposure. Risk is low for women if exposed after age 40.
Respiratory	Linear	Minimum latency of 10 years. Risk decreases with time after exposure. Relative risk for females is twice that for males.
Digestive	Linear	About seven times the risk if exposure occurs at age 30 or less. Risk does not change with time post exposure.
Other	Linear	Contributes significantly to total risk. No age or sex effects have been noted. Insufficient data to permit detailed modeling.

[a] BEIR V does not support the absolute risk model.

The coefficients specifying the BEIR V leukemia model are

$$a_2 = 0.243 \quad a_3 = 0.271$$
$$B_1 = 4.885 \quad B_2 = 2.380$$
$$B_3 = 2.367 \quad B_4 = 1.638 \quad \text{(V.21)}$$

As noted in Table V.2, the BEIR V models have been applied to a variety of cancer types. An example of their application to leukemia and nonleukemia cancers will further illustrate the difficulties encountered by the BEIR III and BEIR V Committees in attempting to assess radiation risk.

BEIR III AND BEIR V COMPARISONS

Leukemia is one of the more likely forms of radiation-induced cancer. Japanese atomic bomb survivors showed an increase in leukemia incidence about 2–3 years following exposure. The incidence increased until about 7 years post exposure, and then it decreased until the leukemia rate reached normal levels in surviving populations during the early 1970s.

While the age at exposure has an important impact on risk, there was no evidence to support earlier beliefs that those exposed before the age of 10 had a greater risk than those exposed between the ages of 10 and 20.

The variation in radiation-induced cancer risks is illustrated by a comparison of the BEIR III and BEIR V estimates. Table V.3 illustrates the variation for both leukemia and nonleukemia cancers. The nonleukemia cancers include

Table V.3 Lifetime Cancer Risk Estimates (Deaths/100,000 Persons)

Cancer Type	Continuous Lifetime Exposure 1 mGy/year (100 mrad/year)		Instantaneous Exposure 0.1 Gy (10 rad)	
	Male	Female	Male	Female
Leukemia				
BEIR III	15.9	12.1	27.4	18.6
BEIR V	70	60	110	80
BEIR V/BEIR III	4.4	5.0	4.0	4.3
Nonleukemia				
BEIR III (absolute)	24.6	42.4	42.1	66.5
BEIR III (relative)	92.9	118.5	192	213
BEIR V (relative)	450	540	660	730
BEIR V/BEIR III (relative)	4.8	4.6	3.4	3.4
BEIR V/BEIR III (absolute)	18.3	12.7	15.7	11.2

respiratory, digestive, breast, and other cancer types. For leukemia, BEIR V leads to a factor of 4–5 greater risk. A similar increase of about 3–5 occurs for nonleukemia cancers if relative risk models are considered.

These factors are considerably larger (11–19) if the BEIR III absolute risk model is compared to BEIR V's relative risk model. Because recent data support the relative risk model, these larger factors do not appear to be reasonable.

DOUBLING DOSE

The qualitative relationship between radiation dose and the probability of a mutation is often described in terms of the doubling dose. The doubling dose is the radiation dose that would lead to a doubling of the mutation rate. Table V.4 summarizes the doubling dose from BEAR and from BEIR I, III, and V.

PROBABILITY OF CAUSATION

Public Law 97-414, the Orphan Drug Act of 1984, directed the Secretary of Health and Human Services to construct radioepidemiological tables providing the probability that certain cancers could result from prior exposure to radiation. The probability of causation (PC) is defined as a number that represents the probability that a given cancer, in a specific tissue, has been caused by a previous exposure or series of exposures to a carcinogenic agent such as ionizing radiation.

The PC tables are based upon the BEIR III (1980) report and are influenced by NCRP 71, which established the groundwork for the PC concept for radiogenic tumors. The current PC tables are somewhat outdated because BEIR III was based on an absolute risk model which has been superseded by BEIR V. An update of these tables, based on BEIR V, is required to ensure consistency with the currently accepted methodology.

The PC has the form

$$PC = R/(1 + R) \qquad (V.22)$$

Table V.4 Doubling Dose

Formulation	Doubling Dose (rem)
BEAR (1956)	5–100
BEIR I (1972)	20–100
BEIR III (1980)	50–250
BEIR V (1990)	<100

where R is the relative excess. In the case of a single exposure of short duration to an individual representative of the U.S. population, the relative excess is given by

$$R = FTK \qquad (V.23)$$

In this equation, F is the exposure factor which characterizes the dependence of R on the radiation dose to the risk organ. The use of whole-body exposure from film or TLD packages is not appropriate because absorbed tissue dose in units of rad is the desired quantity. The appropriate value of F is defined as a function of the absorbed tissue dose (D), measured in rads. The factors T and K are defined in the subsequent discussion.

The specific functional form for F depends on the radiation quality and cancer site. For example, consideration of Ra-224 irradiating the bone and leading to bone cancer results in the simple relationship

$$F_{\text{Bone}} = D \qquad (V.24)$$

for high-LET alpha radiation. For low-LET radiation, the values of F for thyroid, breast, and other cancers are

$$F_{\text{thyroid}} = D \qquad (V.25)$$

$$F_{\text{Breast}} = D \qquad (V.26)$$

and

$$F_{\text{other}} = D + (1/116)D^2 \qquad (V.27)$$

The second factor (T) in the definition of relative excess represents the relative likelihood that a cancer induced at age A_1 will be diagnosed after Y years. For diagnosis times between Y and $Y + 1$ years, Y is utilized in the computation. Under the constant relative risk model, which is used for cancers other than leukemia and bone cancer, T depends only on Y and has a value that increases with Y. For $Y = 0-4$ years, $T = 0$ and it rises to a value of unity for $Y \geq 10$ years. T values of 0.25, 0.5, and 0.75 occur at about $Y = 6, 7,$ and 8 years, respectively.

The constant relative risk model has not been assumed to hold for bone cancer and leukemia. For these two cancer types, T is a conditional probability which assumes that the cancer has been caused by an exposure at age A_1 and will be diagnosed Y years later. For these cases, T is calculated as the lognormal probability that a cancer is detected between years Y and $Y + 1$ after exposure at age A_1. The PC tables compile T for the various forms of cancer.

The final factor defining the relative excess is K, and it provides the dependence of R on age and baseline cancer incidence for persons of age A_2 and sex

(S) for exposure at age A_1:

$$K = K(A_1, A_2, \text{ and } S) \tag{V.28}$$

The reader has by now drawn the conclusion that the PC concept is not precise. A qualitative estimate of the uncertainties of the PC concept may be illustrated by a few examples. If the PC is calculated to be 2% or less, the true PC could be as large as 7% even if we have an accurate knowledge of all the input parameters. If the PC is within the 5–10% range, the true PC could lie within the 1–30% range. Finally, if the PC is calculated to be a least 20%, the true PC could be in the 5–40% range.

A final complication of the PC concept lies in its ties to the BEIR III methodology. The differences between BEIR III and BEIR V suggest that a review of the current PC approach and its underlying assumptions is in order.

INDEX

A-41, 103
Abdomen, 64, 83, 84, 275, 403
Absolute risk, 30, 36, 256, 411–414
Absorbed dose, 13, 84, 143, 200, 288, 289, 354, 368, 369, 385, 394, 398, 399, 410, 415
Absorbed fraction, 85, 289, 387
Absorber, 104, 106
Absorption, 69, 73, 77
Absorption and translocation processes, 69, 392, 393
Absorption cross section, 378
Abundance, 79
Ac-227, 160
Ac-228, 160
Accelerator health physics, 75, 186–208
Accelerator radiation fields, 15, 187, 189–191, 194, 196–198, 360, 376, 377
Accelerator target, 187, 190, 191, 196–198, 201
  Dose equivalent, 196, 197
Accelerator types, 44, 74, 186, 194, 195
  Electron, 44, 186, 188–191, 201–203, 205
  Heavy ion, 45, 75, 186, 188
  Medical, 44, 49, 186, 198, 199, 353
  Neutron, 195
  Particle, 45, 68, 74, 75, 186, 194
  Proton, 45, 186–188, 190, 191, 195, 204, 206
Accelerator voltage, 194, 198
Actinon, 159, 160
Activated charcoal, 337
Activation, 20, 21, 75, 77, 127–129, 132, 135, 153, 190, 191, 195, 238, 239, 323–325, 329, 330, 354, 355, 361, 376, 377, 379
  Air, 190, 191
  Gamma sources, 19–21, 48, 195, 376–379
  Products, 77, 78, 128, 129, 132–135, 137, 138
  Residual accelerator facility activity, 190
  Soil, 190, 191
  Water, 190, 191
Activity, 5
Activity fraction, 113, 300, 301, 405
Acute exposure, 34, 36, 82, 96, 110, 111, 259, 260, 361, 381, 382, 409
Additive risk model, 36, 260
Administrative controls, 87, 98, 104, 266, 278, 293
Adrenals, 28, 249–251, 399
Adults, 165–167

417

INDEX

Aerated conditions, 32
Aerosol, 391, 392
Age, 167, 176, 398, 409–412, 415
Aggregated detriment, 399
Agricultural use of radioisotopes, 74
Airborne dust, 118, 159, 161, 306, 344
Airborne particulates, 80, 159, 306, 341, 342
Airborne radioactivity, 10, 12, 57, 70, 71, 80, 81, 88, 107, 109, 113, 114, 120, 121, 159, 160, 178, 277, 282, 306, 335, 338, 341, 342, 344, 355, 404
Air dose, 143, 167, 171–173, 228
Air energy absorption coefficient, 172, 173
Air ejector, 131
Air ionization chamber, 10, 13, 19
Air sampling, 10–12, 25, 72, 73, 80, 88, 121, 123, 155, 182, 244, 282, 296, 298, 306
   Measurements, 10, 72, 73, 80, 114, 115, 121, 123, 155, 156, 182, 282, 347
Al-27, 238, 354
Al-28, 201, 354
ALARA, 21, 25, 43, 65, 70, 91, 135, 136, 141, 147, 151, 154, 236, 237, 293, 327, 402
   Decay (time), 21, 144, 151–154, 236, 237
   Distance, 70, 151–154
   Shielding, 21, 62, 70, 144, 151–154, 236, 237
Alarming radiation monitor, 152, 153, 328, 329
Alarm set point, 119, 153, 309, 310, 329
ALI (Allowable limit on intake), 7, 27, 96, 121, 214, 252, 312, 390, 391, 403
Alpha counting, 11, 12, 112, 115, 122, 220, 221, 285, 347
Alpha particles, 5, 6, 11, 12, 81, 92, 93, 101, 126, 136, 137, 159–161, 178, 181, 214, 223, 224, 285, 286, 296, 306, 346, 382, 388, 399
Alpha-rays, 81, 415
   Internal dose, 101, 113, 115, 116, 388

Alpha spectroscopy, 326
Alpha track detectors, 336
Alpha track measurement, 163
Am-241, 81–83, 113, 133, 284–286
AMAD, 113, 115, 116, 312, 391, 392
Americium, 103, 302
Angiography, 63
Anisokinetic, 80, 282
Annual incidence, 410
ANSI N13.11, 13, 226
Anterior-posterior, 83, 84, 288, 289
Antineutrino, 215
Aquatic food, 378
Aquatic food pathway, 167–171
Atmospheric diffusion, 27, 247, 338, 406
Atmospheric dispersion, 107, 140, 154, 165, 338
Atmospheric release, 27, 107, 109, 182, 183, 247, 332
Atmospheric stability classes, 107–109, 332
Atmospheric transport, 178, 338
Atomic bomb survivors, 413
Atomic electron, 214, 352
Attenuation coefficient, 272, 292, 355, 374
Attenuation factor, 357
Attenuation length, 195
Au-198, 49
Auger electron, 399
Average absorbed dose, 398
Average lifetime, 9
AVLIS, 100, 101, 117, 118

B-10, 136, 137
B-11, 15
Ba-133, 18
Ba-137, 27
Ba-137m, 28, 249–251
Background count rate, 10, 11, 182, 217–219, 347, 360, 361
Barn, 129, 150, 377, 379
Basal cell layer, 35, 259
Be-7, 77, 190, 191
Be-11, 77
Beam containment, 196, 197
Beam current, 64, 192, 197, 198, 203, 206, 356, 357

Beam dump, see beam stopper
Beam energy, 187, 189, 203, 362
Beam exposure rate, 63, 273, 275, 362
Beam power, 197, 198, 202, 203, 358
Beam pulse, 198
Beam quality, 51, 84
Beam stopper, 190, 191, 196, 197, 201, 354
Beam therapy, 45, 49, 198–200
Beam tube, 307
BEAR, 414
BEIR I, 409, 414
BEIR III, 34, 409, 410, 413, 414
BEIR IV, 36, 337
BEIR V, 34–36, 258, 260, 409, 411–414
Beta air dose, 167, 171–173
Beta counting, 112, 155
Beta decay, 8, 77, 83, 89, 92, 142, 144, 145, 155, 159, 160, 173, 215, 286, 291
Beta energy response, 225
Beta-gamma survey instruments, 13, 145, 225, 226
Beta particle, 6, 91, 112, 214, 226, 306, 385, 388
Beta rays, 70, 85, 88, 89, 91, 92, 101, 126, 132, 142, 145, 171, 214, 225, 292, 298, 299, 353, 382
  Decay, 25, 88, 89, 93, 244, 291
  External dose, 91, 103, 110, 111, 144, 145, 167, 171, 214, 225, 229, 321, 322
  Internal dose, 25, 93, 214, 244, 388
  Range energy, 88, 89, 292, 298
  Skin dose, 132, 322
Betatron, 44
Bi-210, 159
Bi-211, 160
Bi-212, 160
Bi-214, 12, 103, 159, 223, 224, 337
Binding energy, 8, 353, 355
Bioaccumulation factor, 169
Bioassay, 18, 20, 25, 82, 85, 88, 230, 296, 298, 302, 305, 404
Biological effects, 30, 35, 44, 45, 257, 408–416
  Ionizing radiation, 30, 34, 35, 408–416

Breast, 412, 414, 415
Digestive system, 412, 414
Dose response relationships, 409, 411, 412
Doubling dose, 414
Leukemia, 411–415
Marrow, 44
Respiratory system, 412, 414
Skin, 34, 35, 44, 145
Biological elimination, 387
Biological half-life, 26, 50, 58, 85, 116, 165, 179, 245, 259, 289, 385
Biological shield design, 378
Biomedical labeling, 69, 88
Biomedical tracer studies, 68, 69
Bladder, 400
Blood, 46, 48, 69, 361, 389
Body fluids, 393–395
Bone, 26, 31, 45, 46, 48, 49, 141, 257, 415
Bone cancer, 415
Bone marrow, 26, 44, 113, 114, 389, 400
Bone surfaces, 28, 110, 113, 114, 249–251, 301, 390, 400
Boric acid, 137
Boron, 107
Brachytherapy, 43, 48–50, 57–59, 263
Brain, 47, 48, 399
Breast, 28, 114, 249–251, 390, 400, 412, 414, 415
Breast cancer, 412, 414, 415
Breath analysis, 69, 82, 286
Breathing rate, 80, 113, 115, 121, 178, 246–248, 283, 300, 302, 303, 314, 337, 338, 404–406
Bremsstrahlung, 8, 9, 70, 171, 186, 189, 201–203, 273, 355, 356
Broad-beam geometry, 23, 202, 241, 242
Bromine, 136
Bronchial region, 338
Buildup factor, 20–24, 141, 143, 235–237, 240–243, 272, 294, 307, 308, 321, 357, 369, 373–375
Burial loss, 220, 221
Burial site, 103, 179, 180, 236
BWR, 86, 127, 130, 138, 139, 183, 323

C-11, 48, 190, 191, 206, 355, 360
C-13, 19
C-14, 4, 14, 18, 21, 59, 60, 69, 136, 167, 175, 211, 229, 230, 264, 266, 267, 343
Cadmium, 107
Calibration, 15, 78, 182, 227, 228
Cancer, 29, 30, 36, 46, 49, 59, 256, 389, 397–399, 408–415
  Breast, 412, 414, 415
  Fatality rates, 30, 36, 256, 398
  Leukemia, 49, 411–414
  Lung, 161, 412
  Prostate, 59
  Radiation risk, 29, 35, 36, 409–416
  Skin, 35, 259, 322
  Thyroid, 64, 415
Capture gamma sources, 376, 378, 379
Carbon dating, 4
Carcinogen, 44
Cascade, 96–98
Cataract, 31, 389
Ce-141, 137
Ce-144, 133, 137
Cell survival, 31
Centrifuge, 95, 98–100
Cephalocaudal orientation, 60, 268
CERN proton synchrotron, 195, 196
Cf-252, 13, 14, 16–18, 69, 71, 83, 228, 229
Charcoal absorber, 224, 351
Charcoal canister, 336, 337
Charged particle equilibrium, 35, 259
Chemical processing, 95, 101, 103
Chemical reprocessing, 93
Chemical toxicity, 93, 95, 96, 110, 111, 310
Chemical yield, 87, 212, 366, 367
Chest, 45
Children, 165–167
Chimney height, 108
Chronic exposure, 82, 96, 110, 111, 381, 382
Cilia, 393
Cladding, 130, 132, 137, 139, 140, 331, 332
Class D compounds, 28, 97, 101, 110, 115, 116, 120–122, 312, 391
Class W compounds, 97, 101, 111, 113, 115, 116, 122, 391

Class Y compounds, 111, 113, 115, 116, 122, 303, 304, 391
Clearance half-times, 81, 116, 391, 392
Closed window reading, 13, 226
Cm-244, 19
Cm-245, 83
CO_2, 59, 60, 265–267
Co-56, 195
Co-57, 18, 79, 133, 137, 190, 195, 281
Co-58, 103, 133, 137, 138, 190, 195, 377
Co-59, 137, 150, 377
Co-60, 14, 18, 21, 22, 25, 44, 48, 49, 69, 70, 86, 103, 133, 137, 138, 142–144, 150, 151, 190, 195, 199, 229, 237, 240, 290–292, 319–323, 352, 369, 372, 377
Cohort, 410
Collection efficiency, 11, 219–221
Colon, 400
Committed dose equivalent, 25, 27, 82, 116, 162, 180, 249, 251, 252, 302, 304, 305, 312, 313, 346, 390, 392, 395, 396
Committed effective dose equivalent, 27, 50, 59, 121, 123, 163–165, 244, 249, 252, 301, 305, 312, 316, 317, 346
Committed effective dose, 400, 401
Committed equivalent dose, 400, 401
Compartment fractions, 116, 304, 391, 392
Compton scattering, 79, 141, 213, 354
Computed tomography, 47
Concentration factor, 351
Concrete, 106, 130, 189, 191, 194, 196, 202, 204, 205, 237, 240, 262, 263, 356, 357
Concrete activation, 20, 21
Conditional probability, 415
Confidence interval, 10, 12, 218
Constraint, 402
Consumption rate, 170, 176
Contamination, 70, 78, 81, 88, 89, 113, 115, 121, 127, 130, 132, 139, 144, 191, 296, 297
  Control, 70, 75, 81, 88, 89, 130, 136, 137, 191, 278, 296, 329
  Detection, 69, 81, 88, 89, 113, 115, 121, 144, 145, 284, 285, 296, 297

Content, 381
Continuous air monitor, 113, 300
Control dosimeter, 295
Control rod, 107, 132, 139, 151, 152
Conversion electron, 70, 173
Coolant, 126, 127, 130
Core, 126, 127, 129, 132, 137–141, 195, 196
Core damage, 139
Core power, 330
Cosmic radiation, 159, 347
Coulomb barrier, 187, 188, 190
Coulomb force, 352
Counting efficiency, 10, 12, 72, 79, 80, 87, 115, 219, 224, 279–281, 333, 334, 404
Counting statistics, 10, 218, 219
Cow's milk pathway, 351
Cr-51, 103, 133
Criticality, 97, 98, 101, 104, 105, 107, 110, 118, 119, 128, 306–308
  Accident, 98, 101, 104, 118, 119, 307, 308
  Control, 98, 104, 306
  Dose, 98, 101, 110, 118, 119, 306, 307
    Gamma, 118, 119, 306, 307
    Neutron, 118, 119, 306–308
  Fissile nuclide concentration, 104
Critical density, 104–106
Critical mass, 104, 105, 307
Critical organ, 31, 33, 80, 258, 283
Critical parameter, 105
Critical volume, 105, 106
Cross section, 106, 129, 150, 189, 204, 206, 215, 238, 355, 360, 376, 377
  Absorption, 106
  Activation, 129, 377, 378
  Capture, 107, 378
  Neutron removal, 204, 357
  Scattering, 106
Cross wind distance, 107
Crypt cells, 31
Cs-134, 133, 154
Cs-137, 7, 13, 18, 27, 28, 31, 49, 57–59, 81, 82, 120–123, 133, 137, 154, 214, 229, 249–251, 257, 262, 263, 285, 286, 312–315, 317, 318, 369
Cu-64, 195

Cumulated activity, 245, 253, 254, 385, 386
Curium, 83, 103

DAC, 27, 122, 252, 390, 391
Daughter, 391
Daughter activity, 101, 115, 213, 214, 365
DCF, 27, 115, 122, 123, 154, 162, 167, 175, 181, 183, 197, 246–248, 300, 304, 307, 308, 314, 315, 317, 333, 346, 357, 405, 406
Death, 408
Decay, 4, 8, 21, 92, 138, 140, 159, 160, 204, 279, 291, 333, 334, 338, 385, 387
Decay constant, 5, 163, 181, 376, 395
Decay mode, 88, 92, 159, 160
Decommissioning, 20, 158, 236
Decontamination, 297
Decontamination and decommissioning, 20, 21, 236, 237
Deep dose equivalent, 120, 244, 313, 316, 317
Demineralizer activity buildup, 127, 128, 142, 317, 320
Demineralizer beds, 127, 128, 130, 137, 138, 142–144, 317, 320
Detector efficiency, 11, 155, 206, 219, 221, 322, 360
Detector sensitivity, 19, 232
Depleted uranium, 94, 110
Deposition, 74, 179, 264, 303, 337, 381–384, 391, 392
Deposition energy, 81
Deposition factor, 115
Deposition probability, 392
Deposition rate, 165, 170, 179
Deposition velocity, 109
Depth dose, 45, 46, 276
  6 MeV electrons, 45, 46
  18 MeV electrons, 45, 46
  15 MeV x-rays, 45
Derived air concentration (*See* DAC)
Derived investigation level, 26
Detector pulse, 198
Detriment, 397–399
Deuterium, 354
Deuteron, 15, 76, 77
Diagnostic nuclear medicine, 46, 47

Diagnostic radionuclides, 26, 47, 48
  Administration, 26, 48
Diagnostic x-ray, 18, 43, 55
Diffracted beam, 76
Digestive cancer, 412, 414
Disintegration constant, 165, 192, 211, 216, 252, 319, 333, 334, 365–367, 382, 383, 394, 395
Disk source, 121, 127, 129, 319, 322, 368, 370–373
  Thin source, 121, 130, 142, 143, 151, 313, 325–327, 371, 372
  Thick source, 127, 129, 372
Dispersion from continuous source, 107, 109, 154, 338
  Radioactive gas, 107, 108, 141, 154
  Particulates, 109, 141, 154
DNA, 399
DOP testing, 72
Dose, 84, 114, 116, 166, 195, 275, 292, 382, 399, 400, 402, 405–410, 412, 414, 415
  Control, 87, 187, 195, 401, 402
  External, 86, 110, 111, 114, 123, 195, 263
    Beta, 93, 110, 111, 132–135, 146, 214, 228
    Gamma, 20, 86, 110, 111, 118, 119, 123, 143, 213, 228
    Neutron, 5, 6, 20, 110, 111, 118, 119, 195, 196, 204, 205, 213
  Internal, 110, 111, 113–115, 166–178, 380–407
    Alpha emitters, 81, 113, 115, 123, 178, 214, 380, 388
    Beta emitters, 110, 111, 113, 123, 132–135, 146, 173, 174, 214, 244, 380, 388
    Gamma emitters, 110, 111, 118, 123, 171–174, 380, 388
    ICRP-2/10, 381–384
    ICRP-26/30, 114, 116, 121, 293, 389–397
    ICRP-60/61, 397–403
    MIRD methodology, 25, 26, 50, 245, 385–388
  Dose conversion factor (See DCF)
Dose equivalent, 13, 16, 17, 20, 22, 27, 29, 34, 36, 59, 85–88, 114–116, 121, 123, 143, 144, 159, 195–197, 200, 204, 205, 228, 229, 239, 254, 255, 258, 263–265, 289, 294, 295, 312–315, 346, 357, 369, 384, 390, 395, 398, 405
Dose factor, 133, 144, 165, 173, 174, 176, 177, 183, 229, 300, 301, 303, 305, 322
Dose limits, 16, 25, 32, 116, 123, 144–146, 149, 161, 194, 228, 258, 276
  10CFR20, 32, 92, 120, 122, 123, 146, 149, 228, 276, 316, 323, 324
  10CFR50, 166, 167
  10CFR835, 92, 123, 317
  DOE Order 5480.11, 122, 123, 317
  DOE Radiological Controls Manual, 123, 317
  ICRP-2, 33
  ICRP-26, 33, 86, 114, 389
  ICRP-60, 402, 403
  NCRP-91, 34, 121, 313
  NCRP-106, 135, 147
  NRC effluent dose limits, 167, 175
    Any organ from all pathways, 167
    Beta dose in air, 167
    Dose to any organ, 167
    Dose to total body, 167
    Dose to skin, 167
    Gamma dose in air, 167
    Total body from all pathways, 167
Dose rate, 129, 140, 141, 368, 369, 373, 374, 382, 409
Dose rate correction factor, 141
Dose response, 409, 411
  Linear, 411, 412
  Linear quadratic, 411, 412
Dosimetry, 13–18, 32, 75, 76, 87, 88, 225, 227, 298, 305, 361
Doubling dose, 414
Downwind distance, 107, 108
Dry desquamation, 34, 259
Duty factor, 198, 205, 358

Effective dose, 398–400, 403
Effective dose equivalent, 85, 111, 114, 122, 123, 159, 162, 181, 259, 301, 314, 346, 389, 398
Effective half-life, 50, 165, 178, 245, 246, 259, 264, 289, 295, 340, 341, 407

INDEX **423**

Effective energy, 382
Effective height, 107, 109
Effective release height, 107
Effective stack height, 108, 172
Effluent release type, 68, 132, 135–137, 158, 164, 166–168, 183
  Liquid, 137, 166–171, 183
  Noble gas, 136, 166, 167, 171–175, 183
  Particulates, 137, 166, 167, 183
  Radioiodine, 136, 166, 167, 183
  Tritium, 26, 167
Elastic scattering, 187, 195, 213, 352
Electret dosimeter, 225
Electrets, 336, 337
Electromagnetic field, 101, 118, 306, 307
Electron, 44, 70, 117, 186, 188–191, 193, 194, 199, 201–203, 205, 252–254, 388, 399
Electron accelerators, 44, 186, 188–191, 201–203, 205, 354–356
Electron beam, 44, 45, 55, 101, 117, 118, 188, 189, 203, 306, 354
Electron beam current, 55
Electron capture, 8, 9, 70, 77, 214, 215
Electron linear accelerator, 44
Electronic equilibrium, 25, 227, 245
Elimination function, 302, 303, 305
Elution, 4, 366
Embryo, 83–85, 288, 289, 324, 403
Emergency core cooling system, 141, 142
Energy compensated GM probe, 323
Engineering controls, 69, 71, 72, 118, 306, 307
Enrichment, 91, 94–101, 104, 105, 110–112, 117–119, 126, 138, 305, 306
  Stage, 96–99
Entrance skin exposure, 84, 288
Environmental health physics, 74, 91, 158–185
Environmental monitoring program, 74, 158, 164, 179–181
  Assessment of releases, 74, 158, 164–177, 182, 183
  Buildup of activity in ponds, 165, 166, 179
EPA NESHAP, 409
E-perm, 336, 337

Epidemiology, 410, 411, 414
  Meaningful sample sizes, 409, 410
Epithermal neutrons, 17
Equilibrium dose constant, 85
Equilibrium factor, 12, 162, 178, 181, 224, 338
Equivalent dose, 398–400
Erythema, 34, 259
Erythrocytes, 31
Esophagus, 70, 278, 400
Excess cancers, 161, 166, 258, 260, 410, 411
Excess relative risk, 411
Excretion, 394
Excretion function, 341
Expected incidence, 410
Exposure, 10, 22, 52–57, 62–64, 76, 84, 86, 107, 110, 113, 114, 119, 123, 136, 138, 150, 151, 178, 195, 206, 215, 232, 233, 235, 245, 246, 256, 257, 263, 272–275, 277, 307–309, 313, 323, 361, 368, 369, 371–373, 397, 401, 402, 407–410, 412–414
  General public, 136, 141, 155, 158, 161, 164, 167, 183, 398, 400, 403
  Medical, 52, 54, 60, 64, 245, 246, 261, 263, 268–272, 276, 287, 288
    Accelerator, 44, 186–208
    Diagnostic, 49, 60, 245, 246, 268–271, 276, 287, 288
    X-rays, 52, 53, 60, 64, 84, 257, 268–271, 273–276, 287, 288
  Occupational, 20, 21, 110, 113, 121, 141, 147–149, 158, 161, 180, 195, 236, 245, 246, 256–258, 272, 290, 313, 325, 327, 389, 398, 400, 401, 403, 410
  Pregnant women, 83, 84, 276, 287, 288
Exposure control, 20, 21, 65, 78, 98, 236
Exposure factor, 415
External beam therapy, 49
External dose, 8, 25, 30, 43, 64, 118, 119, 123, 126–128, 130, 135, 136, 175, 191, 215, 244, 263, 277, 313, 314, 317, 331, 369

External radiation protection, 78, 235, 277, 368, 369
  Distance, 368–373
    Line source, 127, 129, 150, 151, 368, 370, 371
    Point source, 20, 22, 23, 53, 54, 127, 129, 150, 235, 239, 277, 307, 325, 368, 369
    Slab source, 127, 129, 130, 370
    Thin disk source, 121, 127, 129, 130, 143, 151, 313, 319, 322, 371, 372
    Thick disk source, 127, 129, 372
  Hazard, 93, 101, 110–112, 128
  Shielding, 22, 23, 69, 144, 235
    Beta rays, 69, 134, 189
    Broad-beam, 23, 24
    Gamma rays, 116, 134
    Narrow-beam, 23, 24
    Neutrons, 116, 187, 195, 196
    Primary protective barrier, 53
    Secondary protective barrier, 53
    X-rays, 51–55, 189, 193, 194, 203, 268–270
Extremity dosimetry, 18, 226, 229, 305
Eye, 69, 88, 298, 306
Eye dose, 16, 33, 88, 228, 258

F-18, 48
Facility design, 51, 112, 118, 187, 189, 195–197, 306, 307
Fallout, 347, 348
Fast neutron, 6, 31, 106, 377
Fatal cancers, 34, 36, 398
Fe-54, 137, 377
Fe-55, 190
Fe-58, 137, 377
Fe-59, 103, 133, 137, 195, 377
Fecal sample, 82, 285, 286, 302, 305
Fertile material, 126
Fetal exposure, 64, 83–85, 149, 276, 287–298, 324, 403
Film badge, 15, 18, 70, 85, 227, 230
Filter, 11, 78–80, 112, 127, 128, 132, 136, 138, 141, 151, 153–156, 225, 282, 296, 331, 333, 334, 404
  Activity buildup, 127, 128, 156, 302, 303, 404
Filter type, 11, 220
  Cellulose, 11, 220, 221

Glass-fiber, 11, 153, 220, 221
Membrane, 11, 112, 221
Filtration, 64, 136–138, 142, 220, 221, 287, 306
Fissile material, 102–106, 126
Fission, 104, 106, 107, 118, 119, 126, 127, 130, 136, 307–309, 376
  Fragments, 399
  Gamma dose, 118, 119, 307–309
  Gamma sources, 118, 119, 378, 379
  Neutron dose, 118, 119, 307–309
  Products, 103, 128, 132–137, 139, 140, 365
Fission neutrons, 106, 118, 119, 307–309
Fission product barriers, 130, 155, 331, 332
Fluence, 360
Fluence rate, 129, 132, 240, 330, 358, 374, 376
Fluorescent radiation, 76
Fluoroscopy, 47
Flux, 21, 129, 132, 238, 240, 324, 374, 376
Flux-to-dose conversion factor, 20, 118, 119, 234
Forebrain damage, 85
Fractional yield, 212, 366, 367
Fractionalization, 35, 44, 259, 260, 409
Fuel, 126
Fuel assembly, 126
Fuel bundle, 126
Fuel core, 126, 127, 129, 137, 195
Fuel cycle, 91, 93, 101, 103, 110–112
  Thorium, 91, 101, 102
  Uranium, 91–95, 101, 102, 110–112
Fuel cycle health physics, 91–125
Fuel element, 127, 130, 139
  Cladding failure, 127, 130, 132, 136, 139, 332, 350
Fuel fabrication, 94, 102, 103
Fuel integrity, 128, 139, 332, 350
Fuel pellet, 126
Fuel reprocessing, 118
Fuel rod, 94, 126
Fume hood, 69–73, 299, 306
Fusion energy, 74, 76–78, 186

Ga-67, 48
Gamma camera, 47, 48

Gamma-ray, 137, 196, 199, 201, 205, 206, 234, 291, 292, 299, 353, 354, 358, 360, 362, 376–379, 410
  Air dose, 17, 143, 167, 171–173, 228
  Atomic attenuation coefficient, 141, 240
  Constant, 21, 22, 57, 62, 64, 70, 86, 122, 143, 150, 272, 277, 313, 319, 369, 371–373
  Counting, 78–80, 112, 123
  Energy, 78–80, 85, 92, 143, 234, 240, 273, 279–281, 289, 296, 307, 308, 321, 369, 370, 374, 375, 377, 379
  Energy absorption coefficient, 20, 234, 274
  Exposure, 21, 110, 111, 273, 274, 277, 289, 307–309, 313
  External dose, 20, 21, 91, 92, 110, 111, 118, 123, 126, 140, 141, 143, 150, 191, 225, 226, 307–309, 314, 319
  Fluence rate, 118, 119, 376
  Internal dose, 110, 111, 289, 388
  Linear attenuation coefficient, 22, 24, 143, 374
  Mass absorption coefficient, 24, 234, 369, 370, 374
  Shielding, 52, 235, 307–309, 373–375
  Skin dose, 132
  Spectrum, 78, 79, 201, 307–309, 354
  Yield, 79, 143, 240, 289, 296, 307–309, 321, 369, 370, 375, 377–379
Gamma scan, 343
Gas, 107
Gas channel, 182, 348
Gas filled detector, 18, 19
Gas flow proportional counter, 81, 82, 284, 285
Gas generation, 86, 87, 192, 291, 292
Gaseous diffusion, 95–100
Gaseous effluent pathway, 165, 171–175, 182, 350, 351
Gaseous waste, 103, 137, 138
Gastrointestinal tract, 29, 71, 252–256, 285, 381, 392–395
  Doses, 29, 71, 252–256
Gaussian plume, 107
Ge(Li) detector, 78–80, 112, 191, 279, 285

Geiger–Mueller detector, 63, 81, 112, 285, 323, 358
Genetic background, 409
Genetic effects, 414
Geometry, 98, 104, 105, 129, 225, 368
Geometry effect, 225
Germanium detector, 285
Giant resonance, 189, 356
GI tract model, 29, 394, 395
Glove box, 26, 71–73, 113, 306, 405
Glucose, 59, 60, 264, 265, 267
Gonads, 14, 28, 114, 249–251, 390, 400
Grab sample, 336, 342
Graphite, 106
Gravitational settling, 109, 338
Gross alpha, 343
Gross beta, 343
Ground contact, 109, 338
Ground deposition, 109, 175, 349, 351
Ground-level concentration, 107–109, 174

H-2, 137, 188
H-3, 18, 21, 26, 27, 30, 69, 76–78, 80, 89, 103, 128, 136, 137, 167, 169, 170, 175, 188, 190, 191, 204, 229, 230, 246, 247, 282, 343
Hafnium, 107, 378
Hair, 361
Half-life, 4, 5, 9, 11, 21, 22, 26, 48, 49, 58, 62, 64, 70, 79, 85, 87, 88, 150, 155, 159, 160, 169, 178, 279, 191, 192, 204, 206, 211, 221, 223, 244, 245, 259, 364, 365, 271, 273, 277, 290, 296, 299, 303, 335, 339, 340, 356, 360, 377, 378, 383, 385, 406, 407
Half-value layer, 54–56, 58, 62, 84, 143, 262, 270, 287
He-3, 76, 77
He-4, 76, 77, 188
He-5, 77
Health risk, 408
Heart, 46–48
Heavy ion, 45, 186, 188, 260
Heavy-ion acceleration, 186, 188
Heavy nuclei, 399
Hereditary effects, 389, 397–399
Hf-181, 133

Hg-203, 79, 281, 282
High-level waste, 94, 102, 103, 121–123
High-voltage equipment, 76, 100, 101, 118, 192, 193, 305, 306
Hood, 80
Horizontal standard deviation, 107
Hot particle, 127, 128, 132–135, 144, 145, 322, 323, 329, 331
  Skin dose, 128, 132–135, 145, 322, 323
HTO, 69, 77, 88, 282, 283, 298
Humidity, 225
Hydrogen, 106, 213, 291, 292, 354, 378
Hyperthyroidism, 49
Hypoxic conditions, 32

I-123, 48
I-125, 49, 58, 59, 69, 70, 80, 85, 88, 89, 263, 264, 282, 283, 289, 298
I-126, 7–9, 214, 245
I-129, 86, 87, 103, 291
I-131, 10, 25, 48, 49, 64, 65, 69, 70, 103, 120, 121, 137, 154, 179, 218, 245, 246, 277, 278, 312, 332, 340, 341, 351
ICRP, 163, 181, 251, 252, 258, 290, 322, 381, 389, 391, 395, 397, 401–403
  ICRP-2, 33, 81, 257, 381, 390, 391
  ICRP-10, 25, 26, 246, 381
  ICRP-23, 24
  ICRP-26, 24, 25, 29, 30, 33, 50, 59, 83, 86, 114, 244, 256–258, 286, 381, 389, 397–400
  ICRP-30, 73, 114, 116, 117, 163, 252–257, 293, 304, 381, 389–395, 397, 398, 409
  ICRP-49, 289
  ICRP-60, 73, 324, 381, 397–403, 409
  ICRP-61, 381, 397
Immersion, 174, 175, 350
In-111, 48
In-113, 133
Inconel, 378
Induced activity, 353
Inelastic scattering, 187, 195, 214
Infant, 165–167

Ingestion, 29, 31, 47, 64, 73, 95, 168, 169, 176, 178, 278, 311, 350, 380, 381, 392, 394, 395
Ingestion model, 29, 168, 169, 293, 392
Ingestion rate, 68
Inhalation, 29, 47, 69, 73, 77, 82, 95, 110, 111, 113–117, 120, 122, 161, 175, 176, 191, 283, 286, 302–304, 311, 312, 344, 350, 380, 381, 390, 392, 404–406
Inhalation ALI, 27, 121, 252
Inhalation model, 116, 117
Inhaled particulates, 82, 113–116, 122, 161, 285, 312
Inhaled radioactivity, 31, 82, 110, 111, 113–116, 122, 161, 191, 285
Injection, 47, 265, 278, 380, 381
Insoluble, 303, 323, 393, 395
Intake of radionuclides, 28, 73, 74, 82, 113, 116, 120, 121, 123, 165–167, 253, 260, 277, 283, 284, 300–302, 312, 381, 383, 384, 386, 390, 391, 393–396, 400, 403–405
Intake retention function, 246
Integrated samples, 336
Interlock, 76, 193, 196, 307
Internal dose, 8, 25, 26, 43, 49, 57, 64, 68, 74, 82, 85, 86, 113, 115, 120, 121, 123, 126–128, 130, 135, 136, 167, 178, 215, 244, 264, 286, 296, 300, 301, 316, 317
  Single-uptake, single-compartment model, 74, 82, 85, 113, 115, 123, 282–284, 286, 302, 303, 381–383
  Continuous-uptake, single-compartment model, 178, 179, 286, 339, 340, 383, 384
  qf_2 variation after cessation of intake, 384
Internal dosimetry, 380–407
Internal radiation protection, 68, 74, 78, 81–83, 128, 380–407
  Assessment of dose, 25, 26, 57, 68, 74, 85, 86, 113, 115, 116, 120, 121, 123, 159, 178, 215, 277, 300, 301, 380–407
  Controls, 81
  Radiation hazard, 93, 95, 101, 110–113, 116, 128, 215, 277

INDEX 427

International Commission on Radiological Protection (*See* ICRP)
Intervention, 401, 402
Inverse square law, 288, 309, 329
Investigation level, 26, 246, 402
In vitro, 246
In vivo counting, 182, 246, 302, 305, 347
Iodine, 128, 136, 141, 142, 154, 167, 175, 179, 264, 332
Ion exchange resins, 128, 131, 137, 138, 191
Ionization, 117, 189, 190, 196, 397
Ionization chamber, 10, 13, 19, 63, 81, 83, 112, 132, 135, 198, 225, 231, 275, 282, 284, 336, 343
Ionization detector, 19, 112
Ionizing radiation hazard, 118, 305, 306
Ion pair, 6, 214, 225, 273
Ir-192, 49
Iron, 378
Irradiation cell activity buildup, 192, 203–206
 Toxic gas buildup, 192, 203–205
Irrigated foods pathway, 167, 169–171
Isokinetic sampling, 11, 80, 219, 220
Iostope separation, 95–101
 Centrifuge, 98
 Gaseous diffusion, 95
 Laser techniques, 99–101, 117, 118

K-40, 24, 25, 182, 243, 244, 347, 348
K-capture, 214
Kidney, 31, 93, 110, 257, 399
Klystron, 192, 194
Kr-85, 103, 136, 137, 182, 183, 348, 349
Kr-87, 103, 137
Kr-88, 137, 155, 156, 335
Kr-89, 103
Krypton, 130, 136
Kusnetz method, 336

Laminar flow, 11
Lapel air sampler, 114, 115, 121, 123, 404
Laser, 95, 99, 100, 117, 118, 186, 194, 306, 307

Laser enrichment, 95, 99, 100, 101, 117, 306
Laser isotope separation, 95, 99, 100, 101, 117, 306
Law of Bergonie and Tribondeau, 31
Leakage attenuation factor, 55
Leakage radiation, 51–55, 62, 76, 77, 194, 270
Lens of the eye, 14, 16, 31, 33, 226, 228, 389, 403
Lethal dose, 44, 196
Leukemia, 49, 411–415
Li-6, 15, 77, 137, 227, 360
Li-7, 15, 77, 188, 227
Linear absorption coefficient, 373
Linear accelerator, 18, 48, 186, 201–205, 355, 358
Linear attenuation coefficient, 22, 24, 62, 194, 195, 242, 262, 272
Linear dose response, 30, 411, 412
Linear energy transfer (LET), 32, 35, 36, 186, 259, 260, 352, 377, 415
Linear quadratic dose response, 411, 412
Line source, 127, 129, 150, 151, 290, 291, 325, 326, 368, 370, 371
Liquid effluent pathway, 165, 167–171, 350, 351
Liquid scintillation counting, 69, 82, 191, 283, 285, 286, 299
Liquid waste, 103, 137, 138
Lithium hydroxide, 137
Liver, 31, 47, 59, 113, 265, 301, 400
Loss factor, 57, 265
Lower large intestine, 28, 29, 249–256, 394, 395
Lower limit of detectability, 10–12, 86, 87, 218, 219, 222, 293
Low level counting, 347, 348
Low level waste, 103, 137, 138, 179, 180
Lucas cell, 336
Lung, 28, 31, 45, 46, 48, 69, 81–83, 110, 111, 113–116, 159, 161–163, 181, 249–251, 257, 265, 285, 286, 300, 301, 304, 305, 346, 390–392, 400
Lung model, 82, 113–117, 304, 391–393
 ICRP-30, 116, 117, 303, 304, 391–393

Lymph, 69, 116, 391–393
Lymphocytes, 31

Maintenance, 101, 117, 128
Mammography, 47, 60
Mass attenuation coefficient, 20, 21, 24, 58, 118, 119, 143, 235, 237, 239, 243, 374
Mass defect, 7, 214
Mass energy absorption coefficient, 20, 234, 369, 370
Mass excess, 7
Maximum exposed individual, 166, 167, 177, 180, 311
Mean absorbed dose, 386, 388
Mean dose per unit accumulated activity, 50, 386, 388
Mean dose per unit administered activity, 387
Mean energy emitted per transition, 385, 387
Mean free path, 354
Mean lifetime, 9, 50, 192, 206, 216, 223, 224, 253, 264
Mean residence time, 29, 252, 254, 394, 395
Mean wind speed, 107
Medical exposure, 43, 44, 268–271
  Diagnostic, 43, 60, 268–271
  Therapeutic, 43, 64, 65
Medical health physics, 43–67
Mediolateral view, 60, 61, 268
Memory effect, 225
Mental retardation, 84, 85
Metabolic models, 390
Metabolism, 59, 264, 284
Microwave, 194
Microwave waveguide, 194
Mining and milling, 91, 93–95, 103
Minute volume, 338
MIRD-3, 388
MIRD-11, 26, 387
MIRD theory, 25, 26, 50, 245, 385–388, 395, 396
Mixed radiation fields, 15, 17, 126
Mixing ratio, 168
MLIS, 100
MMAD, 391
Mn-54, 21, 133, 137, 190, 195, 377
Mn-55, 377

Mn-56, 195, 377
Mo-99, 4, 62, 63, 87, 88, 212, 271, 273, 295, 366
Moderator, 104–106
Moist desquamation, 34, 259
Morbidity, 410
Mouth, 278
MPC, 10, 122, 218, 219, 316
Multiple myeloma, 400
Multiplicative risk model, 36, 260
Muons, 75, 187–189, 195, 353, 399
Muscle, 399
Mutation, 260, 414

N-13, 48, 190, 191, 355
N-14, 136
N-16, 19, 128, 188, 323, 350, 351, 379
N-17, 191
Na-22, 133, 190, 191, 195
Na-23, 377
Na-24, 21, 22, 195, 238, 369, 377
NaI detector, 81, 82, 112, 206, 284, 285, 337, 361
Narrow beam, 23, 24, 242
Nasal smear, 82, 113, 115, 121, 285, 286, 296
Nasopharyngeal region, 116, 391–393
National Commission on Radiation Protection (*See* NCRP)
Natural airborne radioactivity, 112, 158, 159, 182, 221
Natural background, 347, 348
Naturally occurring radioactive material, 112, 158, 159, 182, 221, 347
Nb-95, 133, 144
NCRP, 177, 263, 276, 322
  NCRP-37, 56, 58, 261, 262
  NCRP-49, 53–55, 60, 61, 268, 269
  NCRP-71, 414
  NCRP-78, 337
  NCRP-91, 34, 121, 258, 313
  NCRP-93, 158, 159
  NCRP-106, 135, 147, 322, 323
Negative pion therapy, 198–200
Neptunium, 83, 96
Nerve cells, 31
Neutrino, 214, 215
Neutron, 5, 6, 16–18, 20, 21, 68, 75–77, 91, 106, 110, 118, 119, 126, 128, 129, 132, 137, 152, 186–189,

191, 194–197, 199, 202–205, 226, 227, 234, 260, 306–309, 352–358, 362, 365, 370, 376–379, 399
Absorption, 106
Activation, 20, 136, 137, 144, 150, 153, 323–325, 354, 376, 377, 379
Attenuation factor, 118, 119, 307, 308
Capture, 107, 137, 234, 376, 379
Dose, 92, 110, 118, 119, 196, 307–309
Dose equivalent, 16, 17, 20, 119, 204, 205, 229, 234
Dosimetry, 16, 17, 227–229, 362
Fission, 118, 119, 126, 307–309, 365, 376–379
Removal cross section, 204, 357
Scattering, 106, 187, 189, 362
Shielding, 118, 119, 187–189, 194, 195, 203, 307–309, 378
Neutron bubble detectors, 17, 229
Neutron energy, 17, 77, 106, 118, 137, 150, 194, 196, 307, 308, 358, 399
Fast, 31, 77, 106, 137, 150, 194, 196, 213, 227, 399
Thermal, 106, 137, 150, 196, 213, 227, 324, 377, 399
Ni-58, 137, 377
Ni-65, 14
Nitrous oxides, 192, 205, 206, 355
Noble gas, 13, 78, 128, 142, 167, 171–175, 225
Semi-infinite cloud, 140, 141
Nonfatal cancers, 398
Nonionizing radiation, 118, 306, 307
Nonstochastic effects, 7, 33, 110, 114, 214, 252, 258, 290, 302, 305, 389, 390, 397, 408
NORM, 158
Np-237, 83
NRC BRC policy, 409
Nuclear fission, 104, 126
Nuclear medicine, 18, 43, 47
Diagnostic, 47
Therapeutic, 43, 48
Nuclear radiation heat generation, 378
Nuclear reactor (*See* Power reactors)
NVLAP, 226

O-15, 48, 190, 191, 355
O-16, 188, 360, 379

Occupational exposure, 20, 21, 92, 121, 147–149, 180, 236, 245, 256–258, 313, 389, 398, 400, 401, 403, 410
Occupancy factor, 51–53, 61, 62, 161, 178, 181, 268–270, 337, 346
Open window reading, 226
Optical thickness, 321
Optical transport system, 307
Optimization, 402, 403
Orbital electron capture, 7, 8, 214
Organ burden, 50, 264
Organ doses, 26, 29, 50, 84, 114, 141, 167, 168, 175, 176, 255, 290, 300, 380–406
Organ mass, 26, 380, 382, 383, 386
Other cancers, 412, 415
Outages, 128, 136, 138, 139, 149–151
Ovary, 49, 84, 400
Oxygen, 291, 354
Ozone, 192, 205, 206, 355, 359

P-32, 14, 49, 69, 70, 82, 88, 89, 282, 283, 298, 299
Pa-231, 160
Pa-234, 103
Pa-234m, 91, 92, 159
Pair production, 79, 80
Palladium, 284
Pancreas, 399
Parent, 391
Parent activity, 213, 365
Particle production, 187
Particle size, 109, 113, 115, 122, 303, 312, 351
Particulates, 11, 78, 109, 113, 132, 136, 141, 142, 155, 159, 167, 175, 182, 303
Partitioning factor, 154, 332, 333
Pasquill–Gifford equation, 107
Pasquill stability class, 108, 154
Pathways, 120, 164–177, 180, 183, 311, 344, 350, 351, 392–394
Pb-206, 159
Pb-207, 160
Pb-208, 160
Pb-210, 103, 159
Pb-211, 160
Pb-212, 160
Pb-214, 12, 159, 223, 224, 337
PERM, 225

Personnel contamination, 81, 296, 297
Phantom, 17, 228, 361
Photodisintegration, 100
Photoelectric effect, 79, 280
Photoexcitation, 100, 117, 306
Photoionization, 100
Photon, 13, 15, 18, 63, 101, 171, 189–191, 194, 195, 199, 203, 213, 229, 240, 273, 289, 337, 353, 360, 385, 388, 399
Photonuclear, 203, 356
Photon yield, 172, 239, 360
Physical control, 104
Physical half-life, 50, 58, 165, 179, 289, 385
Pions, 75, 187–189, 198–200, 352, 353, 356
Planar imaging, 47, 48
Planned special exposure, 92
Plateout, 284
Plume, 107, 108, 141, 172
Plume centerline, 107, 247, 406
Plutonium, 93, 94, 96, 101, 103, 104, 113, 114, 302
Po-210, 88, 159, 286, 297
Po-212, 160
Po-214, 12, 103, 159, 223, 224, 337
Po-215, 160
Po-216, 160
Po-218, 12, 103, 159, 223, 224, 337
Pocket dosimeters, 14, 226
Point source, 20, 53, 62, 65, 127, 129, 135, 143, 150, 153, 239, 277, 307, 308, 368–370, 372
  With attenuation, 23, 62, 143, 235, 307–309
  Unshielded, 22, 65, 143, 150, 153, 235, 239, 277, 307–309, 320, 321, 325, 368–370, 372
Point source approximation, 22, 54, 65, 129, 143, 150, 196, 239, 240, 291, 294, 307–309, 320, 321, 325, 357, 368, 372
Poison, 106
Polyethylene, 106, 307, 308
Pond activity, 165, 166, 179, 340
Population exposure, 141, 166
Population group, 164–166
Positron, 385, 388

Positron emission, 8, 77, 173, 190, 191, 215, 352, 360
Positron emission tomography (PET), 47, 48, 186
Posterior-anterior, 83, 84, 288, 289
Potable water, 167, 168
Potassium iodide, 89, 299
Potential exposure, 401, 402
Power reactor accidents, 139–141, 154
  Fuel handling accident, 139, 142
  Loss of coolant accident, 139, 142
  Steam generator tube rupture, 139, 142
  Waste gas tank rupture, 139, 140, 142
Power reactor health physics, 126–157
Power reactors, 43, 68, 75, 78, 91, 94, 102, 107, 126–157, 195, 196, 226, 322, 376
  Component activation, 128, 129, 132, 135, 153, 195
  Effluents, 102, 135
  Worker exposure, 17, 19, 128, 134, 135, 141, 147–149
Pr-144, 133
Practice, 401–403
Precipitation removal, 338
Preoperational environmental monitoring, 164, 179–181
Primary beam, 186, 187, 189, 191, 192, 195, 275, 276
Primary beam output, 52, 61, 268, 269
Primary coolant, 126, 127, 130, 131, 136, 137, 139, 140, 142, 150, 154, 332
Primary protective barrier, 53
Primary to secondary leak rate, 130, 131, 332
Probability of causation, 29, 30, 256, 257, 414–416
Product, 96–99, 306
Prompt and delayed fission gammas, 379
Proportional detector, 81, 82, 112, 198, 357
Protection factor, 115, 302, 405
Protective action recommendations, 155, 331, 333
Protective clothing, 114, 115, 134–137, 144
Proton, 45, 75–77, 136, 137, 186, 187,

190, 195, 197–199, 204, 206, 352–354, 357, 358, 360–362, 377, 399
Proton accelerators, 75, 186, 187, 190, 195, 196, 204, 206, 360–362
Proton recoil film badge, 15, 227
Pu-238, 113, 122, 123, 315, 317, 318
Pu-239, 11, 12, 31, 104, 105, 113, 126, 221, 257, 365
Pu-240, 113
Pu-241, 83, 113, 286
Public exposure, 398, 403
Pulmonary clearance, 116, 303, 304, 393
Pulmonary deposition, 115, 116, 303, 304, 393
Pulmonary lung, 46, 116, 303, 304, 392, 393
Pulmonary region, 116, 303, 304, 391–393
Pulse repetition rate, 361
Pulsed radiation fields, 198
PuO_2, 94
PWR, 86, 107, 126, 127, 130, 138, 139, 142, 154, 155

Quality factor, 16, 74, 162, 181, 228, 255, 286, 346, 395, 396, 398
Q-value, 8, 77

Ra-223, 160
Ra-224, 160, 415
Ra-226, 103, 159, 206, 360, 361
Ra-228, 160
Rad, 368, 369, 415
Radiation background measurements, 182
Radiation damage, 86, 87, 293
Radiation exposure, 81, 84, 147–149, 401
Radiation exposure events, 81, 83, 147–149
Radiation oncology, 44
Radiation quality, 409
Radiation survey, 81, 83
Radiation type, 398, 399
Radiation weighting factor, 398, 399
Radiation work permit, 114
Radioactive decay, 4, 5, 9
Radioactive label, 87, 295
Radioactive release, 73, 103, 107–109, 139, 182, 236, 247, 331, 344, 403, 405, 406
  Gas, 103, 107, 108, 344
  Particulate, 103, 109, 344
Radioactive transformation, 365
Radioactive waste, 68, 88, 89, 102, 103, 119–122, 128, 131, 135, 137–139, 299, 344, 397
  Burial facility siting criteria, 179–181, 341–343
  High level, 102, 121, 122
  Low level, 88, 89, 102, 103, 119, 120, 137, 138, 299
  TRU, 103, 306
Radiofrequency equipment, 192–194
Radiofrequency interference, 225
Radiography, 47, 63, 199
Radioisotope administration, 47–49
Radiolytic decomposition, 291, 292
Radiopharmaceutical, 47, 62, 385, 387
Radioprotective chemical, 32
Radiotherapy, 18, 44, 186, 200
Radium, 44, 50, 57, 159, 305
Radium equivalent activity, 50, 58, 261–263
Radon, 12, 13, 91, 112, 123, 158–164, 177, 178, 181, 221, 223–225, 296, 336–338, 345–347, 397, 402
  Measurements, 12, 13, 158, 162, 163, 177, 178, 181, 336–338
  Mitigation in homes, 158, 181, 346, 347
Radon daughters, 12, 13, 112, 159, 161, 162, 177, 178, 181, 221, 223–225, 336–338, 347, 348
Rainout, 338
Range, 187, 200, 353, 354, 387
Rate constant, 394, 395
Rb-88, 155, 156, 333–335
Re-186, 49
Re-188, 49
Reactions, 5
  (α, p), 19
  (γ, n), 190, 354, 355
  (e, e', n), 354
  (n, α), 5
  (n, γ), 5, 17, 213, 228, 377
  (n, p), 5, 213, 377, 379
  (p, n), 204

432 INDEX

Reactor, 126, 377
  Power, 86, 107, 126, 322
    Activation of components, 129, 137, 151
    Health physics concerns, 127, 139, 151–153
    Outages, 127, 136, 138, 139, 149–151
    Primary coolant activity, 86, 130, 132
    Primary coolant system leakage, 128, 130–132, 139
  Research, 74, 75
Reactor operation, 102, 103
Reciprocal dose theorem, 26
Red bone marrow, 14, 28, 249–251, 301, 390, 400
Reference man, 24, 244, 312, 390, 394
Refining, 103
Reflector, 104–106
Relative concentration, 165, 179
Relative excess, 415
Relative risk, 30, 36, 256, 411–415
Relaxation length, 21, 237, 374
Release height, 107, 332
Release rate, 107, 109, 168–173, 332, 333, 347, 351
Rem, 368, 369
Remainder, 114, 251, 252, 390, 399–401
Removal half-time, 116, 304, 391
Representative sample collection, 310
Reprocessing of spent nuclear fuel, 91, 93, 94, 101–103, 121
Research reactor, 74, 75, 78
Residence time, 387
Residual activity, 190
Residual photons, 353
Respiratory cancer, 412, 414
Respiratory protection, 113–115, 136, 244, 355
Respiratory tract, 116, 117, 381, 391–393
Retention function, 26, 116, 286, 303
Retention time, 391
RF cavity, 194
Rh-103, 137
Rh-106, 133, 137
Ring badge, 18, 229

Risk, 29, 30, 34–36, 158, 166, 180, 256, 260, 344, 385, 397, 398, 401, 402, 408–416
Risk coefficient, 166, 180, 256, 258, 260, 344, 398, 408, 409
Risk factors, 398
Risk models, 36, 408–416
  Absolute, 36, 411–414
  Relative, 36, 411–415
Risk population, 410
Rn-219, 160
Rn-220, 160
Rn-222, 12, 31, 103, 159, 160, 163, 178, 181, 223, 224, 257, 336, 337
Roentgen, 143, 368, 369
Rolle method, 336
Room air changes, 59, 60
Ru-103, 137
Ru-106, 63, 133, 137, 273

S-32, 188
Safe industry, 34, 259
Sample counting, 10, 72, 79, 182, 217–219, 347
Sampling efficiency, 18, 219
Sampling system, 11, 12, 72, 219, 220
Saturation activity, 21, 238, 239, 377
Sb-125, 133
Sc-47, 133
Scattered radiation, 51–54, 61, 64, 76, 189, 190, 192, 195, 201, 269, 276
Scattering transmission factor, 54
Scintigraphy, 47
Scintillation counter, 112, 137, 198, 336
S dose factor, 26, 50, 58, 59, 254, 264, 265, 386, 387, 396
Se-75, 79, 281, 282
Secondary beam, 191
Secondary coolant, 127, 139, 183, 332
Secondary protective barrier, 53
Secular equilibrium, 5, 12, 295, 335
Self-absorption, 11, 112, 222
Self-shielding, 285
Semi-infinite cloud model, 140, 141, 171–175
Serial decay relationships, 4, 5, 9, 62, 63, 212, 213, 216, 217, 267, 271, 365–367
Sex, 260, 409, 410, 412, 415

Shielding, 20, 21, 51, 52, 60–62, 70, 76, 77, 87–89, 188, 193–196, 200, 202, 373–375
  Accelerator, 187, 188, 191, 193–196, 200, 202
  Beta, 70, 88, 89
  Design, 51, 87–89, 193–196, 200, 202, 293, 355
  Gamma, 20, 21, 62, 87, 143, 144, 191, 195
  Muons, 187, 188, 195
  Neutron, 77, 187, 188, 195
  Photon, 195
  Structural, 60
  X-ray, 51–53, 60–62, 306
    Primary barrier, 52, 53
    Secondary barrier, 53
Shipping regulations, 294, 295
Shoreline deposits, 167, 169, 350
Single compartment model, 381–384
Skin, 14, 34, 35, 45, 64, 69, 77, 84, 88, 128, 133–135, 167, 174, 175, 229, 245, 259, 275, 288, 298, 306, 400, 403
  Dose assessment, 16, 18, 64, 132–135, 144, 174, 175, 183, 288, 349
  Hot particle, 132–135, 144, 145
  Radiation injury, 34, 35
Skin absorption, 283
Skin damage, 34, 306
Skin pigmentation, 35, 259
Slab source, 127, 129, 130, 368, 370, 373
Sm-153, 49
Small intestine, 28, 29, 249–256, 394, 395, 399
Smoke test, 72
Sn-113, 133
Sodium borotritide, 81, 283
Sodium hydroxide, 141, 142
Sodium iodide, 80, 282
Sodium iodide crystal, 47, 283
Soil, 347
Soil gas, 346
Solid state detector, 112
Solid waste, 103, 137, 138
Solubility, 391
Somatic effect, 410

Source image distance (SID), 64, 83, 84, 275, 287
Source organ, 28, 29, 249–251, 254, 385–388, 395–397
Source strength, 20, 21, 234, 240, 273, 294, 369, 370
Source term, 107, 109, 138, 140, 141, 163–165, 179, 181, 188, 345, 348, 351
Specific absorbed fraction, 386, 388
Specific activity, 4, 9, 21, 62, 96, 132, 138, 144, 211, 216, 271, 272
Specific effective energy (SEE), 28, 249, 250, 396
Specific gamma-ray emission, 7, 57, 143, 150, 214 (*see also* Gamma-ray constant)
SPECT, 47
Spent fuel pool, 101, 128, 130, 137, 139, 151–154, 328–331, 373
Spent power reactor fuel, 93, 94, 101, 103, 128, 137
Spermatocytes, 31
Spleen, 399
Sr-90, 9, 13, 14, 31, 122, 137, 146, 216, 217, 257, 315, 322
Stability classes, 108, 109, 172
Stack, 171, 348
Stack height, 107, 171–174
Stack monitor, 78, 138
Stack sampling, 11, 78, 138, 219, 220
Stainless steel, 106, 137, 142, 378
Standard deviation, 10–12, 218, 222
Standard mortality ratio, 411
Steady-state concentration, 165, 166, 179
Steam generator, 126–128, 130–132, 138, 139, 154, 332
  Surveillance and repair, 128, 136, 138, 139
Stellite, 137
Stochastic effects, 33, 35, 114, 252, 258, 290, 302, 305, 312, 389, 390, 397, 401, 408
Stochastic risk, 30, 35, 83
Stomach, 29, 71, 252–256, 394, 395, 400
Storage ring, 194
Submersion, 350

Sulfur oxides, 192
Surface contamination, 78, 81, 127, 131
Surface deposition rate, 165
Surface dose rate, 13, 45, 63, 84, 88, 93, 275
Systemic blood, 392, 393

Tails, 94, 96–101
Target organ, 28, 29, 114, 249–251, 254, 258, 385, 386, 388, 395–397
Target tissue, 360
Tc-99, 212, 229, 273
Tc-99m, 4, 18, 29, 48, 62, 87, 88, 212, 213, 229, 253–256, 271, 272, 295, 366
Te-125, 133
Te-126, 7, 8, 214, 215
Technique factor, 83
Teenagers, 165–167
Tenth-value layer, 54, 202, 355
Terminal settling velocity, 109
Terminal voltage, 358
Terrestrial radiation sources, 159, 347
Th-227, 160
Th-228, 160
Th-230, 103, 159
Th-231, 92, 160
Th-232, 126, 160
Th-232 series, 160
Th-234, 92, 103, 159
Therapeutic radionuclides, 49, 64, 277
Therapeutic x-rays, 55
Thermal neutron, 5, 17, 106, 377
Thermoluminescent dosimeter, 14–18, 70, 87, 198, 227, 228, 230, 231, 295, 343, 349
Thick source, 127, 129, 372
Thorium, 91, 93, 101–103, 159, 305, 347
Thorium dioxide, 102
Thorium ore, 102
Thorium series, 160, 182, 347, 348, 365
Thoron, 112, 158–160, 296, 347
Thoron daughters, 347, 348
Threshold, 389, 401
Threshold effect, 31
Thymus, 399
Thyroid, 25, 26, 28, 48, 49, 58, 59, 64, 70, 85, 86, 114, 121, 141, 214, 245, 246, 249–251, 263, 264, 277, 283, 289, 312, 313, 331–333, 390, 400, 415
Thyroid cancer, 415
Thyroid counting, 70, 264
Thyroid dose, 26, 49, 64, 85, 86, 121, 155, 245, 289, 290, 332, 333
Tidal volume, 337, 338
Tissue, 45, 173, 174, 199, 200, 206, 214, 252, 263, 385, 389, 398–401, 409
Tissue dose, 5, 6, 83, 143, 171, 213, 361, 398, 415
Tissue energy absorption coefficient, 173
Tl-201, 48
Tl-207, 160
Tl-208, 160
TLD albedo, 15, 17, 227–229
Tomography, 47
Total body, 28, 31, 167, 173, 249–251, 257, 349
Total effective dose equivalent, 50, 244, 265, 290, 302, 313, 316, 317
Toxic gas, 192, 203, 204, 357
Tracer studies, 47, 48
Tracheobronchial region, 116, 159, 162, 178, 181, 338, 346, 391–393
  Radon lung dose, 159, 162, 178, 181, 338, 346
Track etch detector, 224, 336
Transfer rate constant, 252, 253
Transformations in source organ, 249–251
Transient equilibrium, 5, 213
Translocation, 390, 392, 393
Transmission factor, 53, 54, 61, 235, 269, 276, 374
Transportation index, 87, 88, 294, 295
Tritiated thymidine, 30, 31, 257
Tritiated water, 30, 31, 77, 88
Tritium, 30, 76, 78, 80, 128, 136, 137, 169, 170, 175, 191, 204, 246, 247, 283, 343
Tritium gas, 283, 284
Tritium oxide, 80, 81, 283, 284
Tsivoglou method, 336
Tumor, 45–49, 199, 260, 352, 414
Turbine, 127, 131
Two-filter method, 336

U-233, 102, 104, 105, 126
U-234, 92, 93, 115, 159, 304

U-235, 92–96, 98, 100–102, 104–106, 115, 117, 126, 160, 304, 365
U-235 series, 101, 160, 365
U-238, 31, 91–93, 95, 98, 101, 103, 115, 159, 160, 257, 304
U-238 series, 101, 126, 159, 365
UF_4, 93, 95, 111
UF_5, 100
UF_6, 94, 101, 110
$U\text{-}235F_6$, 95, 98–100
$U\text{-}238F_6$, 95, 98–100
Unattached fraction, 178, 337, 338
University health physics, 68–90
  Common radioisotopes, 69
UNSCEAR 88, 399, 400, 409
UO_2, 93–95, 101, 111, 115
UO_3, 101, 110
UO_4, 101, 111
U_3O_8, 93–95, 101, 111
Upper large intestine, 28, 29, 249–256, 394, 395, 399
Uptake, 29, 47, 50, 59, 70, 71, 77, 80, 82, 85, 264, 285, 289, 299, 312, 381–384, 387, 390, 393, 405
Uranium, 14, 23, 91–104, 106, 110–112, 114–119, 126, 159, 188, 305, 306, 347, 378, 379
Uranium dioxide, 94, 126
Uranium fuel cycle, 93–95, 110
Uranium hexafluoride, 91, 95–101
Uranium mine workers, 161
Uranium ore, 91, 94, 95, 102
Uranium series, 101, 159, 182, 347, 348
Uranium vapor, 95, 100, 101, 117
Uranyl fluoride, 96
Urinalysis, 69, 70, 80, 82, 302, 305
Urine, 32, 82, 257, 278, 285, 286
Usage factor, 168
Use factor, 51–53, 60, 61, 268
Uterus, 28, 64, 84, 249–251, 399

Ventilation, 27, 56, 59, 60, 69–71, 80, 160, 161, 182, 192, 246, 247, 266, 277, 284, 306, 347, 405–407
Ventilation hood, 69, 80, 284, 299
Ventilation system removal rate, 27, 57, 59, 60, 160, 161, 192, 246, 247, 266, 407
Vertical standard deviation, 107, 172

Voltage, 54, 55, 64, 76, 193, 197, 198, 305, 306

Washout, 338
Waste management, 86, 87, 91, 137, 138, 179, 180, 236, 278
  Airborne, 180
  Gas, 103, 137, 138
  High-level, 94, 121
  Liquid, 103, 119, 137, 138
  Low-level, 86, 119, 120, 310
  Packaging, 87, 138
  Processing, 128, 138, 310
  Shipping regulations, 87
  Solid, 103, 120, 137, 138
  TRU waste, 103
Waste oil, 119, 120, 310
Water, 106, 163, 164, 168–170, 181, 190, 191, 200, 345, 354, 378
Weapons, 91, 103
Weighting factor, 35, 50, 58, 83, 85, 114, 115, 159, 163, 181, 259, 263, 264, 286, 301, 346, 389, 390, 399–401, 409
Whole body, 14, 265, 283
Whole-body counting, 70, 82, 244, 285, 286, 361, 404
Whole-body dose, 16, 25, 83, 92, 183, 244, 258, 265, 349, 362, 380, 389, 410, 415
Wind speed, 107–109, 154, 172, 332
Working level, 12, 13, 161, 162, 178, 181, 223, 224, 336, 337, 346
Working level concentration, 161, 162
Working level month, 161, 162, 178, 181, 337, 346
Workload, 51–53, 60–62, 267–270

Xe-126, 7, 8, 215
Xe-133, 48, 103, 137, 154
Xe-135, 103, 137, 154
Xe-137, 137
Xe-138, 103
Xenon, 130, 136
X-ray, 8, 9, 13, 31, 43, 47, 51–55, 60–64, 70, 75, 76, 83–85, 100, 101, 118, 171, 189, 193, 194, 203, 214, 215, 229, 267–271, 273–276, 284, 287–289, 295, 305, 306, 410

X-ray (*Continued*)
  Diagnostic, 43, 47, 55, 268–271, 275, 287, 288
  Diffraction, 74, 76
  Shielding, 51, 53, 55, 60–62, 268–271
  Therapeutic, 55
  Production, 118, 189, 193, 194
  Sources, 83, 84, 189, 193, 194, 229, 268–271, 287
X-ray machine, 15, 52–54, 61–63, 68, 83, 84
X-ray tube, 54, 55

Y-90, 9, 13, 14, 49, 123, 137, 216, 217
Yield, 143, 239, 370, 375, 377–379

Zero-threshold dose response, 411
Zirconium, 126, 137, 378
Zirconium-niobium particle, 132–135
Zn-65, 144, 190
Zn-S coated photomultiplier, 81, 82, 285
Zr-94, 137, 377
Zr-95, 133, 137, 144, 377